Profiles of

DRUG SUBSTANCES, EXCIPIENTS, AND RELATED METHODOLOGY

VOLUME **35**

CONTRIBUTING EDITOR

ABDULLAH A. AL-BADR

FOUNDING
EDITOR

KLAUS FLOREY

Profiles of
DRUG SUBSTANCES, EXCIPIENTS, AND RELATED METHODOLOGY

VOLUME **35**

Edited by

HARRY G. BRITTAIN
Center for Pharmaceutical Physics
Milford, New Jersey
USA

Amsterdam • Boston • Heidelberg • London • New York • Oxford
Paris • San Diego • San Francisco • Singapore • Sydney • Tokyo
Academic Press is an imprint of Elsevier

ELSEVIER

ACADEMIC PRESS

Academic Press is an imprint of Elsevier
Linacre House, Jordan Hill, Oxford OX2 8DP, UK
32 Jamestown Road, London NW1 7BY, UK
Radarweg 29, PO Box 211, 1000 AE Amsterdam, The Netherlands
30 Corporate Drive, Suite 400, Burlington, MA 01803, USA
525 B Street, Suite 1900, San Diego, CA 92101-4495, USA

First edition 2010

ISBN: 978-0-12-380884-4
ISSN: 1871-5125 (Series)

For information on all Academic Press publications
visit our website at elsevierdirect.com

Printed and bound in USA
10 11 12   10 9 8 7 6 5 4 3 2 1

CONTENTS

The comprehensive profiling of drug substances and pharmaceutical excipients as to their physical and analytical characteristics remains essential to all phases of pharmaceutical development. It therefore goes without saying that the compilation and publication of comprehensive summaries of physical and chemical data, analytical methods, routes of compound preparation, degradation pathways, uses and applications, etc., have always been and will continue to be a vital function to both academia and industry.

As the science of pharmaceutics grows and matures, the need for information similarly expands along new fronts, and this growth causes an equivalent growth in the repository sources where investigators find the information they need. The content of the *Profiles* series continues to respond and expand to meet this need, and so chapters are published that fall into one or more of the following main categories:

1. Comprehensive profiles of a drug substance or excipient
2. Physical characterization of a drug substance or excipient
3. Analytical methods for a drug substance or excipient
4. Detailed discussions of the clinical uses, pharmacology, pharmacokinetics, safety, or toxicity of a drug substance or excipient
5. Reviews of methodology useful for the characterization of drug substances or excipients
6. Annual reviews of areas of importance to pharmaceutical scientists

As it turns out, the chapters in the current volume are comprehensive in nature, and each provides very detailed profiles of the drug substances involved. The present volume contains a new annual review I have started to cover the exceedingly new and important emerging field of pharmaceutical cocrystals. As always, I welcome communications from anyone in the pharmaceutical community who might want to provide an opinion or a contribution.

Harry G. Brittain

Editor, Profiles of Drug Substances,
Excipients, and Related Methodology

hbrittain@centerpharmphysics.com

CHAPTER 1

Atorvastatin Calcium

Vishal M. Sonje,* Lokesh Kumar,† Chhuttan Lal Meena,‡ Gunjan Kohli,† Vibha Puri,† Rahul Jain,‡ Arvind K. Bansal,† and Harry G. Brittain§

* Department of Pharmaceutical Technology (Formulations), National Institute of Pharmaceutical Education and Research (NIPER), Punjab, India
† Department of Pharmaceutics, National Institute of Pharmaceutical Education and Research (NIPER), Punjab, India
‡ Department of Medicinal Chemistry, National Institute of Pharmaceutical Education and Research (NIPER), Punjab, India
§ Center for Pharmaceutical Physics, Milford, New Jersey, USA

Profiles of Drug Substances, Excipients, and Related Methodology, Volume 35
ISSN 1871-5125, DOI: 10.1016/S1871-5125(10)35001-1

1. DESCRIPTION

1.1. Nomenclature [1,2]

1.1.1. Systematic chemical name

- R-(R^*,R^*)-2-(4-fluorophenyl)-β,δ-dihydroxy-5-(1-methylethyl)-3-phenyl-4-[(phenylamino)carbonyl]-1H- pyrrole-1-heptanoic acid
- (βR,δR)-2-(4-fluorophenyl)-α,δ-dihydroxy-5-(1-methylethyl)-3-phenyl-4-[(phenylamino)carbonyl]-1H-pyrrole-1-heptanoic acid trihydrate

1.1.2. Proprietory names

Lipitor®, Sortis®, Torvast®

1.2. Formulae

1.2.1. Empirical formula, molecular weight, and CAS number

	Formula	Molecular weight	CAS number
Atorvastatin free base	$C_{33}H_{35}FN_2O_5$	558.7	134523-00-5
Atorvastatin calcium	$[C_{33}H_{35}FN_2O_5]_2 \cdot Ca \cdot 3H_2O$	1209.42	134523-03-8

1.2.2. Structural formula

The structural formula of atorvastatin calcium is shown in Fig. 1.1.

1.3. Elemental analysis

The elemental composition (C, H, and N) is given in Table 1.1.

1.4. Appearance

Atorvastatin calcium is a white to off-white crystalline powder.

1.5. Mechanism of action

Atorvastatin is a selective, competitive inhibitor of the 3-hydroxy methyl glutaryl coenzyme A (HMG-CoA) reductase enzyme that is involved in the conversion of HMG-CoA to mevalonate (a precursor of sterols, including cholesterol). A reduction of intracellular cholesterol levels

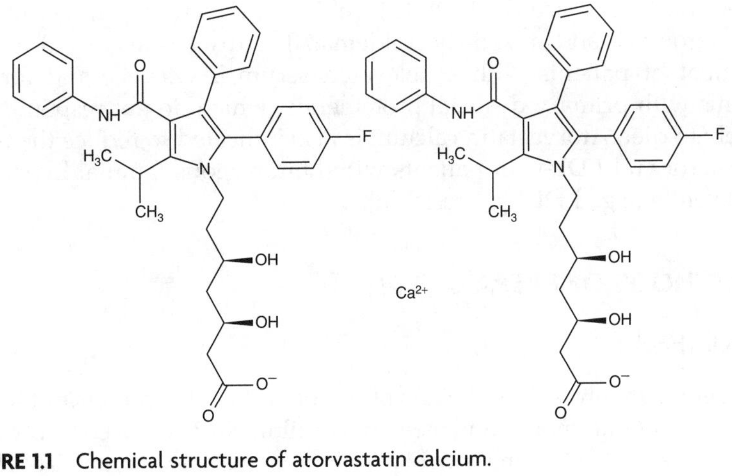

FIGURE 1.1 Chemical structure of atorvastatin calcium.

TABLE 1.1 Elemental analysis of atorvastatin calcium

Element	Theoretical (%)	Actual (%)
Carbon (C)	65.6	65.4
Hydrogen (H)	6.3	6.3
Nitrogen (N)	4.6	4.5
Oxygen (O)	17.2	17.6

promotes an expression of LDL (low-density lipoprotein) receptors on the hepatocyte surface, resulting in an increased extraction of LDL from the blood. As an additional cholesterol-lowering mechanism, HMG-CoA reductase inhibitors also decrease the blood concentrations of VLDLs (very low-density lipoproteins) by inhibiting their synthesis and promoting their catabolism. Atorvastatin calcium also inhibits the cholesterol synthesis in the liver and increases the hepatic LDL receptors on the cell surface to enhance the uptake and catabolism of LDL. The drug also reduces the LDL production and the number of LDL particles.

Atorvastatin calcium possesses an anti-inflammatory property and reduces the accumulation of inflammatory cells in the atherosclerotic plaques. The drug also inhibits the vascular smooth muscle cell proliferation: a key event in the atherogenesis. Atorvastatin also inhibits the platelet function, thereby limiting both the atherosclerosis and the superadded thrombosis; and also improves the vascular endothelial function, largely through the amplification of nitric oxide (NO) generation [3–5].

1.6. Uses and application

Atorvastatin calcium is used as an adjunct to diet to reduce the elevated total-cholesterol, LDL, apolipoprotein B (apo B), and triglyceride (TG) levels, and to increase the HDL-C level in patients with primary hypercholesterolemia and mixed dyslipidemia. The drug is also used for the treatment of patients with an elevated serum TG levels, and for the patients with primary dysbetaliproteinemia, which do not respond adequately to diet. Atorvastatin calcium is also indicated to reduce the total-cholesterol and LDL-C in patients with homozygous familial hypercholesterolemia (e.g., LDL apheresis) [6].

2. METHODS OF PREPARATION

2.1. Method 1

This method involves the reduction of compound **1** in the presence of Raney nickel, aqueous ammonia, and methanol resulting in **2**, followed by condensation of **2** with **3** in the presence of pivalic acid, cyclohexane, and isopropyl alcohol to get **4**—a dioxane derivative (Fig. 1.2). On reaction of **4** with acetonitrile and hydrochloric acid, an ester form of atorvastatin, **5**, is obtained, which on further treatment with sodium hydroxide and calcium acetate gives atorvastatin calcium (patent application no. WO 2006/039441).

2.2. Method 2

As described in US patent 6,777,552, an atorvastatin ester derivative is converted to atorvastatin hemicalcium by mixing the ester derivative with more than 70% excess (molar basis) of calcium hydroxide (see Fig. 1.3). Calcium hydroxide functions as a basic catalyst for the hydrolysis of ester and also supplies calcium ion to form the hemi calcium salt. A significant advantage of this method is that the amount of calcium hydroxide does not have to be as carefully controlled; in contrast to the amount of sodium hydroxide and calcium acetate generally controlled in other processes.

2.3. Method 3

This method entails the conversion of phenylboronate esters of atorvastatin calcium, according to US patent 6,867,306 B2. The method suggests the hydrolysis of phenylboronate ester (see Fig. 1.4), yielding the free acid which is then converted to its ammonium salt by reacting it with methanolic ammonia. The ammonium salt is then reacted with calcium acetate to give atorvastatin calcium. US patent 7,361,772 B2 suggests a process involving the deprotonation of a boronate ester, cleavage of *tert*-butyl ester, and the formation of a calcium salt in one step. The process employs a single, inexpensive calcium oxide reagent without the need for making sodium salt or other intermediates, and thus reduces the number of steps.

A number of other patents, including US patent numbers 4,681,893; 5,003,080; 5,097,045; 5,103,024; 5,124,482; 5,149,837; 5,155,251; 5,216,174;

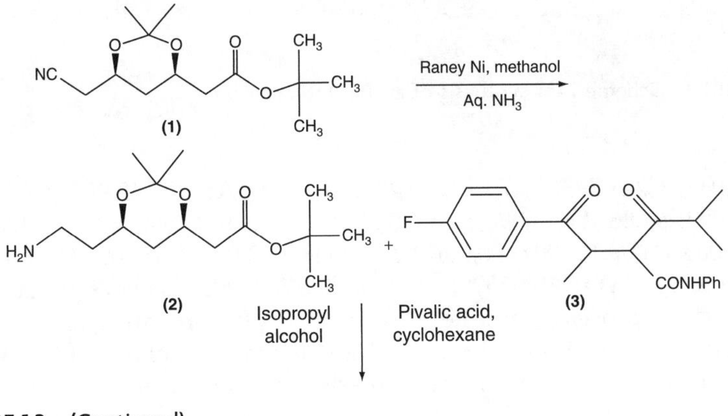

FIGURE 1.2 (Continued)

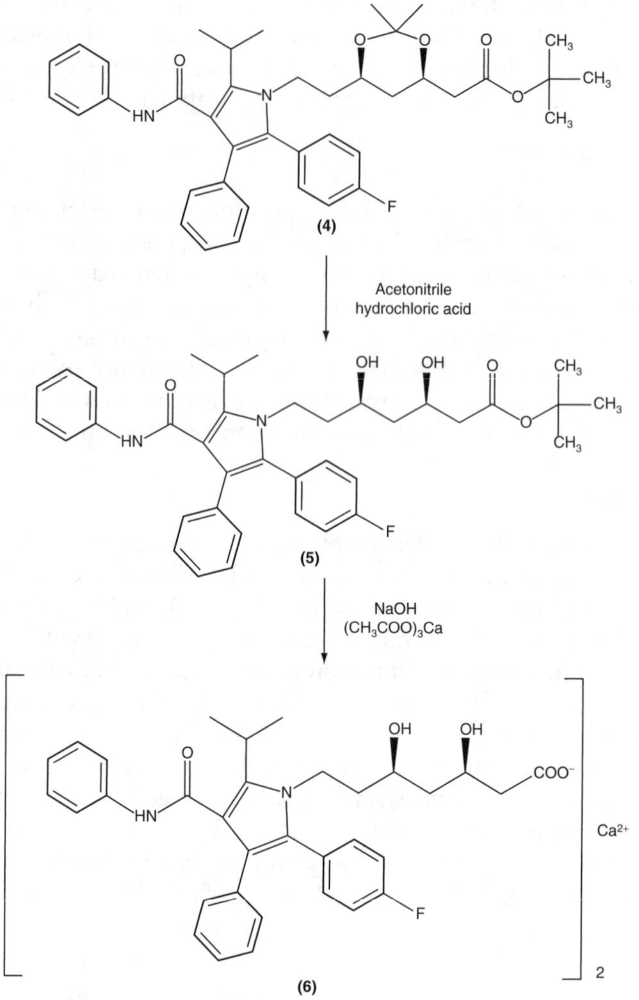

FIGURE 1.2   Scheme 1 for synthesis of atorvastatin calcium.

5,245,047; 5,248,793; 5,273,995; 5,280,126; 5,298,627; 5,342,952; 5,385,929; 5,397,792; 5,686,104; 5,998,633; 6,087,511; 6,126,971; 6,433,213; 6,476,235; 2007/0032662 A1; 2007/0032663 A1; 2007/0032664 A1; 2007/249845 A1; EU 409,281; WO 89/07598; WO 07/057703 A2; WO 08/075165 A1, describe the processes and key intermediates for preparing atorvastatin. US patent 5,273,995 provides procedure for synthesis of one of the stereo-isomer of atorvastatin by chiral synthesis or racemic resolution. US patent

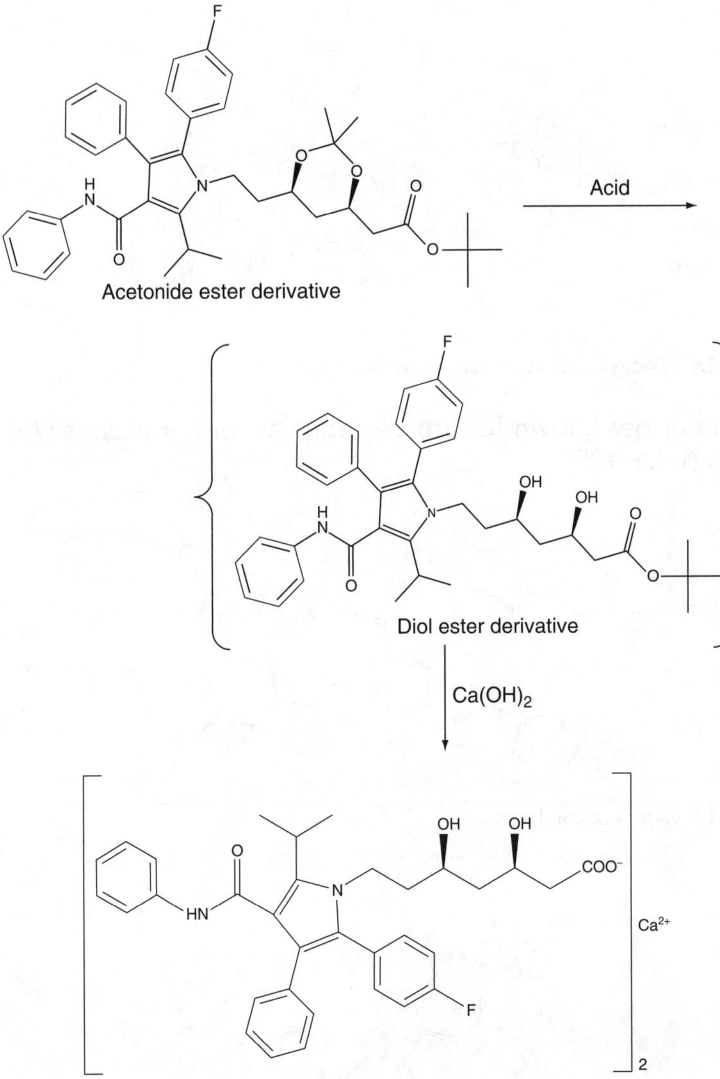

FIGURE 1.3   Scheme 2 for the synthesis of atorvastatin calcium.

numbers 5,003,080; 5,097,045; 5,103,024; 5,124,482; 5,149,837; 2007/
0032663 A1; US 2007/0032662 A1; US 2007/0032664; WO 2008/075165
suggest novel processes and/or superior convergent routes for atorvastatin synthesis.

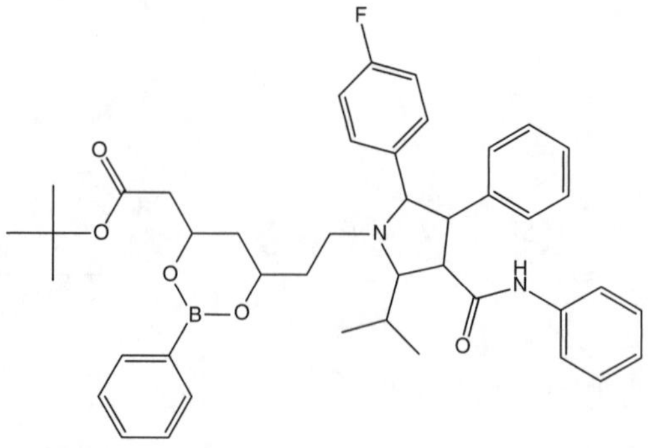

FIGURE 1.4   Phenylboronate ester of atorvastatin.

2.4. Impurities known for atorvastatin calcium processes [7,8] [Figs. 1.5–1.15]

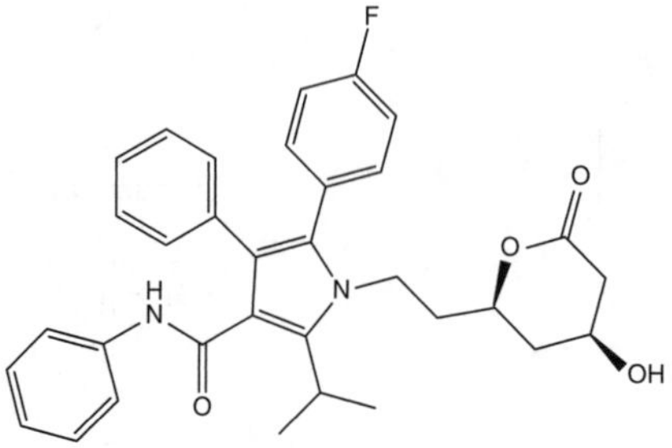

FIGURE 1.5   Atorvastatin lactone.

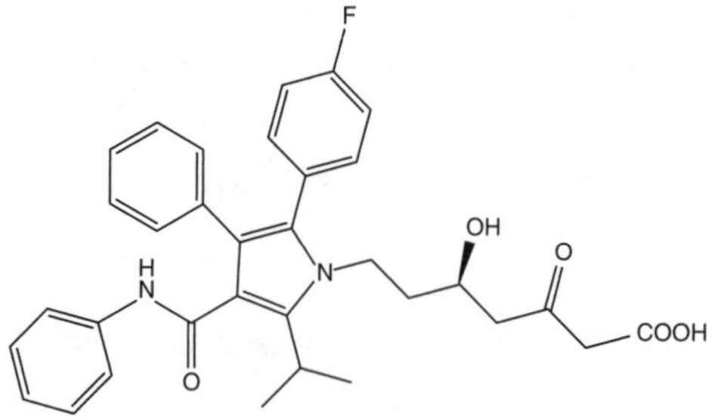

FIGURE 1.6   3-oxo-Atorvastatin.

3. PHYSICAL CHARACTERISTICS

3.1. Ionization constants

Atorvastatin calcium has a pK_a value of 4.46 [1].

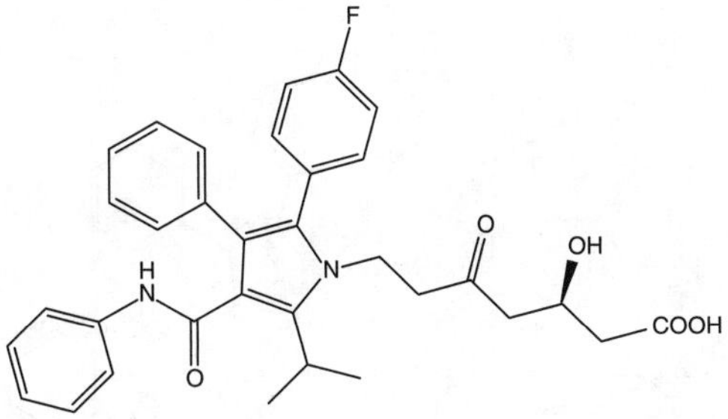

FIGURE 1.7 5-oxo-Atorvastatin.

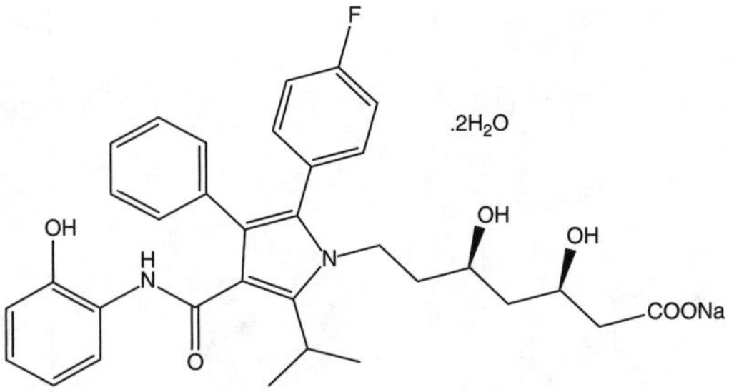

FIGURE 1.8 2-Hydroxy atorvastatin monosodium dihydrate.

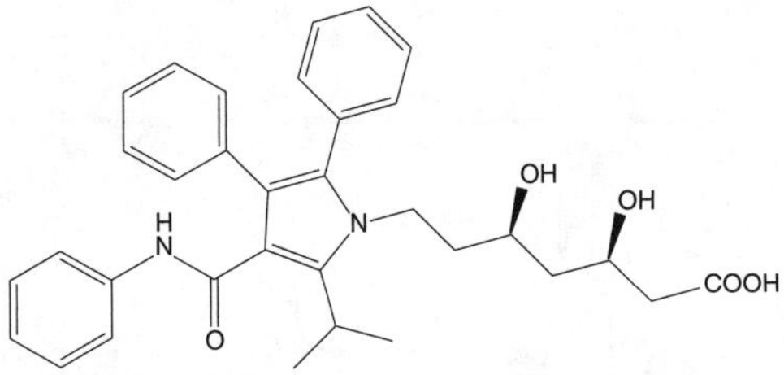

FIGURE 1.9 Desfluoro atorvastatin.

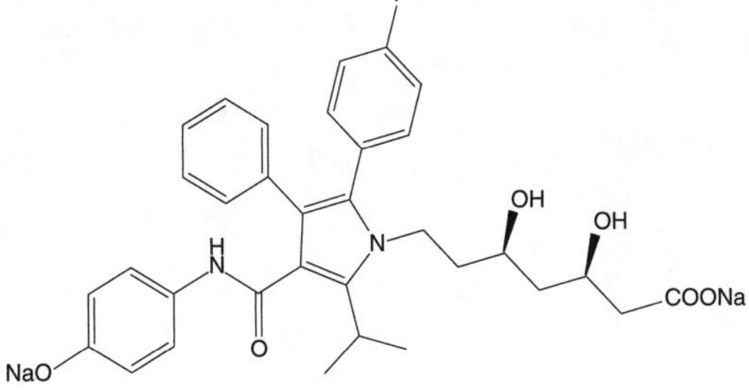

FIGURE 1.10 4-Hydroxy atorvastatin disodium.

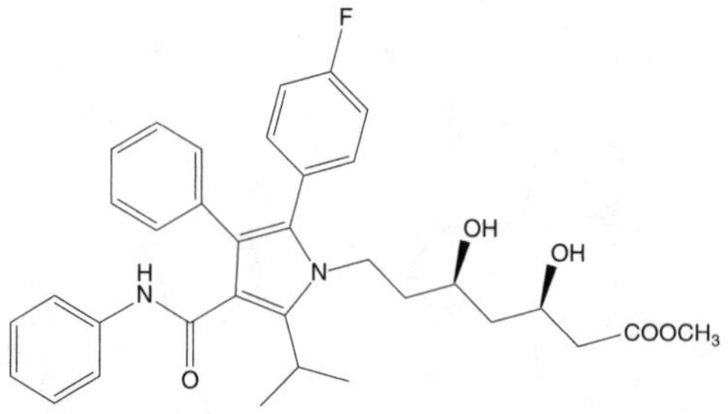

FIGURE 1.11   Atorvastatin methyl ether.

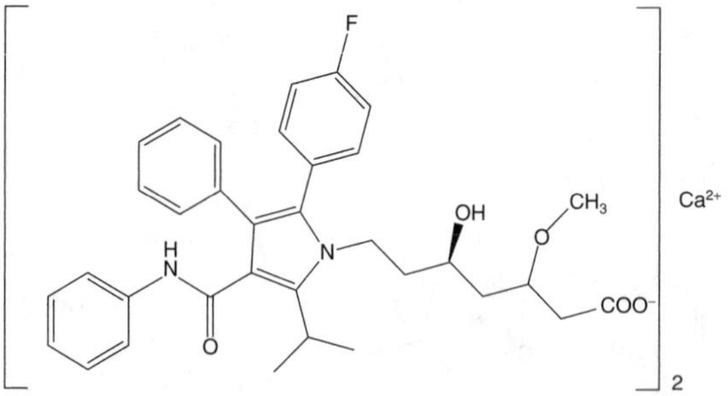

FIGURE 1.12   3-*O*-Methyl atorvastatin calcium.

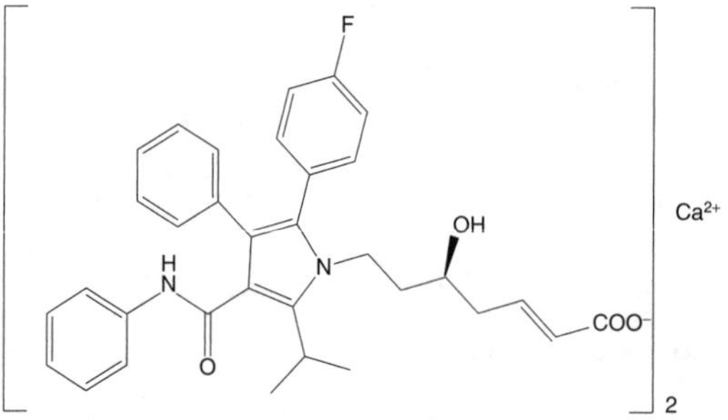

FIGURE 1.13   Anhydro atorvastatin calcium.

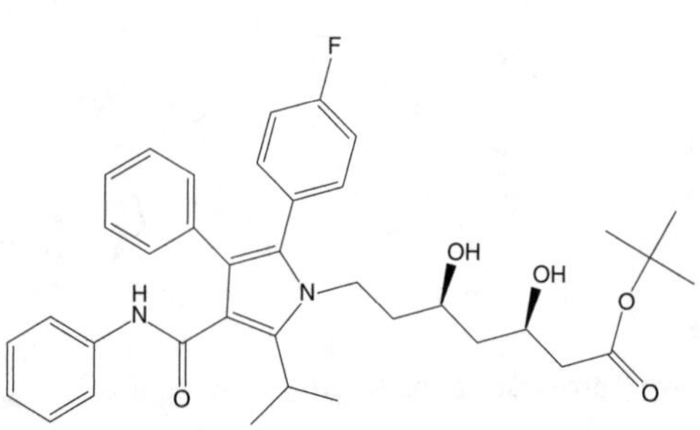

FIGURE 1.14   Atorvastatin *tert*-butyl ester.

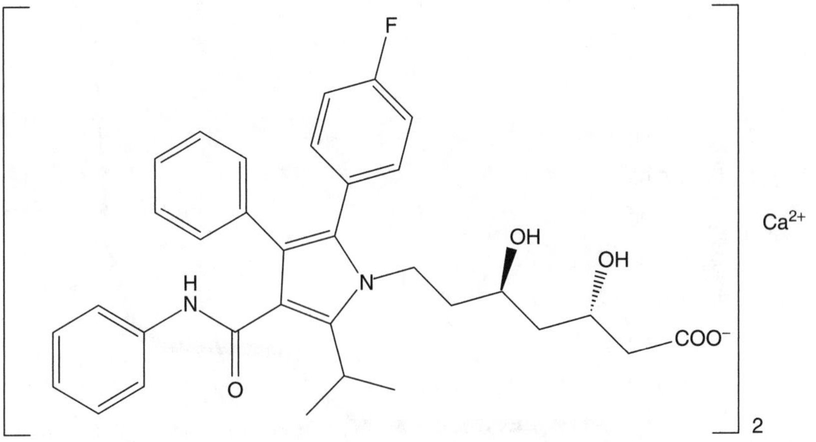

FIGURE 1.15   Atorvastatin diastereomer.

3.2. Solubility characteristics

The solubility of atorvastatin calcium, Form-I, was measured in a variety of solvents (Table 1.2) by taking an excess of atorvastatin calcium in screw-capped vials with 10 ml of the solvent, followed by analysis of the dissolved fraction after 24 h by high-performance liquid chromatography (HPLC). The pH of a saturated solution of atorvastatin calcium in water was found to be 7.12 (Fig. 1.16).

The predicted value of aqueous solubility using the ACD/I-Lab Web service (ACD/aqueous solubility 8.02) was 20 mg/ml at pH 7.12, which, however, was practically obtained to be 0.8 mg/ml.

TABLE 1.2   Solubility of atorvastatin calcium in different solvents

Solvent	Quantity dissolved at 37 °C (mg/ml)
Water	0.8
Methanol	232.8
Ethanol	21.6
Acetone	215.0
Chloroform	216.8
Diethyl ether	0.8
n-Hexane	0.3
Polyethylene glycol 200	32.1
Acetonitrile	14.9
Ethyl acetate	8.8

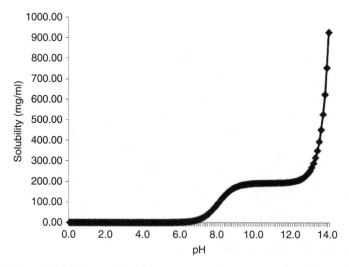

FIGURE 1.16   pH-Solubility profile of atorvastatin calcium, as predicted using the advanced chemistry development program.

3.3. Partition coefficients

The partition coefficient (log P) of atorvastatin has been reported to be 6.36 [9]. Using the ACD/I-Lab Web service (ACD/pK_a 8.03), the predicted partition coefficient was determined to be 4.18. The predicted pH dependence of the atorvastatin distribution coefficient (log D), obtained using the ACD/I-Lab Web service (ACD/log D 8.02), is shown in Fig. 1.17.

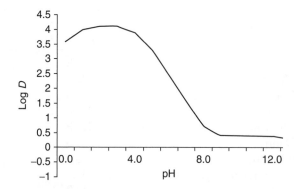

FIGURE 1.17   pH Dependence of log D for atorvastatin calcium, calculated using the advanced chemistry development program.

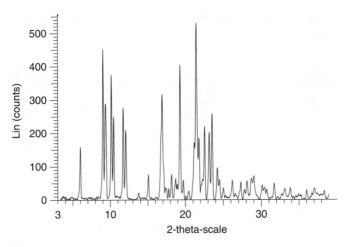

FIGURE 1.18 XRPD pattern of atorvastatin calcium, Form-I.

3.4. Optical activity

The specific rotation of atorvastatin calcium is $-6.0°$ to $-12.0°$ for a 1.0% (w/v) solution in dimethyl sulfoxide [2].

3.5. X-Ray powder diffraction

The X-ray powder diffraction (XRPD) pattern of a sample of atorvastatin calcium, Form-I, was recorded at room temperature on Bruker D8 Advance diffractometer (Karlsruhe, Germany), using nickel-filtered Cu Kα radiation. The sample was mounted in a polymethylmethacrylate sample holder, and analyzed in a continuous mode with a step size of $0.01°$ and a step time of 1 s over an angular range of 3–40° 2θ. The XRPD results are found in Fig. 1.18 and in Table 1.3, being evaluated with the DIFFRAC^{plus} EVA (version 9.0) diffraction software.

Atorvastatin calcium has been obtained in a large number of solid-state crystal forms, most of which have been the subject of patents and patent applications. Figures 1.19–1.58 contain the published diffraction patterns for the described forms, as well as the scattering angles (in units of ° 2θ) and d-spacings (in Å units) of the 10 most intense peaks disclosed for each crystal form.

3.6. Hygroscopicity

The hygroscopicity of atorvastatin calcium, Form-I, was determined by exposing the samples to controlled relative humidity (RH) conditions (21–92% RH) at ambient temperature, and after 7 days the percent weight change was recorded. The substance was found to show less than 0.1%

TABLE 1.3 Scattering angles, interplanar *d*-spacing, and relative intensities in the XRPD pattern of atorvastatin calcium, Form-I

Scattering angle (° 2θ)	*d*-Spacing (Å)	Relative intensity (I/I$_0$)
21.43	4.14	100.00
16.86	5.26	74.40
19.28	4.60	73.70
8.97	9.85	67.80
10.08	8.77	56.00
23.53	3.78	50.80
9.30	9.50	49.90
22.51	3.95	48.30
11.66	7.58	47.70
10.38	8.52	46.30
23.13	3.84	45.60
21.16	4.20	43.40
12.01	7.36	41.50
21.78	4.08	41.40
5.96	14.81	34.50
24.22	3.67	27.70
18.13	4.89	26.10
19.70	4.50	24.30
18.67	4.75	22.80
15.03	5.89	22.70
24.47	3.63	22.30
28.98	3.08	22.10
28.65	3.11	21.70
28.11	3.17	20.90
27.31	3.26	20.80
30.11	2.97	20.80
26.18	3.40	20.10
31.71	2.82	19.60
33.82	2.65	17.90
39.13	2.30	17.70
25.02	3.56	17.50
20.43	4.34	17.50
33.07	2.71	16.30
13.73	6.45	15.80
36.07	2.49	15.50

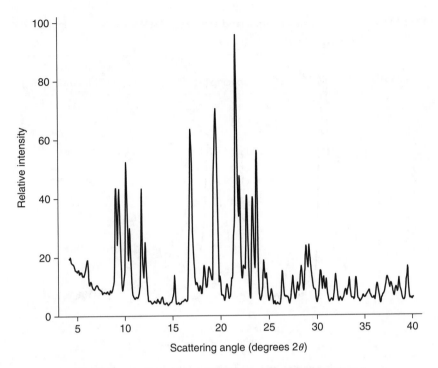

Scattering angles and *d*-spacings of the 10 most intense peaks:

Scattering angle (degrees 2θ)	*d*-spacing (Å)	Scattering angle (degrees 2θ)	*d*-spacing (Å)
9.03	9.780	19.50	4.548
9.39	9.414	21.58	4.115
10.12	8.736	22.69	3.915
11.74	7.530	23.30	3.814
16.89	5.246	23.74	3.745

FIGURE 1.19 XRPD pattern of atorvastatin calcium, Pfizer Form-I, scanned and digitized from the pattern disclosed in US patent 5,969,156.

weight gain at all conditions of RH ranging from 21% to 92% RH, demonstrating its nonhygroscopic nature.

3.7. Thermal method of analysis

3.7.1. Differential scanning calorimetry
The differential scanning calorimetry (DSC) analysis of atorvastatin calcium, Form-I, was performed using a Mettler Toledo 821 DSC system, operating with Star software (version Solaris 2.5.1), and is shown in Fig. 1.59.

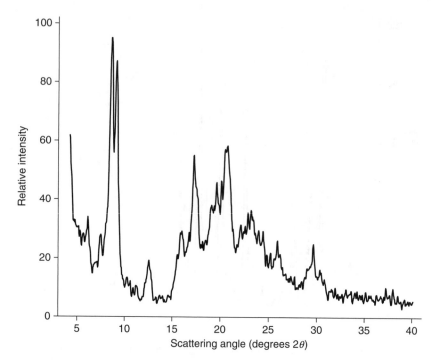

Scattering angles and *d*-spacings of the 10 most intense peaks:

Scattering angle (degrees 2θ)	*d*-spacing (Å)	Scattering angle (degrees 2θ)	*d*-spacing (Å)
6.09	14.491	17.22	5.144
8.58	10.294	20.72	4.282
9.10	9.708	23.19	3.833
12.55	7.049	25.92	3.435
15.99	5.537	29.67	3.009

FIGURE 1.20 XRPD pattern of atorvastatin calcium, Pfizer Form-II, scanned and digitized from the pattern disclosed in US patent 5,969,156.

The temperature scale and the cell constant were calibrated using indium. Samples of approximately 4 mg were heated at 20 °C/min over the temperature range of 25–200 °C, under dry nitrogen purging (80 ml/min) in a pinholed aluminum pan. The melting range of atorvastatin calcium was found to take place over the broad range of 158.4–178.03 °C.

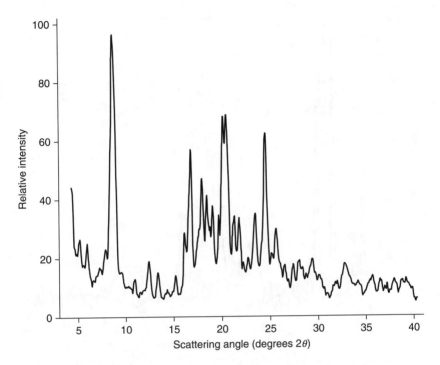

Scattering angles and *d*-spacings of the 10 most intense peaks:

Scattering angle (degrees 2θ)	*d*-spacing (Å)	Scattering angle (degrees 2θ)	*d*-spacing (Å)
8.69	10.172	20.15	4.402
12.45	7.105	20.46	4.336
16.83	5.264	21.78	4.077
17.96	4.934	23.48	3.786
18.48	4.797	24.54	3.625

FIGURE 1.21   XRPD pattern of atorvastatin calcium, Pfizer Form-III, scanned and digitized from the pattern disclosed in US patent 6,121,461.

3.7.2. Thermogravimetric analysis

The thermogravimetric (TG) analysis of atorvastatin calcium, Form-I, is shown in Fig. 1.60, and was performed using a Mettler Toledo 851^e TGA/ SDTA system. The samples were contained in pin-holed aluminum crucibles that were heated at a rate of 20 °C/min from 25 to 200 °C under nitrogen purging (10 ml/min). The initial steps observed from 40 to 160 °C correspond to the loss of hydrate water.

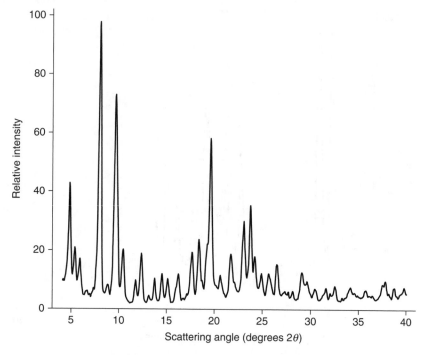

Scattering angles and *d*-spacings of the 10 most intense peaks:

Scattering angle (degrees 2θ)	*d*-spacing (Å)	Scattering angle (degrees 2θ)	*d*-spacing (Å)
4.84	18.245	18.41	4.816
8.00	11.046	19.59	4.529
9.63	9.177	21.73	4.087
12.36	7.156	23.06	3.853
17.69	5.011	23.73	3.746

FIGURE 1.22 XRPD pattern of atorvastatin calcium, Pfizer Form-IV, scanned and digitized from the pattern disclosed in US patent 5,969,156.

3.8. Spectroscopy

3.8.1. Ultraviolet absorption spectroscopy

Atorvastatin calcium was dissolved in methanol at a concentration of 4 μg/ml, and the resulting UV absorption spectrum is shown in Fig. 1.61. The analytically useful wavelength maximum (λ_{max}) of atorvastatin calcium in methanol is approximately 246 nm.

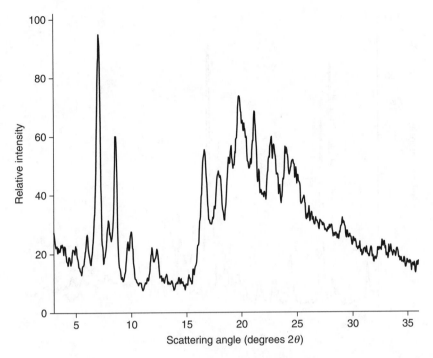

Scattering angles and *d*-spacings of the 10 most intense peaks:

Scattering angle (degrees 2θ)	d-spacing (Å)	Scattering angle (degrees 2θ)	d-spacing (Å)
7.07	12.500	17.89	4.955
8.55	10.338	19.77	4.487
9.99	8.842	21.19	4.190
11.88	7.446	22.70	3.914
16.62	5.328	24.02	3.701

FIGURE 1.23 XRPD pattern of atorvastatin calcium, Pfizer Form-V, scanned and digitized from the pattern disclosed in US patent 6,605,729.

3.8.2. Vibrational spectroscopy

The infrared absorption spectrum of atorvastatin calcium, Form-I, was obtained using a Nicolet FT-IR Impact 410 spectrophotometer equipped with deuterated triglycine sulfate detector, and using Omnic 5.1a software. The sample was contained in a pressed KBr pellet, with each spectrum being derived from 16 single scans over the range of 4000–400 cm^{-1} at a

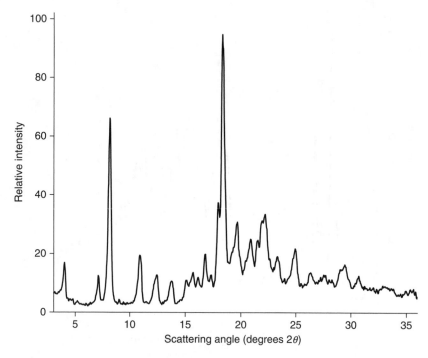

Scattering angles and *d*-spacings of the 10 most intense peaks:

Scattering angle (degrees 2θ)	*d*-spacing (Å)	Scattering angle (degrees 2θ)	*d*-spacing (Å)
4.02	21.973	16.75	5.288
8.17	10.816	18.31	4.843
10.92	8.092	19.70	4.503
12.48	7.089	22.23	3.995
13.78	6.423	24.93	3.569

FIGURE 1.24   XRPD pattern of atorvastatin calcium, Pfizer Form-VI, scanned and digitized from the pattern disclosed in US patent 6,605,729.

spectral resolution of $2 \, cm^{-1}$. The spectrum is shown in Fig. 1.62, and a brief table of peak assignments follows in Table 1.4.

3.8.3. Nuclear magnetic resonance spectrometry
3.8.3.1. ^1H NMR spectrum   The Nuclear magnetic resonance (NMR) spectra of atorvastatin were assigned (Fig. 1.63 and Table 1.5) according to the following numbering system:

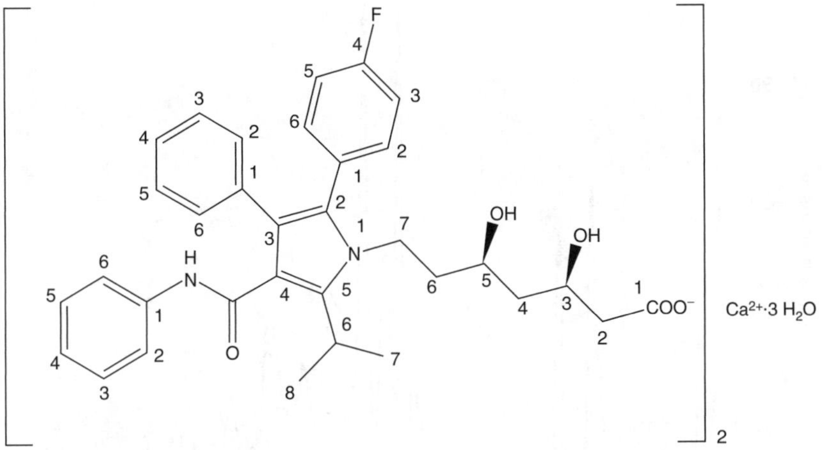

3.8.3.2. ^{13}C NMR spectrum   The ^{13}C NMR spectra of atorvastatin calcium is shown in Fig. 1.64 (Table 1.6).

3.9. Mass spectrometry

Mass spectrometry of mass fragmentation pattern of atorvastatin calcium are shown in Figs. 1.65 and 1.66, respectively.

4. METHODS OF ANALYSIS

4.1. Spectroscopic analysis

4.1.1. Spectrophotometry and colorimetry

Erk developed a spectrophotometric procedure for the assay of atorvastatin, both for the bulk drug substance as well as for pharmaceutical formulations. The procedures are based on the reaction between the drug and bromocresol green, alizarin red, or bromophenol blue, which result in the production of ion-pair complexes (1:1). Beer's law was obeyed over the concentration ranges 5.0–53.0, 7.1–55.8, or 7.5–56.0 $\mu g/ml$ with bromocresol green, alizarin red, and bromophenol blue, respectively. The specific absorptivities, molar absorptivities, Sandell sensitivities, standard deviations, and the percent recoveries were evaluated. Atorvastatin was determined by measurement of its first derivative signal at 217.8 nm.

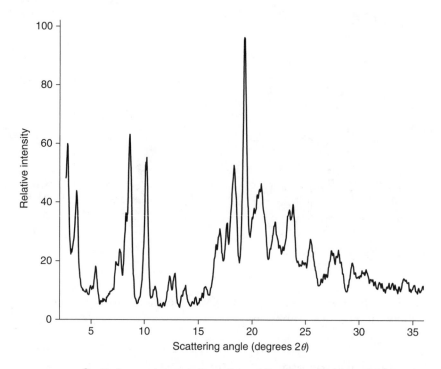

Scattering angles and *d*-spacings of the 10 most intense peaks:

Scattering angle (degrees 2θ)	d-spacing (Å)	Scattering angle (degrees 2θ)	d-spacing (Å)
2.75	32.053	18.27	4.851
3.61	24.447	19.25	4.606
5.45	16.208	20.85	4.258
8.63	10.236	22.19	4.002
10.19	8.671	23.82	3.733

FIGURE 1.25   XRPD pattern of atorvastatin calcium, Pfizer Form-VII, scanned and digitized from the pattern disclosed in US patent 6,605,729.

The calibration curve was established for 4.2–69.0 μg/ml of atorvastatin by first derivative spectrophotometry, and the method showed no interference from the common pharmaceutical adjuvants [11].

Sahu and Patel developed a rapid, simple, accurate, and precise UV spectrophotometric method for the simultaneous determination of amlodipine besylate and atorvastatin calcium in a binary mixture. In the method, absorbance values were measured at 238.2 and 246.6 nm that

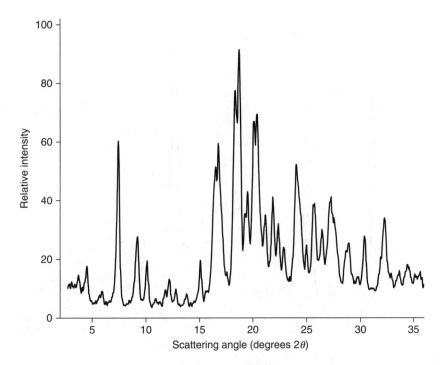

Scattering angles and *d*-spacings of the 10 most intense peaks:

Scattering angle (degrees 2θ)	d-spacing (Å)	Scattering angle (degrees 2θ)	d-spacing (Å)
7.42	11.900	18.70	4.742
9.22	9.580	20.43	4.342
10.12	8.730	21.86	4.062
16.77	5.282	24.04	3.699
18.35	4.830	27.24	3.272

FIGURE 1.26   XRPD pattern of atorvastatin calcium, Pfizer Form-VIII, scanned and digitized from the pattern disclosed in US patent 6,605,729.

corresponded to the respective absorbance maxima of amlodipine besylate and atorvastatin calcium in methanol. Linearity was observed over the concentration range of 5–30 μg/ml for both drugs. The concentration of each drug was obtained by using the absorptivity values calculated at the indicated wavelengths, and the method was statistically validated. A recovery study was performed to confirm the accuracy of the method, and laboratory prepared synthetic mixtures were successfully analyzed using the method [12].

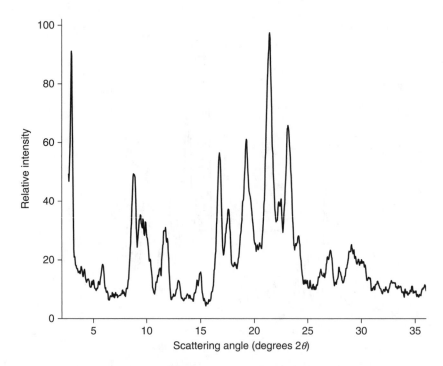

Scattering angles and *d*-spacings of the 10 most intense peaks:

Scattering angle (degrees 2θ)	*d*-spacing (Å)	Scattering angle (degrees 2θ)	*d*-spacing (Å)
2.94	30.042	17.58	5.039
5.86	15.065	19.25	4.608
8.82	10.023	21.43	4.143
11.77	7.513	23.15	3.838
16.75	5.287	27.12	3.285

FIGURE 1.27 XRPD pattern of atorvastatin calcium, Pfizer Form-IX, scanned and digitized from the pattern disclosed in US patent 6,605,729.

4.2. Chromatographic methods of analysis

4.2.1. High-performance liquid chromatography
A number of HPLC methods have been reported in the literature for the analysis of atorvastatin calcium, and the important aspects of these are summarized in Table 1.7.

4.2.2. Thin-layer chromatography
Jamshidi and Nateghi reported a two-step isocratic HPTLC method on silica gel 60F254 plates that used densitometric quantitation at a wavelength of 280 nm for the separation of atorvastatin from plasma

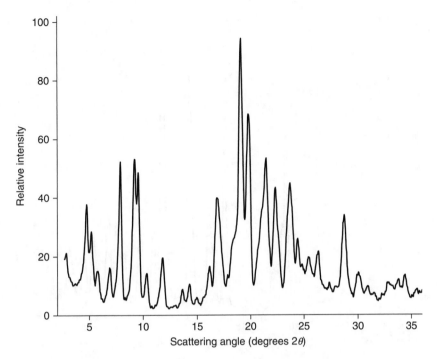

Scattering angles and *d*-spacings of the 10 most intense peaks:

Scattering angle (degrees 2θ)	*d*-spacing (Å)	Scattering angle (degrees 2θ)	*d*-spacing (Å)
4.78	18.472	19.78	4.485
7.92	11.147	21.48	4.134
9.25	9.558	22.36	3.973
16.89	5.246	23.71	3.749
19.12	4.638	28.74	3.103

FIGURE 1.28 XRPD pattern of atorvastatin calcium, Pfizer Form-X, scanned and digitized from the pattern disclosed in US patent 6,605,729.

constituencies. The method used diclofenac sodium as the peak-tracer. The HPTLC method was validated in terms of detection and quantitation limits, linearity, recovery, and repeatability. The method was found to be linear over the range of 101–353.5 ng/zone with an associated correlation coefficient of 0.9969. The limits of detection and quantitation were 30.3 and 101 ng/zone, respectively. The recovery and relative standard deviation obtained from interday analysis were 97.5–103.0% and 1.7–3.4%, respectively [30].

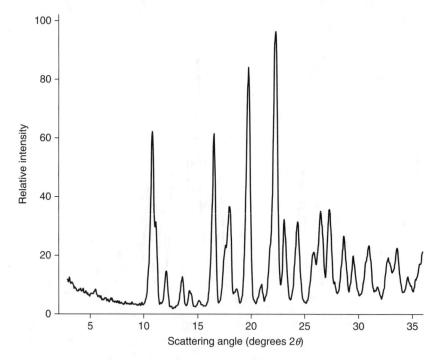

Scattering angles and *d*-spacings of the 10 most intense peaks:

Scattering angle (degrees 2θ)	*d*-spacing (Å)	Scattering angle (degrees 2θ)	*d*-spacing (Å)
10.81	8.179	23.04	3.856
16.52	5.360	24.34	3.654
17.95	4.939	26.47	3.364
19.74	4.494	27.28	3.267
22.30	3.983	28.64	3.115

FIGURE 1.29   XRPD pattern of atorvastatin calcium, Pfizer Form-XI, scanned and digitized from the pattern disclosed in US patent 6,605,729.

5. STABILITY

5.1. Solid-state stability

Crystalline atorvastatin calcium was found to be stable at 80 °C for 8 weeks, and at 40 °C, 75% RH. After 3 months, no significant change in the assay value (initial assay value = 99.3%; final assay value = 104.2%) was observed [31].

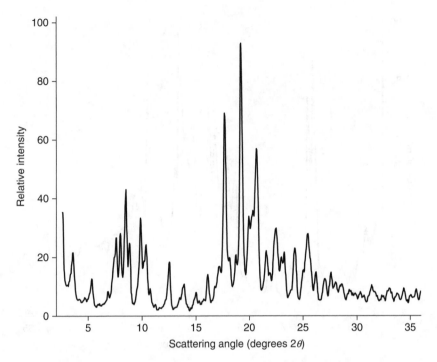

Scattering angles and *d*-spacings of the 10 most intense peaks:

Scattering angle (degrees 2*θ*)	*d*-spacing (Å)	Scattering angle (degrees 2*θ*)	*d*-spacing (Å)
3.56	24.833	17.66	5.018
7.61	11.612	19.17	4.627
8.52	10.372	20.64	4.299
9.87	8.955	22.53	3.943
12.51	7.071	25.45	3.497

FIGURE 1.30   XRPD pattern of atorvastatin calcium, Pfizer Form-XII, scanned and digitized from the pattern disclosed in US patent 6,605,729.

6. METABOLISM AND PHARMACOKINETICS

6.1. Pharmacokinetic profile

The reported pharmacokinetic properties of atorvastatin calcium are summarized in Table 1.8 [3, 32, 33].

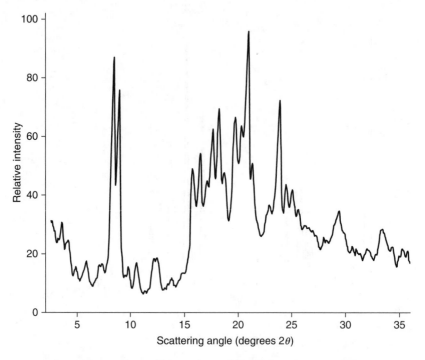

Scattering angles and *d*-spacings of the 10 most intense peaks:

Scattering angle (degrees 2θ)	*d*-spacing (Å)	Scattering angle (degrees 2θ)	*d*-spacing (Å)
8.52	10.367	18.21	4.868
9.03	9.790	19.75	4.492
15.72	5.631	21.01	4.226
16.48	5.375	23.93	3.715
17.64	5.023	29.43	3.032

FIGURE 1.31   XRPD pattern of atorvastatin calcium, Pfizer Form-XIII, scanned and digitized from the pattern disclosed in US patent 6,605,729.

6.2. Absorption

Atorvastatin is readily absorbed after the oral administration. Multiple daily dosages in the form of 2.5–80 mg capsules produce a maximum steady state concentration (C_{max}) of 1.95–252 μg/ml within 1–2 h. The AUC increases in proportion to the dose of atorvastatin, but the increase in C_{max} is greater than for the proportional dose. The low systemic availability is attributed to the presystemic clearance in gastrointestinal mucosa and/or hepatic first pass metabolism. Food significantly

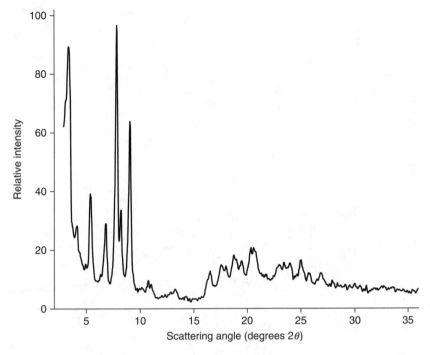

Scattering angles and d-spacings of the 10 most intense peaks:

Scattering angle (degrees 2θ)	d-spacing (Å)	Scattering angle (degrees 2θ)	d-spacing (Å)
3.37	26.228	9.08	9.726
5.39	16.382	16.53	5.357
6.78	13.017	17.56	5.046
7.88	11.206	18.73	4.734
8.23	10.739	20.62	4.304

FIGURE 1.32 XRPD pattern of atorvastatin calcium, Pfizer Form-XIV, scanned and digitized from the pattern disclosed in US patent 6,605,729.

decreases the rate, as well as the extent of absorption; C_{max} and AUC decrease by 48% and 13%, respectively.

6.3. Distribution

Mean volume of distribution (V_d) of atorvastatin is approximately 381 l and is ≥98% protein bound to the plasma proteins.

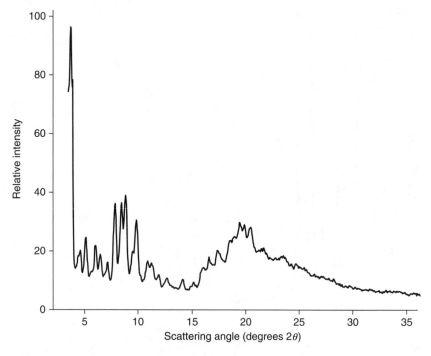

Scattering angles and *d*-spacings of the 10 most intense peaks:

Scattering angle (degrees 2θ)	*d*-spacing (Å)	Scattering angle (degrees 2θ)	*d*-spacing (Å)
3.37	26.230	8.18	10.796
4.81	18.370	8.57	10.309
5.73	15.408	9.54	9.265
6.16	14.333	19.11	4.641
7.58	11.652	20.16	4.401

FIGURE 1.33 XRPD pattern of atorvastatin calcium, Pfizer Form-XV, scanned and digitized from the pattern disclosed in US patent 6,605,729.

6.4. Metabolism

Atorvastatin is extensively metabolized to its *ortho*- and *para*-hydroxylated derivatives, and to various beta oxidation products. At least 70% of HMG-CoA reductase inhibitory activity has been attributed to the active metabolites of atorvastatin.

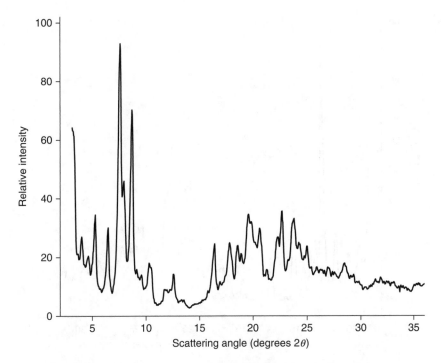

Scattering angles and *d*-spacings of the 10 most intense peaks:

Scattering angle (degrees 2θ)	*d*-spacing (Å)	Scattering angle (degrees 2θ)	*d*-spacing (Å)
5.28	16.728	16.42	5.394
6.44	13.723	17.81	4.975
7.61	11.601	19.57	4.532
8.71	10.147	22.70	3.914
10.27	8.605	23.81	3.733

FIGURE 1.34   XRPD pattern of atorvastatin calcium, Pfizer Form-XVI, scanned and digitized from the pattern disclosed in US patent 6,605,729.

6.5. Elimination

Atorvastatin and its metabolites are primarily eliminated in the bile, and the drug does not undergo any enterohepatic circulation. The mean plasma elimination half-life of atorvastatin is approximately 14 h, but the half-life of its inhibitory activity for HMG-CoA reductase is 20–30 h due to the contribution from the active metabolites. Less than 2% of a dose of atorvastatin is excreted in urine following oral administration [5, 34].

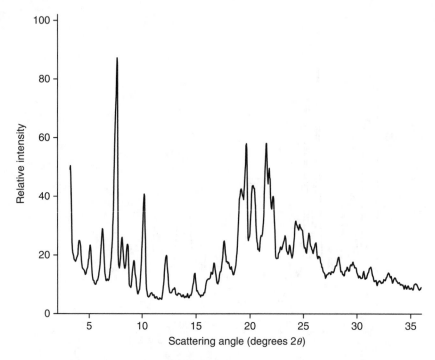

Scattering angles and *d*-spacings of the 10 most intense peaks:

Scattering angle (degrees 2θ)	d-spacing (Å)	Scattering angle (degrees 2θ)	d-spacing (Å)
5.11	17.285	14.90	5.941
6.26	14.116	17.58	5.039
7.66	11.525	19.66	4.511
10.22	8.648	20.23	4.387
12.28	7.203	21.51	4.129

FIGURE 1.35 XRPD pattern of atorvastatin calcium, Pfizer Form-XVII, scanned and digitized from the pattern disclosed in US patent 6,605,729.

6.6. Toxicity

6.6.1. Muscle-related toxicity

Myotoxic side effects, including myopathy or rhabdomyolysis, have been observed with the usage of atorvastatin calcium. Painful myalgia with a significant creatine kinase release (>2000 IU/l) is also associated with the use of atorvastatin [35].

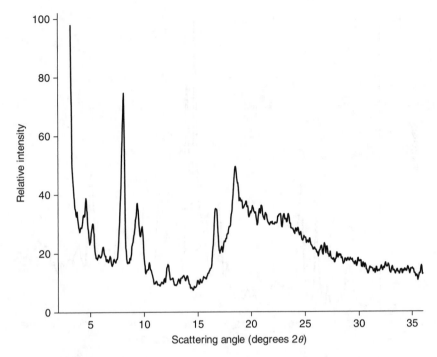

Scattering angles and d-spacings of the 10 most intense peaks:

Scattering angle (degrees 2θ)	d-spacing (Å)	Scattering angle (degrees 2θ)	d-spacing (Å)
4.63	19.073	9.80	9.018
5.26	16.781	10.48	8.437
6.18	14.294	12.20	7.249
8.12	10.879	16.72	5.299
9.36	9.437	18.55	4.779

FIGURE 1.36 XRPD pattern of atorvastatin calcium, Pfizer Form-XVIII, scanned and digitized from the pattern disclosed in US patent 6,605,729.

6.6.2. Gastrointestinal side effects

Gastrointestinal side effects, such as constipation, flatulence, dyspepsia, and abdominal pain, are common with the consumption of the drug [34].

6.6.3. Liver-related toxicity

Hepatic dysfunction due to atorvastatin administration is characterized by a raised serum aspartate (AST) or alanine (ALT) level. The overall frequency of hepatotoxicity with atorvastatin has been estimated at

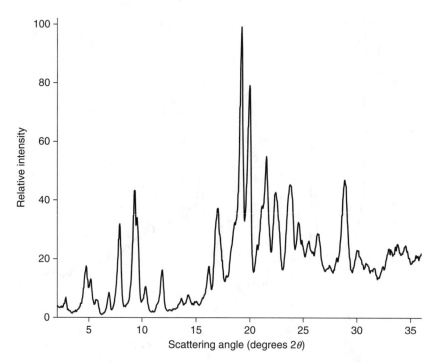

Scattering angles and *d*-spacings of the 10 most intense peaks:

Scattering angle (degrees 2θ)	*d*-spacing (Å)	Scattering angle (degrees 2θ)	*d*-spacing (Å)
4.74	18.622	19.96	4.445
7.97	11.088	21.53	4.125
9.30	9.506	22.39	3.968
16.99	5.214	23.73	3.746
19.22	4.615	28.88	3.089

FIGURE 1.37   XRPD pattern of atorvastatin calcium, Ciba Form-A, scanned and digitized from the pattern disclosed in US patent application 2004/0220255.

approximately 1%. The mechanism of hepatotoxicity seems to be linked to HMG-CoA reductase inhibition. Nausea, bloating, diarrhea, and constipation have been frequently observed with use of atorvastatin [34].

6.6.4. Other side effects
Demyelinating neuropathies have been reported in some individuals [32].

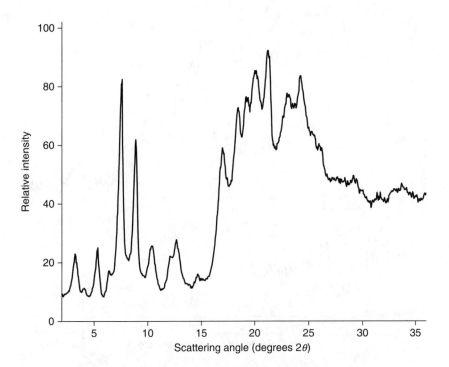

Scattering angles and *d*-spacings of the 10 most intense peaks:

Scattering angle (degrees 2θ)	*d*-spacing (Å)	Scattering angle (degrees 2θ)	*d*-spacing (Å)
3.24	27.269	12.68	6.975
5.34	16.530	16.98	5.218
7.54	11.721	20.00	4.436
8.90	9.924	21.22	4.183
10.41	8.488	24.24	3.668

FIGURE 1.38   XRPD pattern of atorvastatin calcium, Ciba Form-B, scanned and digitized from the pattern disclosed in US patent application 2004/0220255.

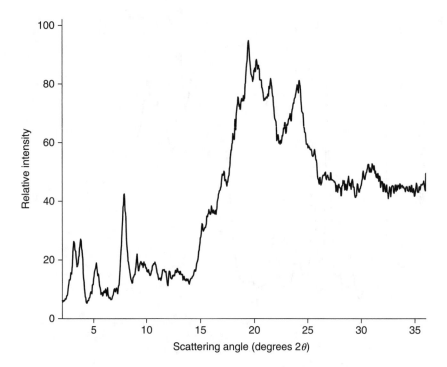

Scattering angles and *d*-spacings of the 10 most intense peaks:

Scattering angle (degrees 2θ)	d-spacing (Å)	Scattering angle (degrees 2θ)	d-spacing (Å)
3.12	28.289	19.43	4.566
3.80	23.237	20.17	4.398
5.23	16.895	21.51	4.127
7.92	11.149	24.19	3.676
9.11	9.703	30.75	2.905

FIGURE 1.39   XRPD pattern of atorvastatin calcium, Ciba Form-C, scanned and digitized from the pattern disclosed in US patent application 2004/0220255.

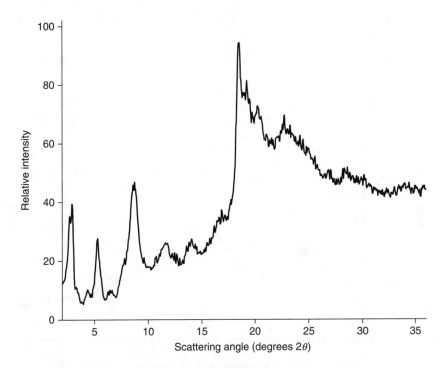

Scattering angles and *d*-spacings of the 10 most intense peaks:

Scattering angle (degrees 2θ)	*d*-spacing (Å)	Scattering angle (degrees 2θ)	*d*-spacing (Å)
2.73	32.321	14.01	6.318
2.92	30.199	18.51	4.790
5.30	16.671	19.24	4.609
8.72	10.135	20.22	4.389
11.79	7.501	22.73	3.909

FIGURE 1.40 XRPD pattern of atorvastatin calcium, Ciba Form-D, scanned and digitized from the pattern disclosed in US patent application 2004/0220255.

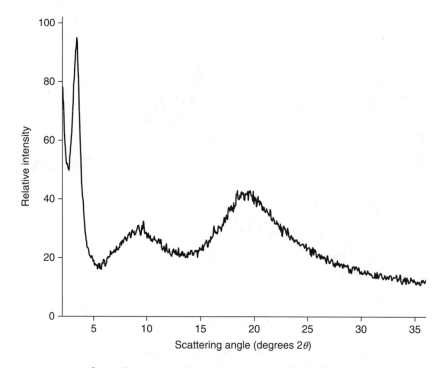

Scattering angles and *d*-spacings of the 3 most intense peaks:

Scattering angle (degrees 2θ)	*d*-spacing (Å)
3.40	25.982
9.24	9.568
19.05	4.654

FIGURE 1.41 XRPD pattern of atorvastatin calcium, Ciba Form-E, scanned and digitized from the pattern disclosed in US patent application 2004/0220255.

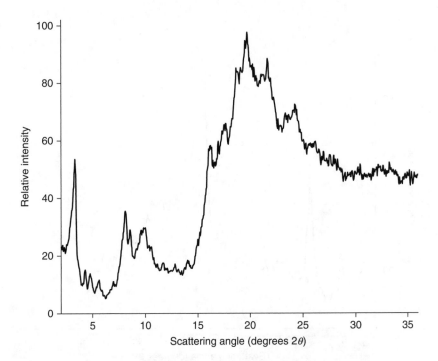

Scattering angles and *d*-spacings of the 10 most intense peaks:

Scattering angle (degrees 2θ)	*d*-spacing (Å)	Scattering angle (degrees 2θ)	*d*-spacing (Å)
3.35	26.370	17.57	5.042
8.11	10.889	18.69	4.744
8.53	10.357	19.71	4.500
9.85	8.970	21.64	4.103
16.20	5.467	24.25	3.667

FIGURE 1.42   XRPD pattern of atorvastatin calcium, Ciba Form-X, scanned and digitized from the pattern disclosed in US patent application 2004/0220255.

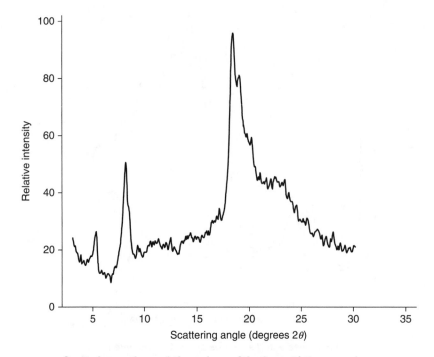

Scattering angles and *d*-spacings of the 6 most intense peaks:

Scattering angle (degrees 2θ)	*d*-spacing (Å)	Scattering angle (degrees 2θ)	*d*-spacing (Å)
5.31	16.638	19.01	4.665
8.20	10.772	20.22	4.388
18.42	4.812	23.40	3.799

FIGURE 1.43   XRPD pattern of atorvastatin calcium, Teva Form-V, scanned and digitized from the pattern disclosed in International patent application WO/2001/36384.

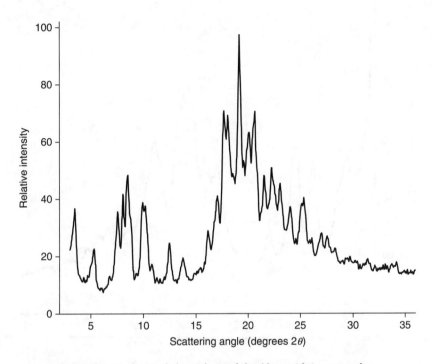

Scattering angles and *d*-spacings of the 10 most intense peaks:

Scattering angle (degrees 2θ)	*d*-spacing (Å)	Scattering angle (degrees 2θ)	*d*-spacing (Å)
3.46	25.487	17.67	5.015
5.33	16.565	19.16	4.629
8.55	10.331	20.69	4.289
9.92	8.906	22.25	3.993
12.45	7.104	25.32	3.515

FIGURE 1.44   XRPD pattern of atorvastatin calcium, Teva Form-VI, scanned and digitized from the pattern disclosed in US patent application 2003/0212279.

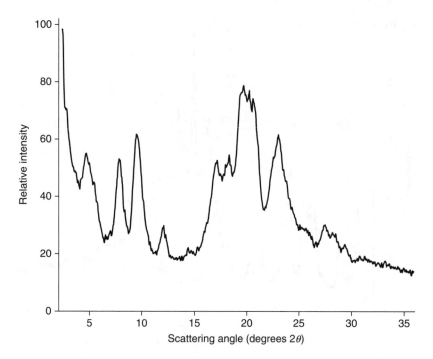

Scattering angles and *d*-spacings of the 10 most intense peaks:

Scattering angle (degrees 2θ)	*d*-spacing (Å)	Scattering angle (degrees 2θ)	*d*-spacing (Å)
4.69	18.820	18.30	4.843
7.85	11.246	19.68	4.508
9.49	9.311	20.57	4.314
12.07	7.328	23.02	3.860
17.15	5.168	27.39	3.253

FIGURE 1.45   XRPD pattern of atorvastatin calcium, Teva Form-VII, scanned and digitized from the pattern disclosed in US patent application 2003/0212279.

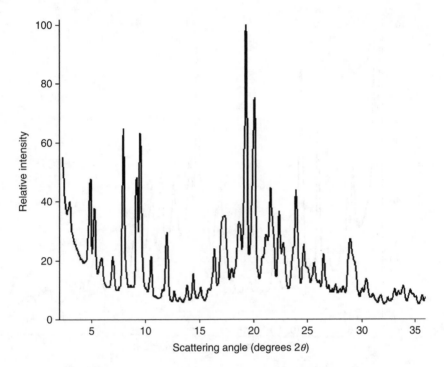

Scattering angles and *d*-spacings of the 10 most intense peaks:

Scattering angle (degrees 2θ)	*d*-spacing (Å)	Scattering angle (degrees 2θ)	*d*-spacing (Å)
4.93	17.903	11.97	7.389
5.30	16.674	19.31	4.594
7.94	11.132	20.12	4.411
9.18	9.628	21.62	4.107
9.52	9.286	23.95	3.712

FIGURE 1.46   XRPD pattern of atorvastatin calcium, Teva Form-VIII, scanned and digitized from the pattern disclosed in US patent application 2003/0212279.

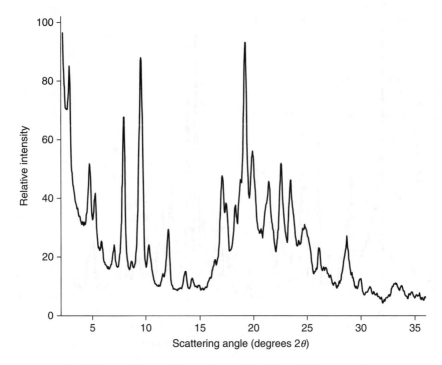

Scattering angles and *d*-spacings of the 10 most intense peaks:

Scattering angle (degrees 2θ)	*d*-spacing (Å)	Scattering angle (degrees 2θ)	*d*-spacing (Å)
4.70	18.772	19.18	4.624
7.97	11.088	19.87	4.465
9.51	9.289	21.38	4.152
12.07	7.324	22.54	3.942
17.04	5.199	23.44	3.792

FIGURE 1.47   XRPD pattern of atorvastatin calcium, Teva Form-IX, scanned and digitized from the pattern disclosed in US patent application 2003/0212279.

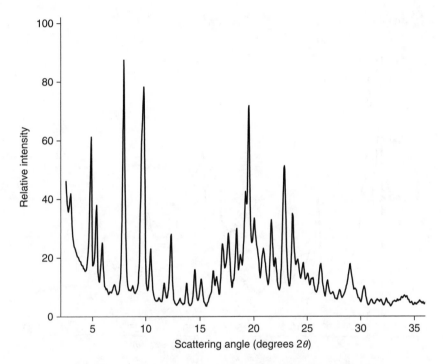

Scattering angles and d-spacings of the 10 most intense peaks:

Scattering angle (degrees 2θ)	d-spacing (Å)	Scattering angle (degrees 2θ)	d-spacing (Å)
4.87	18.149	17.60	5.035
5.38	16.411	19.56	4.535
7.98	11.068	21.62	4.107
9.83	8.992	22.83	3.892
12.37	7.151	23.61	3.765

FIGURE 1.48   XRPD pattern of atorvastatin calcium, Teva Form-X, scanned and digitized from the pattern disclosed in US patent application 2003/0212279.

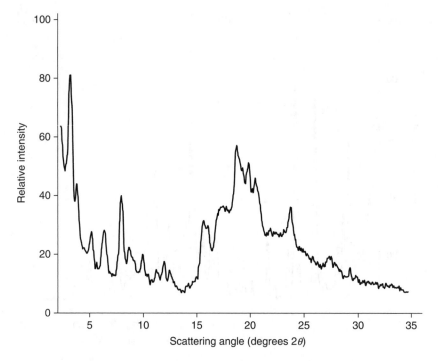

Scattering angles and *d*-spacings of the 10 most intense peaks:

Scattering angle (degrees 2θ)	*d*-spacing (Å)	Scattering angle (degrees 2θ)	*d*-spacing (Å)
3.21	27.533	9.94	8.894
3.79	23.271	15.63	5.664
5.17	17.064	18.74	4.730
6.37	13.875	19.84	4.472
7.95	11.108	23.74	3.745

FIGURE 1.49 XRPD pattern of atorvastatin calcium, Teva Form-XI, scanned and digitized from the pattern disclosed in US patent application 2003/0212279.

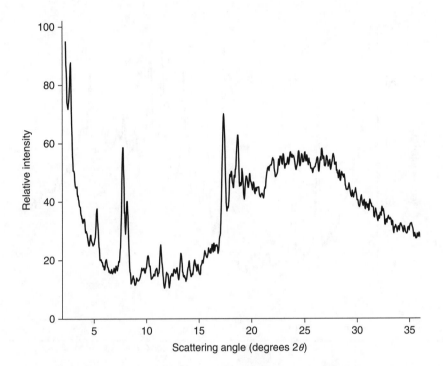

Scattering angles and *d*-spacings of the 10 most intense peaks:

Scattering angle (degrees 2θ)	*d*-spacing (Å)	Scattering angle (degrees 2θ)	*d*-spacing (Å)
2.80	31.491	11.33	7.806
5.26	16.771	13.27	6.665
7.74	11.410	17.35	5.106
8.15	10.843	18.69	4.743
10.16	8.702	22.01	4.035

FIGURE 1.50   XRPD pattern of atorvastatin calcium, Teva Form-XII, scanned and digitized from the pattern disclosed in US patent application 2003/0212279.

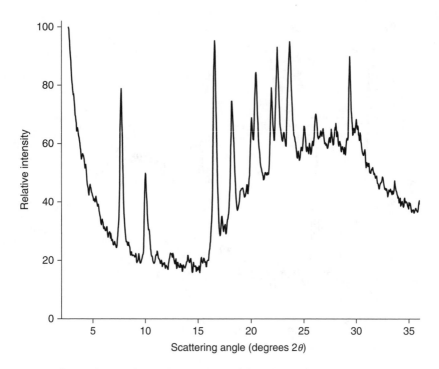

Scattering angles and *d*-spacings of the 10 most intense peaks:

Scattering angle (degrees 2θ)	*d*-spacing (Å)	Scattering angle (degrees 2θ)	*d*-spacing (Å)
7.67	11.524	21.87	4.061
9.99	8.849	22.43	3.960
16.54	5.355	23.66	3.758
18.18	4.876	26.12	3.408
20.43	4.344	29.33	3.043

FIGURE 1.51   XRPD pattern of atorvastatin calcium, Teva Form-XIV, scanned and digitized from the pattern disclosed in US patent application 2003/0212279.

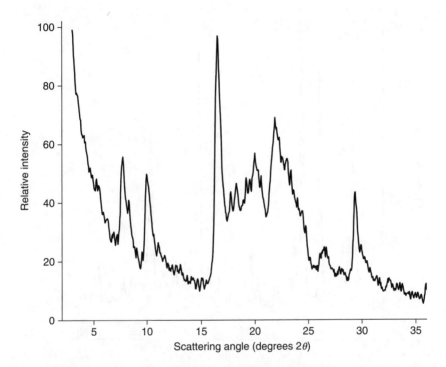

Scattering angles and *d*-spacings of the 10 most intense peaks:

Scattering angle (degrees 2θ)	d-spacing (Å)	Scattering angle (degrees 2θ)	d-spacing (Å)
7.77	11.367	19.98	4.441
9.98	8.852	21.87	4.060
16.55	5.353	26.49	3.362
17.71	5.003	29.36	3.040
18.25	4.857	39.11	2.301

FIGURE 1.52 XRPD pattern of atorvastatin calcium, Teva Form-XVI, scanned and digitized from the pattern disclosed in US patent application 2003/0212279.

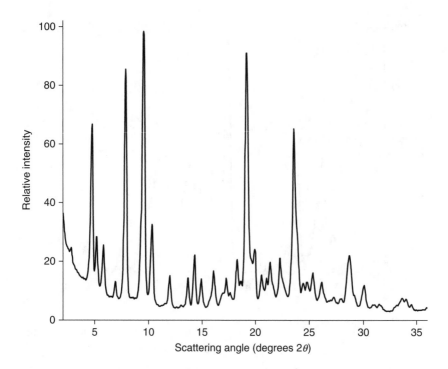

Scattering angles and *d*-spacings of the 10 most intense peaks:

Scattering angle (degrees 2θ)	d-spacing (Å)	Scattering angle (degrees 2θ)	d-spacing (Å)
4.75	18.590	11.96	7.393
5.81	15.206	14.30	6.189
7.90	11.188	19.17	4.626
9.51	9.292	23.57	3.772
10.32	8.566	28.64	3.114

FIGURE 1.53 XRPD pattern of atorvastatin calcium, Teva Form-XVII, scanned and digitized from the pattern disclosed in US patent application 2003/0212279.

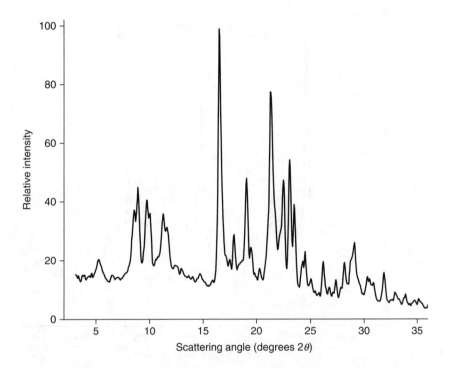

Scattering angles and *d*-spacings of the 10 most intense peaks:

Scattering angle (degrees 2θ)	*d*-spacing (Å)	Scattering angle (degrees 2θ)	*d*-spacing (Å)
8.89	9.938	21.30	4.168
9.72	9.094	22.48	3.952
11.25	7.862	23.05	3.855
16.53	5.359	23.46	3.788
19.07	4.649	29.10	3.067

FIGURE 1.54 XRPD pattern of atorvastatin calcium, Biocon Form-V, scanned and digitized from the pattern disclosed in International patent application WO/2002/057229.

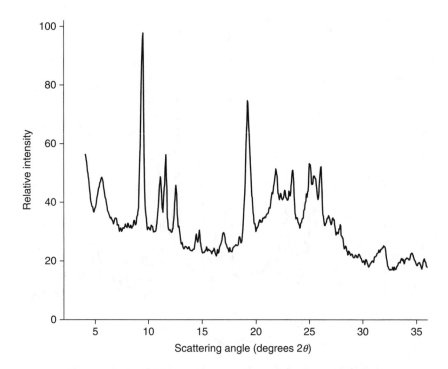

Scattering angles and *d*-spacings of the 10 most intense peaks:

Scattering angle (degrees 2θ)	*d*-spacing (Å)	Scattering angle (degrees 2θ)	*d*-spacing (Å)
5.62	15.710	19.22	4.614
9.45	9.348	21.86	4.063
11.14	7.934	23.42	3.795
11.64	7.598	24.99	3.560
12.59	7.028	26.06	3.416

FIGURE 1.55   XRPD pattern of atorvastatin calcium, Ivax Form-Fa, scanned and digitized from the pattern disclosed in International patent application WO/2003/050085.

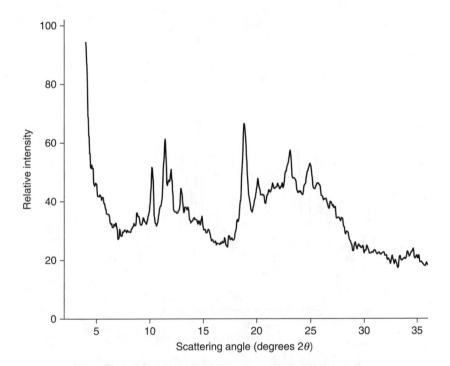

Scattering angles and *d*-spacings of the eight most intense peaks:

Scattering angle (degrees 2θ)	d-spacing (Å)	Scattering angle (degrees 2θ)	d-spacing (Å)
10.23	8.642	18.82	4.711
11.41	7.746	20.09	4.416
11.99	7.377	23.08	3.851
12.89	6.863	24.96	3.564

FIGURE 1.56   XRPD pattern of atorvastatin calcium, Ivax Form-Je, scanned and digitized from the pattern disclosed in International patent application WO/2003/050085.

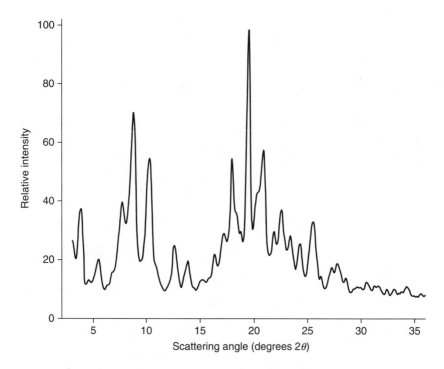

Scattering angles and *d*-spacings of the 10 most intense peaks:

Scattering angle (degrees 2θ)	d-spacing (Å)	Scattering angle (degrees 2θ)	d-spacing (Å)
3.82	23.106	17.94	4.939
5.49	16.074	19.56	4.534
8.81	10.027	20.88	4.251
10.29	8.587	22.58	3.935
12.62	7.007	25.54	3.485

FIGURE 1.57   XRPD pattern of atorvastatin calcium, Morepen Form-VI, scanned and digitized from the pattern disclosed in International patent application WO/2004/022053.

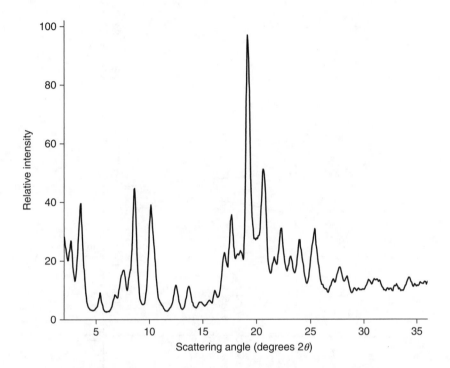

Scattering angles and *d*-spacings of the 10 most intense peaks:

Scattering angle (degrees 2θ)	*d*-spacing (Å)	Scattering angle (degrees 2θ)	*d*-spacing (Å)
2.62	33.634	19.16	4.627
3.55	24.879	20.62	4.305
8.56	10.318	22.29	3.986
10.15	8.711	24.00	3.704
17.71	5.003	25.37	3.508

FIGURE 1.58   XRPD pattern of atorvastatin calcium, Blatter Form-F, scanned and digitized from the pattern disclosed in US patent application 2004/0106670.

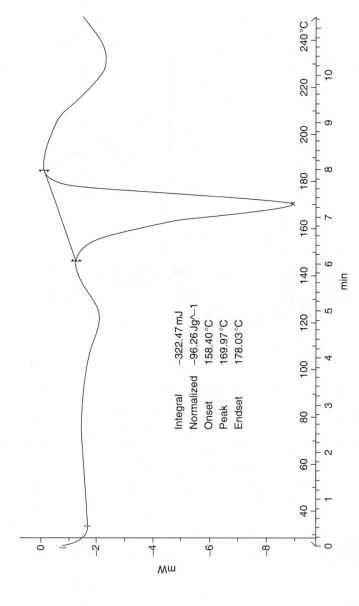

Integral	−322.47 mJ
Normalized	−96.26 Jg^−1
Onset	158.40 °C
Peak	169.97 °C
Endset	178.03 °C

FIGURE 1.59 DSC thermogram of atorvastatin calcium.

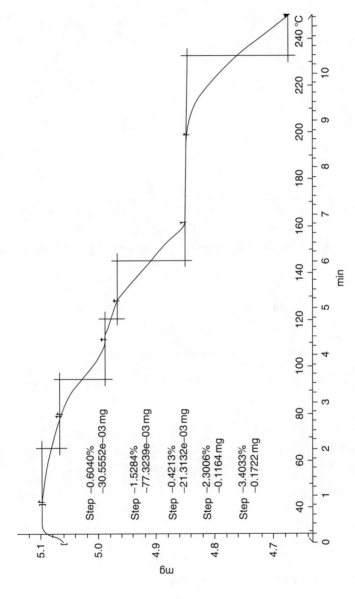

FIGURE 1.60 TG thermogram of atorvastatin calcium.

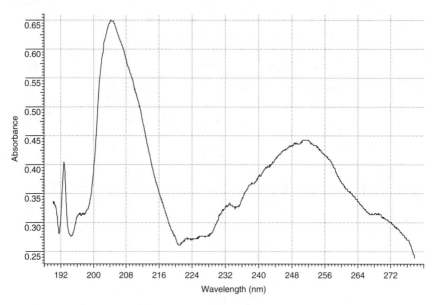

FIGURE 1.61   Ultraviolet absorption spectrum of atorvastatin calcium dissolved in methanol.

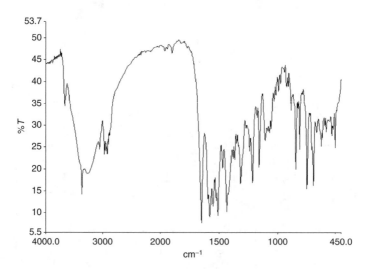

FIGURE 1.62   FTIR spectra of atorvastatin calcium, Form-I.

TABLE 1.4 Assignment for the characteristic infrared absorption bands of atorvastatin calcium

Peak energy (cm^{-1})	Assignment
3365.5	–OH (hydroxyl)
3228.3	–NH
1651.7	–C=O (amide carbonyl)
2903.5	–CH (aromatic)
1595.1	–C=C (aromatic)

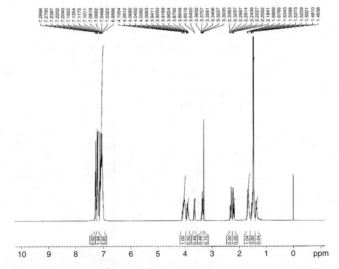

FIGURE 1.63 ^{1}H NMR spectrum of atorvastatin calcium.

TABLE 1.5 Assignment for the resonance bands in the ^{1}H NMR spectrum of atorvastatin calcium (solvent: CH_3OH-d_3)

Chemical shift (ppm)	Number of hydrogen atoms	Position of hydrogen atoms
1.47	6 H	7, 8 (2 × CH_3)
1.50–1.56	2 H	4 (CH_2)
1.63–1.69	2 H	6 (CH_2)
2.18–2.33	2 H	2 (CH_2)
3.35	1 H	5 (CH)
3.65	1 H	6 (CH)
3.86–3.94	1 H	3 (CH)
4.01–4.11	2 H	7 (N–CH_2)
7.03–7.29	14 H	Aromatic protons

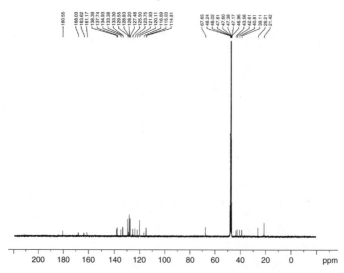

FIGURE 1.64 ^{13}C NMR spectra of atorvastatin calcium.

TABLE 1.6   Assignment for the resonance bands in ^{13}C NMR spectrum of atorvastatin calcium (solvent: $CH_3OH\text{-}d_3$)

Chemical shift (ppm)	Assignment	Chemical shift (ppm)	Assignment
21.42	C7, C8 (isopropyl)	127.48	C2 (pyrrole)
26.21	C6 (isopropyl)	127.48	C1 (4-flurophenyl)
39.11	C4 (heptanoate)	128.20	C3, C5 (phenyl carbamoyl)
40.81	C6 (heptanoate)	128.83	C5 (pyrrole)
42.61	C7 (heptanoate)	129.55	C3, C5 (4-flurophenyl)
43.56	C2 (heptanoate)	137.74	C1 (3-pyrrole phenyl carbon)
67.65	C3,C5	138.39	C1 (phenylcarbamoyl)
114.81	C4 (pyrrole)	161.17	C4 (4-flurophenyl)
120.11	C3, C5 (4-flurophenyl)	168.03	Amide carbonyl carbon
121.93	C6, C2 (phenyl carbamoyl)	180.55	Heptanoate carbonyl carbon
125.50	C3 (pyrrole)		

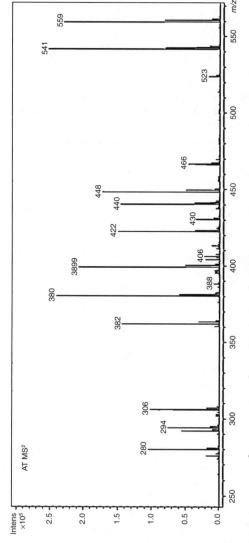

FIGURE 1.65 Mass spectrum of atorvastatin calcium (Reproduced with permission from Wiley; Ref. [10]).

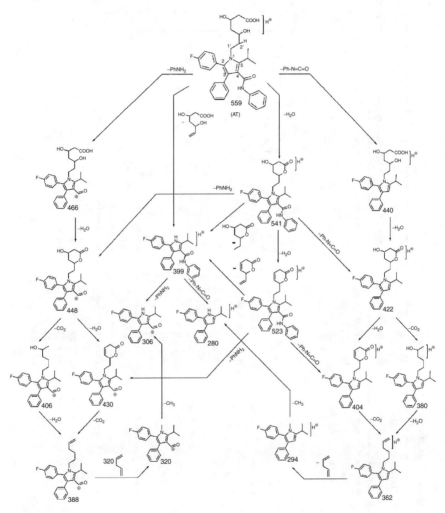

FIGURE 1.66 Mass fragmentation pattern of atorvastatin calcium (Reproduced with permission from Wiley; Ref. [10]).

TABLE 1.7 HPLC methods for the analysis of atorvastatin calcium

Substance	Matrix sample preparation	Stationary phase, analytical column	Mobile phase	Detection	Validation data	Comment	References
AM and AT	Bulk drug and tablets	Perfectsil® Target ODS-3 (5 μm, 250 mm × 4.6 mm i.d.)	Acetonitrile 0.025 M NaH$_2$PO$_4$ buffer (pH 4.5) (55:45, v/v)	UV detection at 237 nm	Linearity: 2–30 μg/ml (r = 0.9994) for AT and 1–20 μg/ml (r = 0.9993) for AM LOD = 0.65 μg/ml (AT) and 0.35 μg/ml (AM) LOQ = 2 μg/ml (AT); 1 μg/ml (AM)	–	[13]
AM and AT	Combination drug product	Lichrospher® 100 C18 (5 mm, 250 mm × 4.0 mm i.d.) column	Acetonitrile and 50 mM potassium dihydrogen phosphate buffer (60:40, v/v) (pH ±3)	UV detection at 254 nm	LOD: 0.4 for AT and 0.6 mg/ml for AM LOQ: 1.0 mg/ml (for both)	–	[14]
AT, FB, and their impurities	Tablets	Acquity UPLC™ BEH C18 column (1.7 μm, 2.1 mm × 100 mm)	Gradient elution of acetonitrile; ammonium acetate (pH 4.7; 0.01 M)	–	Minimum detection limit was at ng/l levels	–	[15]
AT	Aqueous environment	Genesis C column (2.1 mm × 50 mm, 3 μm)	Acetonitrile (gradient from 60% to 100%) and water containing 2 mM methyl amine with 0.1% acetic acid	Electrospray ionization (ESI) tandem mass spectrometry	–	–	[15]
AT	Bulk drug, tablets, and human plasma	RP-Supelcosil C18 (5 μm, 150 mm × 4.6 mm) column	acetonitrile–methanol–water (45:45:10, v/v/v), and a flow rate of 1.0 ml/min	UV detection at 240 nm	Linearity: 0.5–86.0 μg/ml	–	[16]

Analytes	Sample	Column	Mobile phase	Detection	Linearity/LOD	Comments	References
AT, LV, PV, SV, RV	Pharmaceutical formulations, blood, plasma	Intertsil ODS 3V column (4.6 mm × 250 mm, 5 μm)	Ternary gradient elution, 0.01 M ammonium acetate (pH 5.0), acetonitrile and methanol	UV detection at 237 nm	–	Theophylline as an internal standard	[17]
AT, EZ	Pharmaceutical formulations	–	0.01 M ammonium acetate buffer (pH 3.0)–acetonitrile (50:50, v/v)	UV detection at 254 nm	Linearity: 4–400 μg/ml (AT) and 5–500 μg/ml (EZ) LOD:1.25 μg/ml (AT) and 1.48 μg/ml (EZ)	–	[18]
RT, NB, and AT	Aqueous samples, sewage samples	Microbore LC column (100 mm × 1.0 mm i.d.)	Acetonitrile and 10 mM aqueous ammonium acetate	Quantified in the multiple reaction monitoring mode with external standards	LOD: 1–3 pg/ml	–	[19]
AT and its seven related impurities	–	A Zorbax Eclipse XDB C18 Rapid Resolution HT (4.6 mm × 50 mm, 1.8 μm particle size) column	Phosphate buffer (pH 3.5) and a mixture of 10% (v/v) tetrahydrofuran in acetonitrile organic modifier	–	–	Reversed-phase liquid chromatography (RP-HPLC)	[20]
AT and AS	Combined capsule dosage form	Phenomenex Gemini C-18, 5 μm column having 250 mm × 4.6 mm i.d.	0.02 M potassium dihydrogen phosphate–methanol (20:80) adjusted to pH 4 (o-phosphoric acid)	UV detection at 240 nm	Linearity: 0.5–4 μg/ml (AT) and 5–25 μg/ml (AS)	–	[21]
AT and NA	Tablets	Phenomenex Luna C-18, 5 μm column having 250 mm × 4.6 mm i.d.	–	–	Linearity: 2–24 μg/ml (AT) and 60–250 μg/ml (NA)	–	[22]

(continued)

TABLE 1.7 (continued)

Substance	Matrix sample preparation	Stationary phase, analytical column	Mobile phase	Detection	Validation data	Comment	References
AT and FB	Tablets	C18 column	Methanol–acetate buffer, pH 3.7 (82:18, v/v)	UV detection at 248 nm	Linearity: 1–5 μg/ml (AT) and 16–80 μg/ml (FB)	–	[23]
AT and DS (as IS)	Human serum	C18 analytical column	Sodium phosphate buffer (0.05 M, pH 4.0) and methanol (33:67, v/v)	UV detection at 247 nm	LOD: 1 ng/ml (AT) and 4 ng/ml (DS)	–	[24]
AT and HAT	Human serum	–	–	ESI triple–quadrupole mass spectrometer, multiple reaction monitoring (MRM) was used for MS–MS	LOQ: 0.02 ng/ml (AT) and 0.07 ng/ml (HAT) Linearity: 0.10–40.00 ng/ml (AT and HAT) Relative standard deviation (R.S.D.) and percentage deviation less than 8% (AT and HAT)	–	[25]
AT, o-HAT, p-HAT	Human serum	–	Acetonitrile, water, and formic acid	Tandem mass spectrometry operated in the electrospray positive ion mode	Linearity: 0.2–30 ng/ml (AT and p-HAT), and 0.5–30 ng/ml (p-HAT) LOD: 0.06 ng/ml (AT and p-HAT) and 0.15 ng/ml (p-HAT)	Rapid liquid chromatography-tandem mass spectrometry	[26]
AT, o-HAT, p-HAT	Human serum	Reversed-phase C18 column	Isocratic mobile phase	Mass spectrometry	Linearity: 0.1–20 ng/ml LOQ: 100 pg/ml with an R.S.D. of less than 8%	Rosuvastatin as an internal standard	[27]

Analytes	Matrix	Column	Mobile phase	Detection	LOQ / Validation	Method / Internal standard	Ref.
AT, o-HAT, p-HAT (acid and lactone form)	Human serum	–	–	Positive ion electrospray tandem mass spectrometry	LOQ: 0.5 ng/ml	HPLC–ESI-MS; deuterium labeled analog was used as internal standard	[28]
AT, o-HAT, p-HAT	Human plasma	YMC J'Sphere H80 (C-18) (150 mm × 2 mm, 4 μm particle size)	Acetonitrile–0.1% acetic acid (70:30, v/v)	MS-MS	LOQ: 0.250 ng/ml. Interassay precision: based on the percent relative deviation for replicate quality controls for AT, o-HAT, and p-HAT was $\leq$7.19%, 8.28%, and 12.7%, respectively. Interassay accuracy: for AT, o-HAT, and p-HAT was 10.6%, 5.86%, and 15.8%, respectively	Deuterium labeled analog as an internal standard	[29]

Abbreviations: AM, amlodipine; AT, atorvastatin; DS, diclofenac sodium; EZ, ezetimibe; FB, fenofibrate; HAT, hydroxy atorvastatin; IS, internal standard; o-HAT, ortho-hydroxy atorvastatin; p-HAT, para-hydroxy atorvastatin; HPLC–ESI-MS, high-performance liquid chromatography with electrospray tandem mass spectrometry; LV, lovastatin; NA, nicotinic acid; NB, novobiocin; PV, pravastatin; RV, rosuastatin; SV, simvastatin; RT, roxethromycin; UPLC, ultra performance liquid chromatography.

TABLE 1.8   Pharmacokinetic properties of atorvastatin calcium

(1) *Absorption parameters*	
Fraction absorbed %	30
T_{max} (h)	2–3
C_{max} (ng/ml)	27–66
Bioavailability (%)	12
Effect of food (% change in AUC)	13↓
(2) *Distribution*	
Fraction bound (%)	80–90
Lipophilicity	4.06
(3) *Metabolism*	
Hepatic clearance CYP3A4	>70
Systemic metabolites	Active
Clearance (l/h/kg)	0.25
(4) *Excretion*	
$t_{1/2}$ (h)	15–30
Urinary excretion (%)	<2
Fecal excretion	Major route

ACKNOWLEDGMENTS

Lokesh Kumar would like to acknowledge the Department of Science and Technology, Government of India, for providing Senior Research Fellowship (SR/SO/HS-23/2006). Chhuttan Lal Meena would like to acknowledge Rajiv Gandhi National Fellowship [No. F14-2/2006(SA-III)] provided by the University Grants Commission, Government of India. Vibha Puri would also like to acknowledge the financial assistance provided by Indian Council of Medical Research, Government of India, for providing Senior Research Fellowship (45/12/2006/BMS/PHA).

REFERENCES

[1] A.C. Moffat, M.D. Osselton, B. Widdop (Eds.), Clarke's Analysis of Drugs and Poisons in Pharmaceuticals Body Fluids and Postmortem Material, Pharmaceutical Press, London, UK, 2004, pp. 654–655.
[2] Indian Pharmacopoeia, Indian Pharmacopoeial Commission, Ghaziabad, 2007, pp. 749–751.
[3] U. Christians, W. Jacobsen, L.C. Floren, Metabolism and drug interactions of 3-hydroxy-3-methylglutaryl coenzyme A reductase inhibitors in transplant patients: are the statins mechanistically similar? Pharmacol. Ther. 80 (1998) 1–34.
[4] G. Weitz-Schmidt, Statins as anti-inflammatory agents, Trends Pharmacol. Sci. 23 (2002) 482–487.

[5] FDA, Lipitor®—Patient information, 2005.

[6] Physicians' Desk Reference, 61st ed., MDR Thomson, New Jersey, USA, 2007, pp. 2483–2487.

[7] Atorvastatin impurities, http://www.molcan.com/atorvastatin.htm, accessed on: Sep. 16, 2008.

[8] Atorvastatin impurities/metabolites, http://www.alfaomegapharma.com/Atorvastatin-Calcium.php, accessed on: Oct. 7, 2008.

[9] Atorvastatin calcium, http://www.drugbank.ca/drugs/DB01076, accessed on: Sep. 16, 2008.

[10] R.P. Shah, V. Kumar, S. Singh, Liquid chromatography/mass spectrometric studies on atorvastatin and its stress degradation products, Rapid Commun. Mass Spectrom. 22 (2008) 613–622.

[11] N. Erk, Extractive spectrophotometric determination of atorvastatin in bulk and pharmaceutical formulations, Anal. Lett. 36 (2003) 2699–2711.

[12] R. Sahu, V.B. Patel, Simultaneous spectrophotometric determination of amlodipine besylate and atorvastatin calcium in binary mixture, Indian J. Pharm. Sci. 69 (2007) 110–111.

[13] A. Mohammadi, N. Rezanour, M.A. Dogaheh, F.G. Bidkorbeh, M. Hashem, R.B. Walker, A stability-indicating high performance liquid chromatographic (HPLC) assay for the simultaneous determination of atorvastatin and amlodipine in commercial tablets, J. Chromatogr. B 846 (2007) 215–221.

[14] G.B. Chaudhari, N. Patel, P.B. Shah, Stability indicating RP-HPLC method for simultaneous determination of atorvastatin and amlodipine from their combination drug products, Chem. Pharm. Bull. 55 (2007) 241–246.

[15] A.A. Kadav, D.N. Vora, Stability indicating UPLC method for simultaneous determination of atorvastatin, fenofibrate and their degradation products in tablets, J. Pharm. Biomed. Anal. 48 (2008) 120–126.

[16] T.G. Altuntas, N. Erk, Liquid chromatographic determination of atorvastatin in bulk drug, tablets, and human plasma, J. Liq. Chromatogr. Relat. Technol. 27 (2004) 83–93.

[17] Md.K. Pasha, S. Muzeeb, S.J.S. Basha, D. Shashikumar, R. Mullangi, N.R. Srinivas, Analysis of five HMG-CoA reductase inhibitors—atorvastatin, lovastatin, pravastatin, rosuvastatin and simvastatin: pharmacological, pharmacokinetic and analytical overview and development of a new method for use in pharmaceutical formulations analysis and in vitro metabolism studies, Biomed. Chromatogr. 20 (2006) 282–293.

[18] U. Seshachalam, C. Kothapally, HPLC analysis for simultaneous determination of atorvastatin and ezetimibe in pharmaceutical formulations, J. Liq. Chromatogr. Relat. Technol. 31 (2008) 714–721.

[19] X.S. Miao, C.D. Metcalfe, Determination of cholesterol-lowering statin drugs in aqueous samples using liquid chromatography-electrospray ionization tandem mass spectrometry, J. Chromatogr. A 998 (2003) 133–141.

[20] R. Petkovska, C. Cornett, A. Dimitrovska, Development and validation of rapid resolution RP-HPLC method for simultaneous determination of atorvastatin and related compounds by use of chemometrics, Anal. Lett. 41 (2008) 992–1009.

[21] D.A. Shah, K.K. Bhatt, R.S. Mehta, M.B. Shankar, S.L. Baldania, T.R. Gandhi, Development and validation of a RP-HPLC method for determination of atorvastatin calcium and aspirin in a capsule dosage form, Indian J. Pharm. Sci. 69 (2007) 546–547.

[22] D.A. Shah, K.K. Bhatt, R.S. Mehta, M.B. Shankar, S.L. Baldania, RP-HPLC method for the determination of atorvastatin calcium and nicotinic acid in combined tablet dosage form, Indian J. Pharm. Sci. 69 (2007) 700–703.

[23] N. Jain, R. Raghuwanshi, D. Jain, Development and validation of RP-HPLC method for simultaneous estimation of atorvastatin calcium and fenofibrate in tablet dosage forms, Indian J. Pharm. Sci. 70 (2008) 263–265.

[24] G. Bahrami, B. Mohammadi, S. Mirzaeei, A. Kiani, Determination of atorvastatin in human serum by reversed-phase high-performance liquid chromatography with UV detection, J. Chromatogr. B 826 (2005) 41–45.

[25] V. Bořek-Dohalský, J. Huclová, B. Barrett, B. Němec, I. Ulč, I. Jelínek, Validated HPLC–MS–MS method for simultaneous determination of atorvastatin and 2-hydroxyatorvastatin in human plasma-pharmacokinetic study, Anal. Bioanal. Chem. 386 (2006) 275–285.

[26] M. Hermann, H. Christensen, J.L.E. Reubsaet, Determination of atorvastatin and metabolites in human plasma with solid-phase extraction followed by LC–tandem MS, Anal. Bioanal. Chem. 382 (2005) 1242–1249.

[27] R.V.S. Nirogi, V.N. Kandikere, M. Shukla, K. Mudigonda, S. Maurya, R. Boosi, Simultaneous quantification of atorvastatin and active metabolites in human plasma by liquid chromatography–tandem mass spectrometry using rosuvastatin as internal standard, Biomed. Chromatogr. 20 (2006) 924–936.

[28] J. Mohammed, O. Zheng, C.C. Bang, T. Deborah, Quantitation of the acid and lactone forms of atorvastatin and its biotransformation products in human serum by high-performance liquid chromatography with electrospray tandem mass spectrometry, Rapid Commun. Mass Spectrom. 13 (1999) 1003–1015.

[29] W.W. Bullen, R.A. Miller, R.N. Hayes, Development and validation of a high-performance liquid chromatography tandem mass spectrometry assay for atorvastatin, *ortho*-hydroxy atorvastatin, and *para*-hydroxy atorvastatin in human, dog, and rat plasma, J. Am. Soc. Mass Spectrom. 10 (1999) 55–66.

[30] A. Jamshidi, A.R. Nateghi, HPTLC determination of atorvastatin in PLASMA, Chromatographia 65 (2007) 763–766.

[31] http://www.fda.gov/ohrms/DOCKETS/dockets/05p0452/05p-0452-cp00001-01-vol1.pdf, accesed on: Sep. 16, 2008.

[32] A.S. Wierzbicki, R. Poston, A. Ferro, The lipid and non-lipid effects of statins, Pharm. Ther. 99 (2003) 95–112.

[33] A. Corsini, S. Bellosta, R. Baetta, R. Fumagalli, R. Paoletti, F. Bernini, New insights into the pharmacodynamic and pharmacokinetic properties of statins, Pharmacol. Ther. 84 (1999) 413–428.

[34] A.P. Lea, D. McTavish, Atorvastatin: review of its pharmacology and therapeutic potential in the management of hyperlipidaemias, Drugs 53 (1997) 829–847.

[35] Y. Shitara, Y. Sugiyama, Pharmacokinetic and pharmacodynamic alterations of 3-hydroxy-3-methylglutaryl coenzyme A (HMG-CoA) reductase inhibitors: drug–drug interactions and interindividual differences in transporter and metabolic enzyme functions, Pharm. Ther. 112 (2006) 71–105.

CHAPTER 2

Clopidogrel Bisulfate

Maria L.A.D. Lestari,* **Suciati,***
Gunawan Indrayanto,* and **Harry G. Brittain**[†]

Contents

* Faculty of Pharmacy, Airlangga University, Dharmawangsa Dalam, Surabaya, Indonesia
† Center for Pharmaceutical Physics, Milford, New Jersey, USA

Profiles of Drug Substances, Excipients, and Related Methodology, Volume 35
ISSN 1871-5125, DOI: 10.1016/S1871-5125(10)35002-3

1. DESCRIPTION

1.1. Nomenclature [1]

Clopidogrel bisulfate is the hydrogen sulfate salt of clopidogrel, for which the free compound is known by the following systematic names:
Systematic chemical name

- (αS)-α-(2-Chlorophenyl)-6,7-dihydrothieno[3,2-*c*]pyridine-5(4*H*)-acetic acid methyl ester

Alternate systematic chemical names

- Methyl-(+)-(*S*)-α-(*o*-chlorophenyl)-6,7-dihydrothieno[3,2-*c*]pyridine-5(4*H*)-acetate
- (+)-Methyl-α-5-[4,5,6,7-tetrahydro[3,2-*c*]thienopyridyl]-(2-chlorophenyl)acetate

1.2. Formulae [2]

1.2.1. Empirical formula, molecular weight, CAS number

Clopidogrel	$C_{16}H_{16}ClNO_2S$	321.82	[113665-84-2]
Clopidogrel bisulfate	$C_{16}H_{18}ClNO_6S_2$	419.90	[135046-48-9]

1.2.2. Structural formula

The structural formula of clopidogrel is:

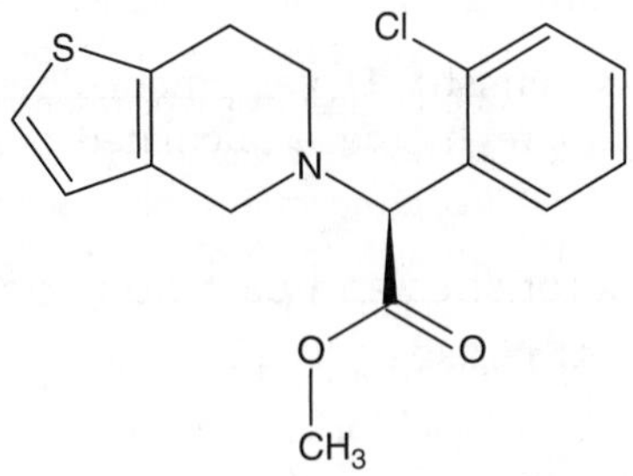

The site of interaction for salt formation is at the pyridine nitrogen, which is capable of forming salts only with extremely strong acids.

1.3. Appearance

Clopidogrel is a colorless oil under ambient conditions, while the bisulfate sale is obtained in the form of white crystals [1].

1.4. Elemental analysis [1]

	C (%)	HX (%)	Cl (%)	N (%)	O (%)	S (%)
Clopidogrel	59.71	5.01	11.02	4.35	9.94	9.96
Clopidogrel bisulfate	45.77	4.32	8.44	3.34	22.86	15.27

1.5. Uses and applications

Clopidogrel contains a center of dissymmetry, and hence is capable of being resolved into its two mirror image compounds. It has been found that only the (S)-enantiomer (which corresponds to the dextrorotatory form) has antithrombotic activity and that the (R)-enantiomer (which corresponds to the levorotatory form) does not exhibit antithrombotic activity. Moreover, in animal studies the (R)-enantiomer triggered convulsions at high doses [4, 5]. Consequently, (R)-clopidogrel bisulfate is considered to be one of the impurities in (S)-clopidogrel bisulfate bulk drug substance.

2. PHYSICAL CHARACTERISTICS

2.1. Ionization constant

Using the Advanced Chemistry Development Physical Chemistry program, the pK_a of clopidogrel has been calculated as 4.56 ± 0.20.

2.2. Solubility characteristics and partition coefficient

Clopidogrel bisulfate has been reported to be freely soluble in methanol, sparingly soluble in methylene chloride, and practically insoluble in ethyl ether and in water at neutral pH. At pH 1, the compound is freely soluble in water [3], which is one of the reasons why the hydrogen sulfate salt is the preferred form of the drug substance.

The partition coefficient in octanol/water of clopidogrel bisulfate has been reported to be about 3.9 at pH 7.4 [3].

2.3. Crystallographic properties

Although clopidogrel bisulfate has been found to crystallize in six different polymorphic forms [9, 10], only Form-I and Form-II are used in pharmaceutical products. Although the density of orthorhombic Form-II is less dense (1.462 g/cm^3) than is the monoclinic Form-I (1.505 g/cm^3), the two crystal forms appear to be fairly equivalent in their relative stability. However, Form-II is more compact and much less electrostatic than is Form-I, making it more useful for secondary processing [9].

The methods for the preparation of clopidogrel bisulfate Form-I and Form-II, as described in United States patent 6,504,030 [9], have been reproduced. X-ray powder diffraction (XRPD) patterns of these products were obtained using a Rigaku MiniFlex powder diffraction system, equipped with a horizontal goniometer operating in the $\theta/2\theta$ mode. The X-ray source was nickel-filtered Kα emission of copper (1.54184 Å). Samples were packed into the sample holder using a back-fill procedure, and were scanned over the range of 3.5–40° 2θ at a scan rate of 0.5° 2θ/min. Using a data acquisition rate of 1 point per second, the scanning parameters equate to a step size of 0.0084° 2θ. Calibration of the diffractometer system pattern was effected by scanning an aluminum plate, and using the aluminum scattering peaks having d-spacings of 2.338 and 2.024 Å to verify both the angle and scan rate. The intensity scale for all diffraction patterns was normalized so that the relative intensity of the most intense peak in the pattern equaled 100%. The patterns obtained for the two polymorphic forms of clopidogrel bisulfate are shown in Figs. 2.1 and 2.2, along with tables containing the angles and calculated d-spacings of the most intense scattering peaks.

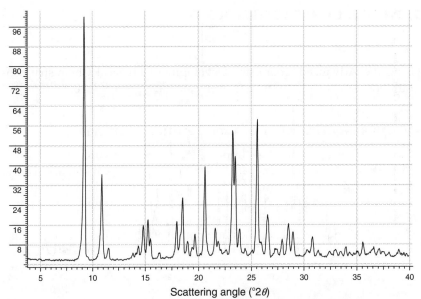

Scattering angles and *d*-spacings of the 10 most intense peaks

Scattering angle (°2θ)	d-Spacing (Å)	Scattering angle (°2θ)	d-Spacing (Å)
9.22	9.580	20.63	4.302
10.89	8.117	23.23	3.825
15.27	5.798	23.47	3.788
17.95	4.937	25.59	3.479
18.50	4.793	26.54	3.356

FIGURE 2.1   X-ray powder diffraction pattern of clopidogrel bisulfate, Form-I.

2.4. Thermal properties

2.4.1. Melting range
Polymorphic Form-I has been reported to exhibit a melting point range spanning 198–200 °C, while Form-II exhibits a melting point range spanning 176–178 °C [11].

2.4.2. Differential scanning calorimetry
Measurements of differential scanning calorimetry (DSC) were obtained on clopidogrel bisulfate Form-I and Form-II using a TA Instruments 2910 thermal analysis system. Samples of approximately 1–2 mg were accurately weighed into an aluminum DSC pan, and then covered with an aluminum lid that was inverted and pressed down so as to tightly contain

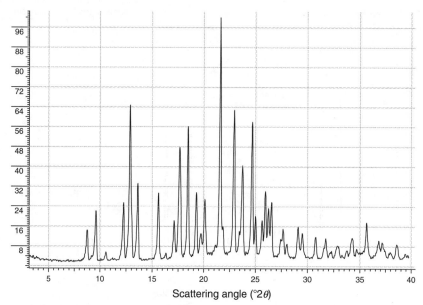

Scattering angles and *d*-spacings of the 12 most intense peaks

Scattering angle (°2θ)	*d*-Spacing (Å)	Scattering angle (°2θ)	*d*-Spacing (Å)
9.61	9.199	18.47	4.800
12.28	7.204	19.29	4.597
12.89	6.864	21.61	4.109
13.62	6.497	22.95	3.872
15.62	5.670	23.76	3.742
17.66	5.018	24.71	3.601

FIGURE 2.2 X-ray powder diffraction pattern of clopidogrel bisulfate, Form-II.

the powder between the top and bottom aluminum faces of the lid and pan. The samples were then heated over the temperature range of 20–200 °C, at a heating rate of 10 °C/min.

The thermograms obtained for the two clopidogrel bisulfate polymorphic forms are shown in Figs. 2.3 and 2.4. While the temperature maximum associated with the melting endotherm of Form-II (180.7 °C) agreed reasonably well with the melting point value, the temperature maximum associated with the melting endotherm of Form-I (184.1 °C) was significantly less than the melting point value.

The endothermic melting transitions were integrated, enabling a determination of the enthalpies of fusion for the two forms. For Form-I it was found that $\Delta H_f = 70.6$ J/g, while for Form-II it was found that

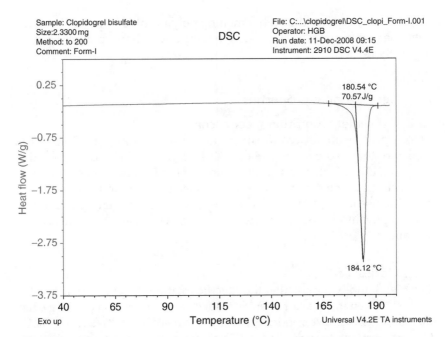

FIGURE 2.3 Differential scanning calorimetry thermogram of clopidogrel bisulfate, Form-I.

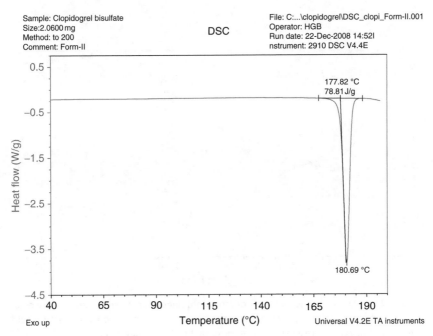

FIGURE 2.4 Differential scanning calorimetry thermogram of clopidogrel bisulfate, Form-II.

$\Delta H_f = 78.8$ J/g. Since the higher melting polymorphic form exhibited the lower heat of fusion, it is concluded that the two polymorphs are enantiotropically related.

2.5. Spectroscopic properties

2.5.1. Ultraviolet absorption spectroscopy

Clopidogrel bisulfate was dissolved at a concentration of 5.5 μg/ml in methanol that had been acidified with sulfuric acid. The UV absorption spectrum of this solution was obtained using a Beckman model Lambda 3B spectrophotometer, and the resulting spectrum is found in Fig. 2.5. The most intense absorption band was observed at a wavelength of 220 nm, while (as the inset shows) two very weak peaks were also noted at 270 and 278 nm.

2.5.2. Optical activity

The Merck Index reports that the specific rotation ($[\alpha]_D^{20}$) of clopidogrel as $+51.52°$, when the measurement is made at a concentration of 1.61 g/dl in methanol. It was also reported that for the bisulfate salt, $[\alpha]_D^{20} = +55.10°$ ($c = 1.891$ in methanol). The latter value is consistent with the data

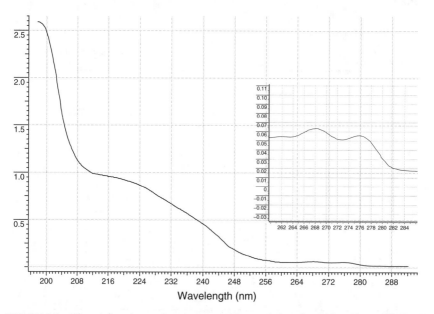

FIGURE 2.5 Ultraviolet absorption spectrum obtained clopidogrel bisulfate, dissolved at a concentration of 5.5 μg/ml in methanol that had been acidified with sulfuric acid. The inset spectrum represents an expansion of the absorbance scale.

published in the patent literature, namely $[\alpha]_D^{20} = +55.16°$ ($c = 1.68$ in methanol) for Form-I and $[\alpha]_D^{20} = +55.10°$ ($c = 1.68$ in methanol) for Form-II [11].

2.5.3. Infrared absorption spectroscopy

The infrared absorption spectra of clopidogrel bisulfate Form-I and Form-II were obtained at a resolution of 4 cm^{-1} using a Shimadzu model 8400S Fourier-transform infrared spectrometer, with each spectrum being obtained as the average of 40 individual spectra. The data were acquired using the attenuated total reflectance sampling mode, where the samples were clamped against the ZnSe crystal of a Pike MIRacle^{TM} single reflection horizontal ATR sampling accessory.

As shown in Figs. 2.6–2.9, the infrared absorption spectra of the two polymorphic forms of clopidogrel bisulfate are quite different. These differences are summarized in Table 2.1, which lists the energies of the main observed peaks and which contrasts the energies of peaks assigned to the same vibrational mode.

2.5.4. Raman spectroscopy

The Raman spectra of clopidogrel bisulfate Form-I and Form-II were obtained in the fingerprint region using a Raman Systems model R-3000HR spectrometer, operated at a resolution of 5 cm^{-1} and a laser

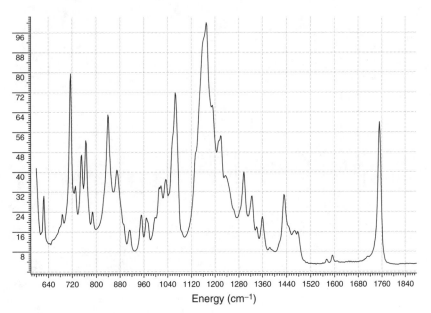

Energy (cm^{-1})

FIGURE 2.6   Infrared absorption spectrum obtained in the fingerprint region for clopidogrel bisulfate, Form-I.

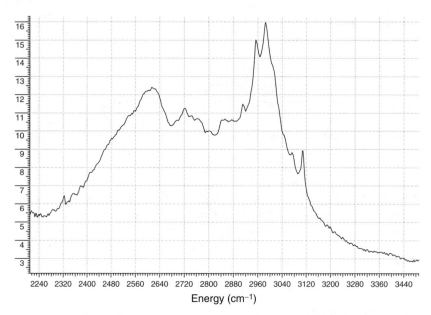

FIGURE 2.7 Infrared absorption spectrum obtained in the high-frequency region for clopidogrel bisulfate, Form-I.

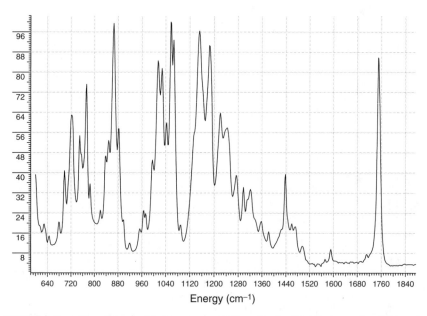

FIGURE 2.8 Infrared absorption spectrum obtained in the fingerprint region for clopidogrel bisulfate, Form-II.

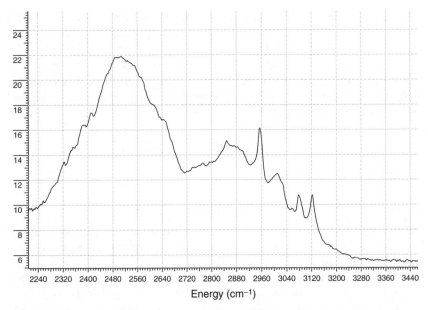

FIGURE 2.9 Infrared absorption spectrum obtained in the high-frequency region for clopidogrel bisulfate, Form-II.

TABLE 2.1 Energies of corresponding bands in the infrared absorption spectra of clopidogrel bisulfate, Forms I and II

Energy, Form-I band (cm⁻¹)	Energy, Form-II band (cm⁻¹)	Energy, Form-I band (cm⁻¹)	Energy, Form-II band (cm⁻¹)
624.9			1151.4
	698.2	1170.7	
715.5	721.3		1184.2
752.2	750.3	1218.9	1220.9
765.7	773.4		1244.0
839.0			1272.9
869.8	866.0	1298.0	1298.0
912.3		1325.0	1321.1
950.8		1359.7	
968.2		1431.1	1438.8
1018.3	1014.5	1477.4	
1033.8	1028.0		1591.2
	1056.9	1751.2	1749.3
1066.6	1066.6		

wavelength of 785 nm. The data were acquired using front-face scattering from a thick powder bed contained in an aluminum sample holder.

As shown in Figs. 2.10 and 2.11, the Raman spectra of the two polymorphic forms of clopidogrel bisulfate are substantially different. These differences are summarized in Table 2.2, which lists the energies of the main observed peaks and which contrasts the energies of peaks assigned to the same vibrational mode.

3. METHODS OF ANALYSIS

3.1. Known impurities of clopidogrel

Clopidogrel bisulfate has several related compounds that are to be impurity species [2, 6, 7]. The molecular structures of these impurities are shown in Fig. 2.12.

Related compound (**A**) has the systematic name (+)-(*S*)-(*o*-chlorophenyl)-6,7-dihydrothieno[3,2-*c*]pyridine-5-(4*H*)-acetic acid [2], and is a hydrolysis product of the ester group of clopidogrel. This impurity is formed as a result of the combined effects of moisture and temperature [2, 6, 7].

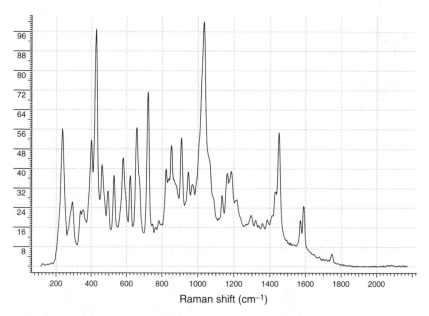

FIGURE 2.10 Raman spectrum obtained in the fingerprint region for clopidogrel bisulfate, Form-I.

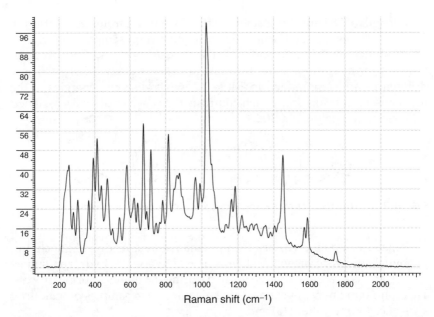

FIGURE 2.11   Raman spectrum obtained in the fingerprint region for clopidogrel bisulfate, Form-II.

TABLE 2.2   Energies of corresponding bands in the Raman spectra of clopidogrel bisulfate, Forms I and II

Raman shift, Form-I band (cm^{-1})	Raman shift, Form-II band (cm^{-1})	Raman shift, Form-I band (cm^{-1})	Raman shift, Form-II band (cm^{-1})
235.4	254.4		781.9
	278.9	822.0	814.2
292.3	303.5	852.2	
351.4	364.8		875.6
	390.4	907.9	
399.4		947.0	964.8
	412.8		989.3
426.1	435.1	1032.8	1025.0
460.7	469.6	1133.2	
495.3		1162.2	1163.3
528.7	538.8	1185.6	1183.4
580.0	581.2		1221.3
620.2	621.3	1449.9	1449.9
657.0	641.4	1571.5	1570.4
	673.7	1589.3	1588.2
720.6	716.1		1746.6

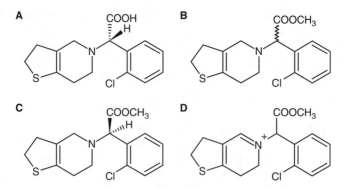

FIGURE 2.12   Structures of four known impurities of clopidogrel bisulfate.

Related compound (**B**) has the systematic name methyl-(±)-(*o*-chloro-phenyl)-4,5-dihydrothieno[2,3-*c*]pyridine-6-(7*H*)-acetate, hydrogen sulfate salt [2], and is a racemic residue formed during the manufacturing process. This compound may appear as a racemic mixture in samples of bulk drug substance as impurities (**1a**) and (**1b**) [6, 7].

Related compound (**C**) is the (*R*)-enantiomer of clopidogrel bisulfate, and therefore has the systematic name methyl-(−)-(*R*)-(*o*-chlorophenyl)-6,7-dihydrothieno[3,2-*c*]pyridine-5(4*H*)-acetate, hydrogen sulfate salt [2, 6, 7].

Related compound (**D**) has been reported to result from an oxidation process, and has the systematic name 5-[1-(2-chlorophenyl)-2-methoxy-2-oxoethyl]-6,7-dihydrothieno[3,2-*c*]pyridine-5-ium [8].

3.2. Compendial methods of analysis

3.2.1. Identification

USP32-NF27 [2] sets forth the use of infrared absorption spectroscopy (method ⟨197 K⟩), a chromatographic method, and the sulfate test (method ⟨191 K⟩) for identification of clopidogrel bisulfate bulk drug substance. The IR absorption method requires mixing the assayed substance with potassium bromide, and then recording the spectra of the test specimen and the USP reference standard over the range of 2.6–15 μm (3800–650 cm^{-1}). The chromatographic method uses the liquid chromatography (LC) method of the assay section, with the monograph specifying that the retention time of the major peak be similar with that of the reference standard. Furthermore, treatment of a sample with barium chloride TS should result in the formation of a white $BaSO_4$ precipitate which is insoluble in hydrochloride acid and nitric acid. As a confirmatory test, addition of lead acetate TS to a sample will result in the formation of a white $PbSO_4$ precipitate that is insoluble in ammonium acetate TS.

For identification of clopidogrel bisulfate in pharmaceutical preparations, USP32-NF27 [2] recommends use of a UV absorption and a chromatographic method. The test solution required for performance of the UV absorption method is obtained as for the dosage form uniformity test, where one tablet containing clopidogrel bisulfate is dissolved in 0.1 N HCl. The spectrum is recorded over 250–300 nm, where clopidogrel bisulfate exhibits a wavelength maximum at 270 nm. The chromatographic method to be used for identification of clopidogrel bisulfate in pharmaceutical preparations is the same as used for assay of the bulk drug substance.

3.2.2. Impurity analysis

USP32-NF27 [2] requires use of a liquid chromatographic method to determine three clopidogrel bisulfate-related compounds, and this method is the same as the chromatographic method used to determine clopidogrel in bulk substances. Stock solutions of clopidogrel bisulfate and each of its related compounds are prepared by dissolution in methanol, and these are subsequently diluted with methanol to obtain concentrations of 20 μg/ml (for clopidogrel bisulfate RS), 40 μg/ml (for related compound **A**), 120 μg/ml (for related compound **B**), and 200 μg/ml (for related compound **C**). Five milliliters of each of these solutions is then diluted with mobile phase to obtain final concentrations of 0.5, 1, 3, and 5 μg/ml, respectively, for clopidogrel and the three impurities.

Test solutions of analytes are prepared by dissolving 100 mg of sample with 5 ml methanol, and then diluting to volume with mobile phase. With the flow rate set adjusted to 1.0 ml/min, the relative retention times will be approximately 0.5 for related compound **A**, 0.8 and 1.2 for the two enantiomers of related compound **B**, 1.0 for clopidogrel, and 2.0 for related compound **C**. The resolution between clopidogrel and the first enantiomer of related compound **B** must be greater than 2.5. The monograph specification of USP32-NF27 is that not more than 0.2% of related compound **A** is found, not more than 0.3% of the first enantiomer of related compound **B** is found, and not more than 1.0% of related compound **C** is found. In addition, the monograph requires that not more than 0.1% of any other impurity is found, and that the total impurity content be less than 1.5%. The concentrations of all clopidogrel-related compounds are to be expressed as bisulfate salts, obtained by using bisulfate salt equivalents as stated on the USP reference standard labels.

3.2.3. Assay methods

A liquid chromatographic method is utilized for the determination of clopidogrel bisulfate in samples of the bulk drug substance. The method uses a column (L57 column size 15 cm × 4.6 mm) packed with ovomucoid (a chiral-recognition protein) that is chemically bonded to silica particles of 5 μm diameter and a pore size of 120 Å. Both the reference standard and the sample to be analyzed are dissolved in methanol, and then diluted with mobile phase. The mobile phase is 75:25 0.01 M phosphate buffer/acetonitrile, and the flow rate is adjusted to 1.0 ml/min. Observation is made on the basis of the UV absorbance at 220 nm, and the clopidogrel peak has a relative retention time about 1.0 min.

The assay method for determination of clopidogrel bisulfate content in tablets uses the same LC method as used for the bulk drug substance. Here, not less than 20 tablets are finely powdered, and a quantity of the powder (equivalent to about 75 mg of clopidogrel base) is dissolved in and diluted with methanol. The quantity of clopidogrel base (in mg units) is calculated using a gravimetric factor based on the molecular weight of clopidogrel base and the molecular weight of clopidogrel bisulfate. The calculation also takes into account the peak responses of the standard and the sample:

$$1000 \times \left(\frac{321.82}{419.90}\right) C \left(\frac{r_u}{r_s}\right) \tag{1}$$

where C is concentration of clopidogrel bisulfate USP reference standard, while r_u and r_s are the respective peak responses obtained from the sample and the standard.

3.3. Spectroscopic methods of analysis

3.3.1. Ultraviolet absorption spectroscopy

Sankar *et al.* [12] reported a spectrophotometric method for the analysis of clopidogrel and repaglinide in their pure forms and in their combination tablet. In this method, water was used as the solvent, and clopidogrel was determined on the basis of its absorbance at a wavelength of 225 nm. The regression curve was linear over the range of 10–60 μg/ml.

Another spectrophotometric method for determining clopidogrel bisulfate in aspirin combination tablets was reported by Mishra and Dolly[13]. In this method, the absorbance additivity technique and the graphical determination of absorbance ratios methods were applied. Stock solutions of clopidogrel bisulfate and aspirin were prepared in methanol, and working solutions were then prepared by mixing 1 ml of the stock solutions with 1 ml of H_2SO_4 and heating in a water bath for

30 min. Linearity in the calibration curve was obtained over the range of 4–18 μg/ml for both methods, and ranged from 97.0% to 101.4%.

Rajput *et al.* reported another method to determine of clopidogrel bisulfate and aspirin in combination tablets, developing and validating a spectrophotometric assay based on chemometric methods [14]. Methanol was used to prepare standard solutions of clopidogrel bisulfate and aspirin, as well as to extract analytes from the tablets. In this method, chemometric methods based on inverse least square (ILS) and classical least square (CLS) were applied. For the purpose of establishing a calibration set, 12 mixture compositions of aspirin (0–20 μg/ml) and clopidogrel bisulfate (0–30 μg/ml) were employed, and the UV spectra of each mixture was recorded over the spectral region of 200–310 nm (16 wavelength point absorbance measurements were obtained with a 2-nm interval spanning 220–250 nm). In order to validate the method, calibration models developed based on ILS and CLS methodologies were used to determine clopidogrel bisulfate and aspirin in mixtures. Ten combination mixtures were used as a validation set and subjected to recovery studies. It was found that the recovery based on the ILS method was 97.33–103.33%, while the recovery obtained using the CLS method was 95.55–103.50%.

A spectrophotometric method was developed by Zaazaa *et al.* [15] to analyze clopidogrel bisulfate and its alkaline degradation, where the method used two spectrophotometric derivatization methods and a spectrophotometric method based on bivariate calibration. The first method used second derivative spectrophotometry, in which the derivative curves were recorded at $\Delta\lambda = 4$ with a scaling factor of 100. The wavelength values used for the calibration curve were 219.6, 270.6, 274.2, and 278.4 nm, which corresponds to the zero-crossing points of the degradant. This second derivative spectrophotometric method was able to determine clopidogrel bisulfate in the presence of 65% degradant, with a recovery ranging from 99.32% to 100.51% and linearity over the 4–37 μg/ml concentration range. In addition, this method was also employed to study the kinetics of the alkaline degradation of clopidogrel bisulfate at 278.4 nm.

The second spectrophotometric method developed by Zaazaa *et al.* [15] used the first derivative of ratio spectra, with peak amplitudes being recorded at 217.6 and 229.4 nm ($\Delta\lambda = 4$ nm, scaling factor $= 10$). The absorption spectrum of clopidogrel bisulfate was then divided by the absorption spectrum of a 15 μg/ml solution of the degradant over the range of 5–38 μg/ml. By using this latter method, clopidogrel bisulfate could be determined in the presence of up to 70% of its degradant.

The third spectrophotometric developed by Zaazaa *et al.* [15] used a bivariate calibration technique, which is simple and does not require performance of a derivatization procedure. In addition, no full spectrum information is required, eliminating the need for extensive data processing. In this method, each substance (clopidogrel bisulfate and an alkaline analyte) are observed at two different wavelengths (210 and 225 nm), and calibration curves were established for each wavelength. The bivariate calibration technique could enable the determination of clopidogrel bisulfate in the presence of up to 70% of its alkaline degradant with recoveries ranging from 98.0% to 101.10%. Linearity in the clopidogrel bisulfate calibration curve was obtained over the range of 5–38 μg/ml, while linearity for its alkaline degradant extended over the range of 5–25 μg/ml.

3.3.2. Vibrational spectroscopy

Two infrared absorption methods (i.e., FTIR) and a Raman spectroscopic method were used to quantify polymorphic clopidogrel bisulfate Form-I and Form-II [16, 17]. In addition, qualitative analysis of these polymorphs was also conducted using FTIR [16], where each sample was scanned over in the spectral region of 450–4000 cm^{-1} at a resolution of 4 cm^{-1}. The sampling procedure used KBr pellets, loaded to contain approximately 3% of analyte. It was found that absorption bands associated with C–H and C–O bonds were stronger for Form-II relative to Form-I, and that unique absorption bands for Form-I and Form-II were observed at 841 and 1029 cm^{-1}, respectively. These absorption bands were reported to be useful in the quantitative or qualitative analysis of clopidogrel polymorphs.

For quantitative analysis, it was necessary to sieve both polymorph forms prior to analysis in order to obtain some degree of uniformity in particle size. The sieving method was chosen owing to the possibility of polymorphic interconversion during grinding [16]. The quantitative analysis of Form-I was performed on the basis of absorption intensities at 2987, 1175, and 841 cm^{-1}, and at 1497, 1187, and 1029 cm^{-1} for Form-II. Quantification of Form-I in a binary polymorph mixture was performed at 841 cm^{-1} due to the sensitivity of this band, and the readily interpretable spectral difference. Moreover, this band was not affected by pressure effects associated with preparation of the KBr pellet. The range of the method was evaluated between 10% and 90% of Form-I in samples of Form-II. However, the method was found to be limited to low levels of Form-I in Form-II since the characteristic Form-II bands were not visible for mixtures containing less than 30% of Form-I. The recovery of the method ranged from 100.0% to 101.0% for Form-I and in Form-II percentages of 20%, 50%, 60%, and 80% [16].

In another study, samples for polymorphic analysis were processed using gentle grinding after mixing [17]. Here, a combination of FTIR and Raman spectroscopy with multivariate analysis was utilized to quantify Form-II in mixtures with Form-I. Multivariate analysis using the PLS method gave the best results when compared to other methods such as CLS and PCR. For the Raman spectroscopic method, the respective LOD and LOQ values were determined to be 1.0% and 3.0% when using PLS multivariate analysis, while for the FTIR method the respective LOD and LOQ values were found to be 0.7% and 0.2%. Both methods were able to quantify down to 3% of clopidogrel bisulfate Form-II in mixtures with Form-I. The accuracy of the method was characterized by RSD values smaller than 10% over the analysis range of 3–15% (w/w). It was concluded that by employing PLS multivariate analysis for both FTIR and Raman spectroscopy, quantification of lower amounts of clopidogrel bisulfate Form-II polymorph was possible [17].

3.3.3. Nuclear magnetic resonance

An NMR spectroscopic method was utilized by Badorc and Fréhel [5] to determine the enantiomeric purity of the dextrorotatory and levorotatory enantiomers of clopidogrel bisulfate. The chiral lanthanide shift reagent, Eu(tfc)$_3$ [tfc = 3-(trifluoromethyl-hydroxymethylene)-d-camphorato], was added to a CDCl$_3$ solution containing the racemic drug substance, whereupon the enantiomeric purity could be determined using 60 MHz ^{1}H NMR spectroscopy. In the absence of the chiral shift reagent, NMR analysis of the racemate showed that the hydrogen attached to the asymmetric center in the α position to the ester function appeared as a singlet, characterized by a chemical shift $\delta = 4.87$ ppm in CDCl$_3$. After the addition of the shift reagent, the singlet became separated into two well-resolved singlets (separation of about 6 Hz) which corresponded to the protons of the two enantiomers. It was determined that the band having the smaller chemical shift corresponded to the proton of the dextrorotatory enantiomer, while the band having the larger chemical shift was due to the levorotatory enantiomer. It was reported that this method was able to detect more than 5% (w/w) of one enantiomer in the presence of the other.

Mohan et al. [8] used an NMR spectroscopic method to characterize impurity D in samples of clopidogrel bisulfate. The method entailed ^{1}H NMR (at 400.13 MHz) and ^{13}C NMR (at 100.62 MHz), with a sample concentration of 1 mg/ml in DMSO-d_6 (this solvent also served as an internal chemical shift standard). It was found that the ^{1}H NMR spectrum of impurity D exhibited one hydrogen band than did that of clopidogrel, while the ^{13}C NMR and DEPT135 NMR spectra indicated the presence of one methyl carbon, two methylene carbons, eight methine carbons, and five quaternary carbons. This corresponded to a similar structure as for clopidogrel, but with one less methylene carbon and one more methine carbon.

3.4. Potentiometric method of analysis

Saber *et al.* [19] reported the development and validation of a potentiometric method to determine clopidogrel bisulfate in pharmaceutical preparations. In this method, a polyvinyl chloride membrane reference electrode was developed by using two plasticizers having different dielectric constants. The reference electrode consisted of a 7:63:30 (w/w/w) mixture of tetrakis (*p*-chlorophenyl) borate, *o*-nitrophenyl octyl ether or dioctyl phthalate as plasticizers, and PVC. This reference electrode was soaked in 0.1 M clopidogrel solution for a minimum of 2 days prior to use, and maintained in the same solution when not in use. An EIL-type RJ 23 calomel reference was used as working electrode, while an Ag/AgCl combination electrode was used for pH measurement, with the internal solution being silver–silver chloride in a 0.1 M clopidogrel solution.

It was found that the limit of detection was 1.0×10^{-5} M for both plasticizers and that linearity and Nernstian response was obtained over the range of 1.0×10^{-5}–1.0×10^{-2} mol/l (pH range of 1.5–4.0). This potentiometric method was validated for the determination of clopidogrel over the concentration range of 4.2 μg/ml to 4.2 mg/ml with an average recovery of 100.65%. Both sensors showed stable potential readings and calibration slopes for extensive time periods.

3.5. X-Ray powder diffraction methods of analysis

Quantitative and/or qualitative XRPD methods have been reported to determine the polymorphic content of clopidogrel bisulfate samples, and these have been summarized in Table 2.3. Koradia *et al.* [16] reported the qualitative analysis of clopidogrel bisulfate in both active pharmaceutical ingredients and tablet dosage forms. Based on the interplanar distances (*d*-spacing) associated with each polymorph, it was concluded that the molecular packing in Form-I was more dense than that of Form-II, indicating that Form-II would be less stable relative to Form-I. This result was similar with that reported by Bousquet [9].

Uvarov and Popov [20] reported the quantification of polymorphic forms I and II using two different XRPD methods. One method was based on the single peak intensity or direct method, and the second used whole-powder-pattern decomposition (WPPD) method. The Form-I nonoverlapping peaks suitable for analysis work were found at *d*-spacings 8.13, 5.98, and 4.32 Å, while the useful peaks for Form-II had *d*-spacings of 7.24 and 6.88 Å. It was shown that use of the WPPD method yielded better

TABLE 2.3 Summary of X-ray powder diffraction methods used to analyze clopidogrel bisulfate

Radiation source	Scan range (° 2θ)	Step size (° 2θ)	Tube voltage (kV)	Current (mA)	Type of analysis	References
Cu Kα	2–40	Not reported	Not reported	Not reported	Qualitative	[9]
Cu Kα	5–40	0.02	35	20	Qualitative	[16]
Cu Ni	1–40	0.02	60	300	Qualitative	[54]
Cu Kα	6–36	0.02	40	40	Quantitative	[20]
Not reported	8.99–9.30	Not reported	Not reported	Not reported	Quantitative	[21]
Cu Kα	2–40	0.013	40	40	Quantitative	[17]

reproducibility as compared to the direct method, although the limits of detection were comparable for both methods.

Alam *et al.* [21] reported an XRPD method for quantification of the two polymorphic forms of clopidogrel bisulfate. In this method, particle size uniformity of both polymorphs was obtained by ball milling samples to obtain mean particle sizes of 9–17 μm. The highest peak intensity of clopidogrel bisulfate using the highest intensity scattering peak of Form-I (9.14° 2θ), linearity was characterized by an R^2 value of 0.9885, LOD and LOQ values of 0.29% and 0.91%, respectively, and recoveries in the range of 97–102%. This method was shown to be able to detect low concentration of Form-I (down to levels of 1%) in mixtures of Form-I and Form-II. It was also noted that the presence of excipients changed the peak intensities and caused baseline shifting. These problems must be considered in any quantitative analysis of polymorph of clopidogrel bisulfate in pharmaceutical dosage forms by means of XRPD [16].

3.6. Thermal methods of analysis

Koradia *et al.* [16] have used hot stage microscopy (HSM), DSC, and Thermogravimetric analysis (TGA) to study Form-I and Form-II of clopidogrel bisulfate. For the HSM method, samples were heated on the hot stage and simultaneously observed with both normal and polarized illumination. It was found that Form-I exhibited an irregular plate morphology and that agglomeration was characteristic for Form-II. Under polarized light examination, Form-I appeared as alternately dark and bright when rotated by 45°, while Form-II appeared as fully bright and turned to dark at a rotation of 90°. In addition, half extinction was exhibited for Form-II at rotation angles of 30° and 60°.

As detailed by Koradia *et al.* [16], the DSC and TGA studies were run at a heating rate of 10 °C/min, and a nitrogen purge was set at 80 ml/min for the DSC work and at 20 ml/min for the TGA work. The DSC thermogram of Form-I consisted of three endothermic transitions, with temperature maxima at 181, 186, and 190 °C. The DSC thermogram of Form-II also consisted of three endotherms, characterized by temperature maxima at 177, 179, and 182 °C. During the TGA analysis, no temperature-induced loss of mass was observed for either polymorph, indicating that the two polymorphic crystal forms were anhydrous and nonsolvated.

Alam *et al.* [21] reported that the endothermic transition of Form-I was observed at 179.78 °C, and that the endothermic transition of Form-II was observed at 177.34 °C.

4. CHROMATOGRAPHIC METHODS OF ANALYSIS

4.1. Thin-layer chromatography

Table 2.4 contains a summary of some of the thin-layer chromatography (TLC) methods that have been developed and validated for the analysis of clopidogrel bisulfate. High-performance TLC methods have been shown to be able to separate clopidogrel from its degradants in accelerated degradation studies that used stressing by acid–base, hydrogen peroxide, heat degradation, and photochemical means [15, 22]. In the acid–base degradation study (performed using both 1.0 N HCl and 1.0 N NaOH), two additional peaks appeared in front of the main clopidogrel peak [22].

Similar results were obtained in another study of the alkaline degradation of clopidogrel that used either 0.5 N NaOH or 1.0 N NaOH, with one degradant peak appearing before the elution of the main peak [15]. Infrared spectroscopic and mass spectrometric structural elucidation of the alkaline degradants using showed the presence of the carbonyl group and the hydroxyl group of carboxylic acids [15]. An additional peak was also found eluting after the main peak for both the peroxide and dry heat (100 °C) degradation studies. In this work, no degradant peaks of photochemical origin were noted when clopidogrel bisulfate was exposed to direct sunlight for 24 h, indicating that clopidogrel bisulfate was photochemically stable under these conditions [22].

4.2. Gas chromatography

Kamble and Venkatachalam [23] reported a gas chromatography (GC) method for the determination of clopidogrel in tablet dosage forms. This method used a DB-17 capillary column of 30 m length and 0.25 mm internal diameter, equilibrated at an oven temperature of maintained at 250 °C. Dioctyl phthalate was used as an internal standard, and clopidogrel was detected at 4.1 min by means of a flame ionization detector. Linearity in the method ranged from 0.5 to 5.0 mg/ml, with a recovery of 99.89%.

4.3. Liquid chromatography

High-performance liquid chromatography (HPLC) methods represent the most commonly used analytical technology for the determination of clopidogrel bisulfate in either its bulk drug substance form or its pharmaceutical dosage forms. HPLC methods are most preferred in pharmaceutical compendia [2]. A summary of HPLC methods for the analysis of clopidogrel bisulfate is provided in Table 2.5. Since clopidogrel bisulfate contains a center of dissymmetry, some HPLC methods are

TABLE 2.4 Summary of TLC methods used to analyze clopidogrel bisulfate

Stationary phase	Mobile phase (v/v)	Solvent	Wavelength	TLC preparation	Sample	Limit of detection (LOD), limit of quantitation (LOQ), and recovery (Rec)	References
Silica gel 60F-254 HPTLC plate	Carbon tetrachloride: chloroform: acetone = 6:4:0.15	MeOH	230 nm	• Prewashing plate with MeOH and heating at 60 °C for 5 min • Saturating chamber with mobile phase for 30 min	Drug powder and tablet	LOD: 40 ng/spot LOQ: 120 ng/spot Rec: 99.13–101.30%	[22]
Silica gel 60F-254	n-Heptane: tetrahydrofuran = 1:1	MeOH	• Spot detected at 254 nm (UV light) • Quantitative analysis at 230 nm	• Saturating chamber with mobile phase for 15 min	Drug powder and clopidogrel impurity (SR 25334)	Clopidogrel bisulfate (impurity level of SR 26334: 0.028%) LOD: 0.024 mg/ml LOQ: 0.014 mg/ml Rec: 98.5–100.3% SR 25334 (impurity level of SR 0.095%) LOD: 0.079 LOQ: 0.047 Rec: 94.5–107.1%	[55]

| Silica gel 60F-254 HPTLC plate | Acetone:MeOH:toluene: glacial acetic acid = 5.0:1.0:4.0:0.1 | MeOH | 235 nm | • Saturating chamber with mobile phase for 30 min | Drug powder and tablet | LOD: 112.66 ng/ml LOQ: 375.55 ng/ml Rec: 99.42–100.86% | [12] |
| Silica gel plate | Hexane:MeOH:ethyl acetate = 8.7:1:0.3 | MeOH | 248 nm | n/a | Tablet | LOD: 0.04 μg/band LOQ: 0.4 μg/band Rec: 98.90–102.17% | [15] |

n/a: not available.

TABLE 2.5 Summary of HPLC methods used to analyze clopidogrel bisulfate in bulk drug substance and pharmaceutical dosage forms

Column	Mobile phase	Column temperature (°C)	Detection (nm)	Solvent	Limit of detection (LOD), limit of quantitation (LOQ), and recovery (Rec)	References
BDS C$_8$ semimicro column (250 mm × 2.1 mm i.d.)	0.010 M disodium hydrogen phosphate (pH 3):ACN = 35:65	Ambient	235	Standard: stock solution in MeOH then diluted with mobile phase Sample: dissolved in ACN then with water then with mobile phase	LOD: 0.12 μg/ml LOQ: 0.39 g/ml Rec: 99.62%	[31]
Chromolith Performance 18e (100 mm × 4.6 mm i.d.)	ACN:50 mM phosphate buffer = 50:50 (pH 3)	n/a	235	MeOH	LOD: 0.97 μg/ml LOQ: 3.53 μg/ml Rec: 99.1%	[32]
Phenomenax C$_{18}$ (250 mm × 4.6 mm i.d.)	ACN:MeOH:20 mM phosphate buffer (pH 3) = 50:7:43	20	240	Mobile phase	LOD: n/a LOQ: n/a Rec: 100.20%	[26]
Phenomenax Gemini C$_{18}$ column	0.1% (v/v) triethylamine (pH 4.0):ACN = 25:75	n/a	225	MeOH	LOD: 0.03 μg/ml LOQ: 1.0 μg/ml	[29]
Phenomenex Luna 5 m C$_{18}$ 100A column	ACN:50 mM potassium dihydrogen phosphate buffer:MeOH = 50:30:20 (solution adjusted to pH 3.0)	Ambient	240	Mobile phase	Rec: 100.34%	[30]
Nucleosil C$_8$ SS column (150 mm × 4.6 mm i.d.)	ACN:phosphate buffer = 55:45 (solution adjusted to pH 3.0)		235	MeOH	LOD: 0.047 μg/ml LOQ: 0.157 μg/ml Rec: 99.34-100.94%	[28]
Waters C$_{18}$ (150 mm × 3.9 mm)	0.05 M ammonium formate (pH 4.0):ACN = 40:60	Ambient (+25 °C)	225	Stock solution dissolved in MeOH	LOD: 0.3 μg/ml LOQ: 1.0 μg/ml	[19]

Column	Mobile phase		Detection	Sample preparation	Recovery / Application	Reference
Hypersil BDS C$_8$ column	Solvent A: ACN:potassium phosphate buffer (pH 2.3; 10 mM) = 20:80 (v/v) Solvent B: ACN:potassium phosphate buffer (pH 2.3; 10 mM) = 80:20 (v/v) Time gradient program of T (min)/%B (v/v): 0.01/0, 5/0, 15/15, 40/30, 45/060/0	n/a	220	then standard series/samples diluted with mobile phase	Rec: 99.20–99.88% Qualitative analysis of degradation sample	[8]
Hypersil BDS C$_8$ column	Solvent A: ACN:potassium phosphate buffer (pH 2.3; 10 mM) = 20:80 (v/v) Solvent B: ACN:potassium phosphate buffer (pH 2.3; 10 mM) = 80:20 (v/v) Time gradient program of T (min)/%B (v/v): 0.01/0, 5/0, 15/15, 40/30, 45/060/0	n/a	LC–MS/MS operated in electrospray ionization positive and negative mode. Fragmentation pattern: m/z: 321/323 m/z: 212/214 m/z: 183/185 m/z: 125/127 m/z: 155/157	n/a	Characterization of clopidogrel bisulfate degradation products	[8]

(continued)

TABLE 2.5 *(continued)*

Column	Mobile phase	Column temperature (°C)	Detection (nm)	Solvent	Limit of detection (LOD), limit of quantitation (LOQ), and recovery (Rec)	References
ULTRON ES-OVM column (4.6 mm × 150 mm i.d.)	ACN:0.01 M potassium dihydrogen phosphate = 25:75	17	220	Tablets were dissolved with 5.0 ml and further diluted with mobile phase	Hydrolysis product: LOD: 0.5 ng LOQ: 0.15 ng Rec: n/a R-enantiomer: LOD: 1.5 ng LOQ: 4.9 ng Rec: n/a	[24]
ULTRON ES-OVM column (4.6 mm × 150 mm i.d.)	ACN:0.01 M potassium dihydrogen phosphate = 22:78	25	220	Tablets were dissolved with 5.0 ml and further diluted with mobile phase	Hydrolysis product: LOD: 0.5 ng LOQ: 0.15 ng Rec: n/a R-enantiomer: LOD: 1.5 ng LOQ: 4.9 ng Rec: n/a	[6]
CHIRAL-AGP (4.0 mm × 100 mm)	ACN:0.015 M di-potassium hydrogen phosphate (pH 6.5) = 40:60	30	220	MeOH	Clopidogrel: LOD: 194.6 ng/ml LOQ: 589.7 ng/ml Rec: 99.8–101.2% Impurity **1a**: LOD: 1.9 ng/ml LOQ: 5.7 ng/ml Rec: 98.9–101.6% Impurity **1b**: LOD: 1.79 ng/ml LOQ: 5.44 ng/ml Rec: 99.6–101.2%	[7]

Column	Mobile phase	Detection (nm)	Sample	Validation data	Reference
ChiraDex® column (4 mm × 250 mm)	ACN:MeOH:0.01 potassium dihydrogen phosphate = 15:5:80	220	Mobile phase	Impurity 2: LOD: 1.4 ng/ml LOQ: 4.2 ng/ml Rec: 98.9–100.9% R-enantiomer LOD: 1.04 ng/ml LOQ: 3.17 ng/ml Rec: 99.9–101.0%	[25]
Chiral-AGP (100 mm × 4.0 mm)	ACN:1 mM DMOA in 10 mM ammonium acetate (pH 5.5) = 16:84	230	n/a	S-Clopidogrel: LOD: n/a LOQ: n/a Rec: 99.9–101.2% R-enantiomer: LOD: 0.25 µg/ml (impurity level 0.07%) LOQ: 0.75 µg/ml (impurity level 0.2%) Rec: 104.6–105.2% S-acid: LOD: 0.031 µg/ml (impurity level 0.008%) LOQ: 0.09 µg/ml (impurity level 0.024%) Rec: 104.0–105.0%	[59]

conducted using chiral chromatography in the determination of clopido-grel bisulfate and its impurities [6, 7, 24, 25], since the use of chiral columns enables separation of the (S)- and (R)-enantiomers, as well as separation from other impurities and degradants.

However, most of the HPLC methods use C_{18} column or C_8 column technology [8, 19, 26–28], including the determination of clopidogrel bisul-fate in an aspirin combination tablet [29, 30]. Sippel *et al.* [27] also reported an HPLC method using a C_{18} column to determine clopidogrel bisulfate in coated tablets, and evaluation of its degradation under acid, alkaline, and peroxide conditions, as well as its photostability. Alkaline conditions induce rapid degradation of clopidogrel bisulfate, as has been shown by HPLC analysis of the alkaline degradation products of clopidogrel bisulfate [27, 31].

Mohan *et al.* [8] reported a successful separation of clopidogrel bisul-fate and its degradants using a C_8 HPLC column, followed by prepara-tive-HPLC collection of the degradants for their characterization by LC–MS/MS. The preparative-HPLC isolation of the impurities was con-ducted by using an Xterra MS C_{18} ODB HPLC column. Degradation studies were conducted in acid, base, and peroxide conditions by moist-ening clopidogrel bisulfate with water and maintenance of the samples in an oven at 120 °C for 24 h. In this study, the presence of related compound **D** was found when the drug substance was exposed to oxidative condi-tion, while this compound was not found for the other degradation conditions.

The use of semimicro column and monolithic columns for the HPLC determination of clopidogrel bisulfate have also been reported [31, 32]. The use of semimicro or small-bore columns is advantageous when com-pared to conventional columns, since the reduction in diameter enables a reduction in solvent usage owing to the lower flow rates and a concomi-tant reduction in run times. Analysts must consider, however, the possi-bility that higher back pressures might be encountered when using semimicro columns [33]. Monolithic columns have also been used to obtain shorter run times, but here higher flow rates (i.e., 4.0 ml/min) are used [32]. Due to their internal structure, monolithic silica columns pro-vide higher separation efficiencies, as well as higher permeability. These qualities enable running the columns at higher flow rates with little effect on sample separation or interfering back pressures [34, 35]. In these works, the retention times for clopidogrel bisulfate were 6.28 min (semimicro column) and 1.13 min (monolithic column).

4.4. Capillary electrophoresis

One capillary zone electrophoresis method was reported by Fayed *et al.* [18] for the separation and determination of clopidogrel and its impuri-ties. In this method, an uncoated fused silica capillary was used at 20 °C, with an applied voltage of -12 kV and sample detection at 195 nm. Optimum separation of clopidogrel and its impurities (related compounds **A**, **B**, and **C**) was achieved by using 5% (m/v) of the sodium salt of sulfated β-cyclodextrin in 10 mM (pH 2.3) buffer to achieve chiral selection. The buffer composition consisted of equal volumes of triethylamine/phosphoric acid, and was pH adjusted with 1.0 M HCl. The reported limits of detection were 0.40, 1.00, 0.50, and 0.25 μg/ml for clopidogrel and related compounds **A**, **B**, and **C**, respectively. The reported limits of quantification for clopidogrel, and related compounds **A**, **B**, and **C** were 0.13, 0.33, 0.16, and 0.08 μg/ml, respectively. The average recovery for clopidogrel in this method was 99.45%, and it was successfully used to analyze clopidogrel and its impurities in two commercial bulk samples.

5. DETERMINATION IN BIOLOGICAL SAMPLES

Clopidogrel is extensively metabolized *in vivo* by carboxylesterase hydrolysis on the ester function, resulting the formation of clopidogrel carboxylic acid (CCA) as the inactive metabolite of clopidogrel. In addition, small amounts of clopidogrel are converted to a pharmacologically active metabolite (AM) via the intermediate metabolite inactive 2-oxoclopidogrel which is then converted to an AM by a two-step cytochrome P450 oxidation process [36, 37].

Due to the instability of clopidogrel AM and the abundant availability of the more stable CCA in human plasma ($\pm$85%), CCA is used to indirectly determine the pharmacokinetics of clopidogrel [36–38]. Furthermore, there is also possibility that (S)-clopidogrel undergoes an *in vivo* chiral inversion into the other clopidogrel enantiomer, which becomes hydrolyzed to (R)-CCA [36]. Metabolic pathways and potential *in vivo* chiral inversions of clopidogrel are described in Fig. 2.13. Until recently, only chromatographic methods were used to determine clopidogrel in biological samples.

5.1. Liquid chromatography

Most of the liquid chromatographic methods use CCA as a standard in the analysis of clopidogrel metabolites, facilitating the acquisition of pharmacokinetic information for clopidogrel [31, 39–43]. Although the analysis of clopidogrel AM in plasma is difficult due to its instability [44], some workers have reported the successful development of LC methods to determine AMs of clopidogrel [37, 38, 44–46]. Despite previous reports of very small levels of unchanged clopidogrel being detected in plasma [47, 48], a number of successful determinations of unchanged clopidogrel

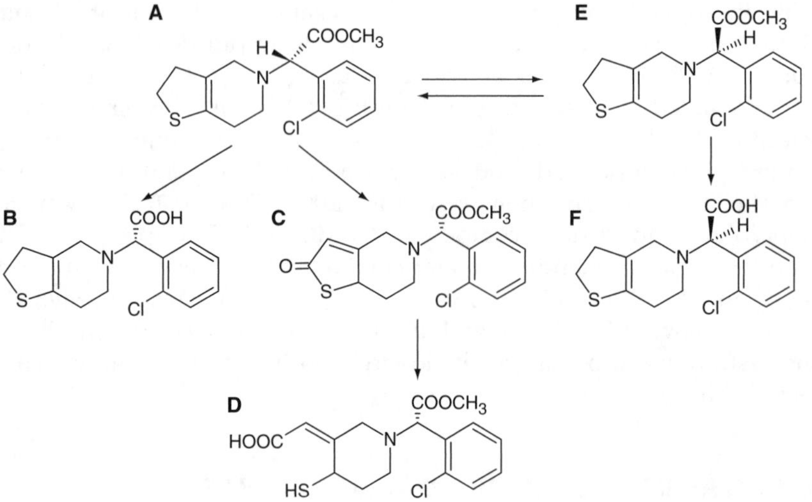

FIGURE 2.13 Metabolic pathway and possible *in vivo* chiral inversion of clopidogrel, modified from references [37, 38]. Clopidogrel (**A**) is metabolized into inactive metabolite (**B**) (clopidogrel carboxylic acid), and an inactive 2-oxoclopidogrel (**C**). The latter compound can undergo a two-step cytochrome P450 process, and become converted to the active metabolite of clopidogrel (**D**). The *in vivo* process results in a chiral inversion, changing clopidogrel into its other enantiomer (**E**), which then is metabolized into the clopidogrel (*R*)-acid (**F**).

in biological samples have been reported that use LC methods [38, 46, 49–51]. Summaries of the LC methods for analysis in biological samples of clopidogrel, its active metabolite, and its inactive metabolite are detailed in Table 2.6.

For the determination of CCA in biological samples, methods not based on LC–MS/MS technology [39, 41–43] and methods that used LC–MS/MS [40, 52] have been reported. Most of the sample extraction methods used liquid–liquid extraction (LLE) technology, since this extraction method is simpler and able to minimize matrix effects. Consequently, LLE methods are considered to provide cleaner samples as compared to solid phase extraction (SPE) methods. Since LC–MS/MS methodology uses nonvolatile solvents or a combination of nonvolatile and volatile solvents, difficulties in the evaporation process and associated interferences when samples are injected onto the system can arise [51]. However, Bahrami as well as Souri [42, 43] applied a combination of nonvolatile and volatile solvents in which the nonvolatile solvents were acidic buffers (pH 5 or less). Analytes eluted from SPE prepared samples did not undergo evaporation as applied commonly encountered in extraction procedures [37, 45].

TABLE 2.6 Summary of HPLC methods used to analyze clopidogrel bisulfate and its metabolites in biological samples

Analyte(s)	HPLC conditions	Sample	Preparation of standard, sample extraction and clean up	Limit of detection (LOD), limit of quantitation (LOQ), and recovery (Rec)	References
CCA and sulfafurazole (i.s.)	Column: Hypercarb PGC (50.0 mm × 3.0 mm) at ambient temperature Detector: LC–MS operated in electrospray ionization positive mode SIM monitoring: Carboxylic acid metabolite: m/z 322 Clopidogrel: m/z 308 Sulfafurazole: m/z 268 Mobile phase: 70% methanol in water containing 0.1% (v/v) trifluoroacetic acid	Human plasma	Standard: MeOH Sample: SPE: Hypercarb cartridge Precondition: 1.0 ml MeOH and 1.0 ml water Column wash: 0.5 ml of 10% MeOH in water Elution: 0.8 ml mixture of 70% ACN in water containing 0.1% TFA	Carboxylic acid metabolite LOD: 28 ng/ml LOQ: 93 ng/ml Rec: 73.0–75.2%	[52]
CCA and atorvastatin (i.s.)	Column: Kromasil ODS (250 mm × 4.6 mm i.d.) at 30 °C Detector: UV/Vis 220 nm Mobile phase: Solvent A (0.05% TFA in water)/solvent B (ACN) run at gradient program started from 90% solvent A and 10% solvent B then decreased to	Rat plasma	Standard: Water:MeOH:ACN = 40:40:20 Sample: LLE: ethyl acetate: dichloromethane = 80:20 Reconstitution of dry residue with similar solvent as for standard	Carboxylic acid metabolite LOD: 75 ng/ml LOQ: 125 ng/ml Rec: 85.8–88.5%	[41]

(continued)

TABLE 2.6 (continued)

Analyte(s)	HPLC conditions	Sample	Preparation of standard, sample extraction and clean up	Limit of detection (LOD), limit of quantitation (LOQ), and recovery (Rec)	References
	10% solvent A and 90% solvent B which was held constant for 3 min then increased to 60% solvent A and 40% solvent B at time 16 min held constant for 2. At time 18 min increased to 90% solvent A and 10%, held for 2 min until time 20 min, then baseline				
CCA and repaglinide (i.s.)	Column: Luna 3μ C$_{18}$ (75 mm × 4.6 mm i.d.) at 35 °C	Human plasma	Standard: SR25990C: MeOH	Carboxylic acid metabolite LOD: n/a LOQ: 20 ng/ml Rec: 85–90%	[40]
	Detector: LC–MS operated in electrospray ionization SIM monitoring: SR25990C: m/z 308.00 Repaglinide: m/z 453.55 Mobile phase: ACN:water: formic acid = 60:40:0.1		Repaglinide: stock solution in MeOH then diluted with water Sample: SPE continued with LLE: Sample added with 250 ml water and 500 ml 0.1 ammonium acetate, then loaded on 1 ml Chem Elut extraction cartridge. LLE performed with 4 ml dicholorometane, dried, and reconstituted with 200 μl of mobile phase		

Analyte	Sample matrix	Method	Sample preparation	Results	Reference
CCA and ticlopidine (i.s.)	Human plasma	Column: Nova-pack® C_8 4 μm (250 mm × 4.6 mm) Detector: UV/Vis 220 nm Mobile phase: 30 mM K_2HPO_4: THF:ACN = 79:2:19 (pH mobile phase = 3)	Standard: MeOH Sample: LLE: Sample added with 200 μl of 300 mM K_2HPO_4 (pH 5) and 4 ml chloroform, then extracted. Dry residue obtained from the organic layer was reconstituted with 100 μl mobile phase	Carboxylic acid metabolite LOD: 0.02 μg/ml LOQ: 0.2 μg/ml Rec: 77.4–83.2%	[42]
CCA, aspirin, and salicylic acid	Human plasma	Column: semimicro column ODS (250 mm × 1.5 mm) Detector: n/a Mobile phase: 10 mM phosphate buffer (pH 2.5) and ACN. Composition of mobile phase was not reported	Standard: n/a Sample: LLE using 10% n-hexane/ethyl acetate	Carboxylic acid metabolite LOD: n/a LOQ: 80.7 ng/ml Rec: n/a	[39]
CCA and phenytoin (i.s.)	Human serum	Column: Shimpack CLC-ODS (150 mm × 4.6 mm i.d.) at 50 °C Detector: UV/Vis 220 nm Mobile phase: 0.05 M sodium phosphate buffer (pH 5.7): ACN = 56:44	Standard: MeOH Sample: LLE using 2.0 N HCl and extracted with ethyl acetate. Residue reconstituted in mobile phase	LOD: 0.02 μg/ml LOQ: 0.05 μg/ml Rec: 96.1–101.8%	[43]

(continued)

TABLE 2.6 (continued)

Analyte(s)	HPLC conditions	Sample	Preparation of standard, sample extraction and clean up	Limit of detection (LOD), limit of quantitation (LOQ), and recovery (Rec)	References
CCA and clopidogrel	Column: C_{18}-ether HPLC analytical column at 50 °C. Detector: LC–MS/MS operated in electrospray ionization (ESI) interface at positive ion mode. MRM monitoring; Clopidogrel: m/z 322.0663 → 212.0478. CCA: m/z 308.0506 → 198.0322. Mobile phase: Gradient from 10% to 100% of ACN:0.1% formic acid	Human plasma	Standard: n/a. SPE	Clopidogrel: LOD: 0.25 μg/l. LOQ: n/a. Rec: n/a. CCA: LOD: 25 μg/l. LOQ: n/a. Rec: n/a	[58]
Purified analyte of clopidogrel AM	Column: Lichrocart 60RP8E (125 mm × 4 mm). Detector: LC–MS operated in electrospray ionization positive mode. SIM monitoring; m/z 356.5, m/z 358.5. Mobile phase: MeOH:water:ACN:diethylamine = 40:60:0.2:0.1	Fraction H obtained from incubation of (7S)-2-oxoclopidogrel with human microsomes	HPLC preparative at UV detection 234 nm. Elution was done using ACN/10 mM ammonium acetate (pH 6.5) gradient (10–24%) using Ultrabase UB225 column	LOD: n/a. LOQ: n/a. Rec: n/a	[44]

Analyte	Matrix	Method	Extraction	Results	Ref.
Clopidogrel (S)-acid and (R)-acid	Rat plasma	Column: Hypersil ODS (250 mm × 4 mm) Detector: spectrofluorometric with excitation wavelength = 280 nm and emission wavelength = 330 nm Mobile phase:ACN: triethylammonium acetate buffer 0.01 M (pH 3.3) = 55:45	SPE and LLE: First elution with hexane (discarded) then analytes Plasma sample was then derivatized with (S)-(−)-α-(1-naphtyl)ethylamine with the aid of HOBT and EDAC. Extraction was performed with water and hexane. Organic phase was then evaporated and reconstituted with mobile phase	LOD: n/a LOQ: n/a Rec: n/a Linearity of (S)-acid: 0.60–40 mg/l Linearity of (R)-acid: 0.60–4.0 mg/l	[36]
Clopidogrel AM, MPB derivative of clopidogrel AM (MP-AM) (i.s.)	Human plasma	Column: Inertsil ODS-3 column (2.1 mm × 50 mm) at 40 °C Detector: LC-MS/MS operated in ESI interface at positive ion mode MRM monitoring: Clopidogrel AM: m/z 356 → 155 MP-AM: m/z 504 → 354 Mobile phase: MeOH:1% formic acid = 70:30	Standard solution: derivatization of clopidogrel AM in ACN with 3′-methoxyphenacyl bromide (MPB) in ACN and ammonium chloride buffer (pH 9). Volume was then made with ACN Sample: Derivatization as for standard solution followed by extraction Extraction: SPE prewashed with MeOH and water in equal. After sample application, SPE then washed with 1% formic acid followed by 50 mM ammonium acetate equivolume. Analyte eluted with MeOH followed by 50 mM aqueous ammonium acetate. Both eluates then mixed and injected	LOD: n/a LOQ: n/a Rec: 88.9–98.6%	[37, 45]

(continued)

TABLE 2.6 *(continued)*

Analyte(s)	HPLC conditions	Sample	Preparation of standard, sample extraction and clean up	Limit of detection (LOD), limit of quantitation (LOQ), and recovery (Rec)	References
Free clopidogrel base, CCA, clopidogrel AM, and diltiazem (i.s.)	Column: Aquasil C$_{18}$ column (100 mm × 3 mm) Detector: LC–MS/MS operated in electrospray ionization positive mode SRM monitoring: Unchanged clopidogrel: *m/z* 322 → 212 Carboxyl metabolite: *m/z* 308 → 198 Active metabolite: *m/z* 356 → 212 Diltiazem: 415 → 178 Mobile phase: gradient of ACN/ 0.1% formic acid (10–90%, v/v)	Human plasma	Standard: n/a Sample: dissolved in ACN	n/a	[46]
Free clopidogrel base, CCA, clopidogrel AM, and 1-methyl-4-phenylpyridinium bromide (i.s.)	Column: Kromasil C$_8$ column (100 mm × 3 mm) Detector: LC–MS/MS operated in electrospray ionization positive mode SRM monitoring: Unchanged clopidogrel: *m/z* 322 → 212	Human plasma	Standard: n/a Sample: n/a	n/a	[56]

	Matrix	Sample preparation	Method	Validation	Ref.
			Carboxyl metabolite: m/z 308 → 198 Active metabolite: m/z 356 → 212 1-Methyl-4-phenylpyridinium bromide: 170 → 127 Mobile phase: ACN/0.1% formic acid:water/0.1% formic acid = 90:10		
Clopidogrel and CCA	Human plasma	Standard: Sample: LLE using ethyl acetate: pentane = 1:1 (pH 4). Dry extract was reconstituted in ACN	Column: Varian Monochrom silica 3u (100 mm × 4.6 mm) Detector: positive MRM Other details were not available Mobile phase: ACN:ammonium acetate 2 nM = 500:140	LOD: n/a LOQ: n/a Rec: n/a	[57]
Clopidogrel bisulfate and ticlopidine HCl (i.s.)	Human plasma	Standard: stock standard in MeOH then diluted with MeOH:water = 50:50 Sample: LLE: sample extracted with extraction mixture diethyl ether/n-hexane (8:2, v/v) then dried organic residue was reconstituted with 250 μl mixture of 5 mM ammonium formate/MeOH (20:80)	Column: Waters symmetry C$_8$® (150 mm × 4.6 mm i.d.) at 30 °C Detector: LC–MS/MS operated in electrospray ionization MRM monitoring; Clopidogrel: m/z 322.2 → 212.1 Ticlopidine: m/z 264.1 → 154.2 Mobile phase: 5 mM ammonium formate:MeOH = 5:95	LOD: not reported LOQ: 5 pg/ml Rec: 88.2–99.5%	[51]

(continued)

TABLE 2.6 (continued)

Analyte(s)	HPLC conditions	Sample	Preparation of standard, sample extraction and clean up	Limit of detection (LOD), limit of quantitation (LOQ), and recovery (Rec)	References
Clopidogrel and $^2H_3$-clopidogrel (i.s.)	Column: Luna C_8 (50 mm × 2.0 mm) at 50 °C Detector: LC–MS/MS operated in atmospheric pressure ionization positive mode MRM monitoring: Clopidogrel: m/z 322.07 → 212.15 $^2H_3$-clopidogrel: m/z 327.00 → 217.10 Mobile phase: Solvent A (ACN containing 0.1% (v/v) formic acid) and solvent B (water containing 0.1%, v/v formic acid) was run in gradient mode. 35% solvent A and 65% solvent B were held for 0.8 min. At 1.40 min, solvent A% was 70% and solvent B 30%, held until 1.60 min. At 1.70 min, composition was 35% solvent A and 65% solvent B, held until time 3.0 min then baseline	Human plasma	Standard: stock standard in MeOH then diluted with MeOH:water = 50:50 Sample: LLE: sample added with 500 µl of 50 mM ammonium acetate buffer (pH 6.8), then with 2 ml diethyl ether. Dried organic residue was reconstituted in 50 µl of 0.1% (v/v) formic acid in ACN then with 50 µl of 0.1% formic acid in water	LOD: not reported LOQ: 10 pg/ml Rec: 61.5–68.6%	[50]

Clopidogrel, ticlopidine HCl (i.s.)	Human plasma	Column: Hypersil GOLD C$_{18}$ column (2.1 mm × 150 mm) Detector: LC–MS/MS operated in turbo ion spray ionization MRM monitoring: Clopidogrel: m/z 322.2 → 211.9 Ticlopidine: m/z 264.1 → 125.10 Mobile phase: ACN:10 mM ammonium acetate in water = 85:15	Standard: MeOH Sample: LLE using pentane. Dried organic residue was dissolved in mobile phase		[49]
Clopidogrel	Human plasma	Column: Zorbax SB-C$_8$ Detector: LC–MS/MS operated in electrospray ionization Mobile phase: MeOH: ammonium formate (5 mM/l, pH 6.0)	LLE	LOD: n/a LOQ: n/a Rec: 103.1–109.3%	[47]

According to Ksycinska *et al.* [40], buffering plasma samples with 0.1 M ammonium acetate at pH 4.0 helped to weaken drug that had become bound to plasma, therefore making easier the partitioning of drug from an aqueous environment into the organic solvent. In contrast with the common preference for extracting CCA from biological sample, Mitakos and Panderi [52] reported an SPE method for extracting CCA from human plasma samples that did not require acidification. The extraction efficiency obtained using this SPE method ranged from 73.0% to 75.2%. However, this value is seen to be lower relative to other values obtained using the LLE method, which uses either buffer/volatile solvent mixtures or volatile solvents only [40–42].

In the biological samples (blood or plasma), AMs of clopidogrel contain a thiol group, which is reactive and causes degradation of the AM. To overcome the instability of clopidogrel AM, a derivatization method was used, where an alkylating reagent (3-methoxyphenacyl bromide) served to block reactivity of the thiol group. Since this method enabled a successful stabilization of the thiol group, higher amounts of clopidogrel AM in human plasma samples were reported when compared to samples which did not undergo the derivatization reaction [37, 45].

Due to the instability of unchanged clopidogrel, Taubert *et al.* [46] obtained standards of clopidogrel free base by extracting the substance from crushed clopidogrel bisulfate tablets with methanol, followed by extraction using potassium carbonate/cyclohexane, and finally purified over Celite™ 545. However, other workers have reported [49–51] obtaining the free base of clopidogrel by dissolving clopidogrel bisulfate in methanol without the use of potassium carbonate/cyclohexane extraction.

In the analysis of biological samples, the use of an internal standard is needed to ensure that the method employed is specific and selective for the analyte. Isotopically labeling of the analyte is preferable. However, this type of internal standard is not commercially available, and an alternative is to choose an internal standard that has a relevant structure and similar retention behavior to the target analyte [51]. Several drugs have been used as internal standards, with ticlopidine being commonly used due to its close similarity with clopidogrel [42, 49, 51]. However, it should be noted that ticlopidine was observed to have a similar retention time to CCA, so high percentages of aqueous content in the mobile phase and lower flow rates should be used to increase the resolution of ticlopidine and clopidogrel. However, this can result in longer run times and in the reduction of sensitivity [42, 43].

Stability of samples prior to and during analysis is an important consideration when developing and validating an analytical method. For analysis of CCA, Souri *et al.* [42] reported that the stability of CCA in rat plasma samples was up to 48 days, or 3 cycles of freeze–thaw, when stored at -70 ± 5 °C. When stored at ambient temperature (20–25 °C), samples can withstand up to 24 h of exposure, and after reconstitution, samples were stable up to 35 h in an autosampler maintained at 15 °C.

Ksycinska as well as Bahrami [40, 43] reported that clopidogrel in human plasma was stable for 1 month storage at −20 °C, and 60 days for human serum stored at −40 °C. Rapid degradation of clopidogrel AM can be overcome by derivatization, with it being reported that clopidogrel AM in plasma was stable for up to 4 months [45]. For the analysis of unchanged clopidogrel in plasma, Robinson et al. [50] reported that clopidogrel was only stable for 4 h at room temperature. In contrast, Shin as well as Nirogi [49, 51] reported that clopidogrel in human plasma samples did not show significant loss when stored at room temperature for as long as 24 h.

5.2. Gas chromatography

Only one gas chromatographic method has been reported to determine CCA in human plasma samples [53]. The standard metabolite used was the hydrochloride salt of CCA, while an analogous hydrochloride salt of the carboxylic acid was used as internal standard. Extraction by LLE method followed by SPE, and subsequent derivatization, was used to extract the metabolite of clopidogrel from human plasma and serum since this procedure could minimize matrix effects. In the LLE procedure, formic acid and diethyl ether were used to extract the analyte, followed by an SPE extraction of the residue in methanol using a C_{18} SPE column.

Derivatization was conducted by the addition of a 10% n-ethyl-diiso-propylethylamine solution and α-bromo-2,3,4,5,6-pentafluorotoluene. Sample obtained from the derivatization procedure were dissolved in ethyl acetate prior to injection in splitless mode using a DB-1 capillary column. Helium was used as the mobile phase, and the injector temperature was set at 290 °C with a transfer line temperature of 270 °C. Sample detection used ion trap MS for detection, with the detector being set at negative chemical ionization with $m/z = 262$ (for CCA) and $m/z = 286$ (for the internal standard). The limit of quantitation was 5 ng/ml, and the average recovery ranged from 92.0% to 114%. In addition, the extraction efficiency ranged from 48.2% to 55.6% for concentrations of 5, 50, and 250 ng/ml. Samples were reported to be stable for up to 6 months when stored at −18 °C.

REFERENCES

[1] S.C. Sweetman (Ed.), Martindale: The Complete Drug Reference, 35th ed., The Pharmaceutical Press, London, 2007 (CD ROM).

[2] United States Pharmacopoeia 32 National Formulary 27, The United States Pharmacopoeial Convention, Rockville, MD, 2009, 1992–1993.

[3] Product monograph Plavix, http://www.sanofi-aventis.ca/products/en/plavix.pdf (May 30, 2009).
[4] S.J. Gardell, Perspect. Drug Discov. Des. 1 (1993) 521–526.
[5] R. A. Badorc, D. Fréhel, US Patent no. 4847265, 1989.
[6] M.-H. Ki, M.-H. Choi, K.-B. Ahn, B.-S. Kim, D.S. Im, S.-K. Ahn, H.-J. Shin, Arch. Pharm. Res. 31 (2008) 250–258.
[7] R. Petkovska, C. Cornett, A. Dimitrovska, Maced. J. Chem. Chem. Eng. 27 (2008) 53–64.
[8] A. Mohan, M. Hariharan, E. Vikraman, G. Subbaiah, B.R. Venkataraman, D. Saravanan, J. Pharm. Biomed. Anal. 47 (2008) 183–189.
[9] A. Bousquet, B. Castro, J. Saint-Germain, Polymorphic form of clopidogrel hydrogen sulfate, United States Patent 6,504,030, issued January 7, 2003.
[10] R. Lifshitz-Liron, E. Kovalevski-Ishai, S. Wizel, S.A. Maydan, R. Lidor-Hadas, US Patent no. US 2003/0114479 A1, 2003.
[11] M.S.J. Mukarram, Y.A. Merwade, R.A. Khan, US Patent no. US 7291735 B2, 2007.
[12] D.G. Sankar, S.K. Sumanth, A.K.M. Pawar, P.V.M. Latha, Asian J. Chem. 17 (2005) 2022–2024.
[13] P. Mishra, A. Dolly, Indian J. Pharm. Sci. 68 (2006) 365–368.
[14] S.J. Rajput, R.K. George, D.B. Ruikar, Indian J. Pharm. Sci. 70 (2008) 450–454.
[15] H.E. Zaazaa, S.S. Abbas, M. Abdelkawy, M.M. Albedrahman, Talanta 78 (2009) 874–884.
[16] V. Koradia, G. Chawla, A.K. Bansal, Acta Pharm. 54 (2004) 193–204.
[17] Z. Német, Á. Demeter, G. Pokol, J. Pharm. Biomed. Anal. 49 (2009) 32–41.
[18] A.S. Fayed, S.A. Weshahy, M.A. Shehata, N.Y. Hassan, J. Pharm. Biomed. Anal. 49 (2009) 193–200.
[19] A.L. Saber, M.A. Elmosallamy, A.A. Amin, A.H.M.A. Killa, J. Food Drug Anal. 16 (2008) 11–18.
[20] V. Uvarov, I. Popov, J. Pharm. Biomed. Anal. 46 (2008) 676–682.
[21] M. S. Alam, S.B. Patel, A.K. Bansal, AAPS Annual Meeting, Atlanta, Georgia, 2008.
[22] H. Agrawal, N. Kaul, A.R. Paradkar, K.R. Mahadik, Talanta 61 (2003) 581–589.
[23] N. Kamble, A. Venkatachalam, Indian J. Pharm. Sci. 67 (2005) 128–129.
[24] Y. Gomez, E. Adams, J. Hoogmartens, J. Pharm. Biomed. Anal. 34 (2004) 341–348.
[25] K. Nikolic, B. Ivković, Ž. Bešović, S. Marković, D. Agbaba, Chirality, 2009, early view, doi:10.1002/chir.20681.
[26] K. Anandakumar, T. Ayyappan, V.R. Raman, T. Vetrichelvan, A.S.K. Sankar, D. Nagavalli, Indian J. Pharm. Sci. 69 (2007) 597–599.
[27] J. Sippel, L.L. Sfair, E.E.S. Schapoval, M. Steppe, J. AOAC Int. 91 (2008) 67–72.
[28] R.B. Patel, M.B. Shankar, M.R. Patel, K.K. Bhatt, J. AOAC Int. 91 (2008) 750–755.
[29] M. Gandimathi, T. Ravi, Indian J. Pharm. Sci. 69 (2007) 123–125.
[30] P. Shrivastava, P. Basniwal, D. Jain, S. Shrivastava, Indian J. Pharm. Sci. 70 (2008) 667–669.
[31] A. Mitakos, I. Panderi, J. Pharm. Biomed. Anal. 28 (2002) 431–438.
[32] H.Y. Aboul-Enein, H. Hoenen, A. Ganem, M. Koll, J. Liq. Chromatogr. 28 (2005) 1357–1365.
[33] J.C. Spell, J.T. Stewart, J. Pharm. Biomed. Anal. 18 (1998) 453–460.
[34] N. Tanaka, H. Kobayashi, N. Ishizuka, H. Minakuchi, K. Nakanishi, K. Hosoya, T. Ikegami, J. Chromatogr. A 965 (2002) 35–49.
[35] K. Cabrera, J. Sep. Sci. 27 (2004) 843–852.
[36] M. Reist, M.R.-D. Vos, J.-P. Montseny, J.M. Mayer, P.-A. Carrupt, Y. Berger, B. Testa, Drug Metab. Dispos. 28 (2000) 1405–1410.
[37] M. Takahashi, H. Pang, K. Kawabata, N.A. Farid, A. Kurihara, J. Pharm. Biomed. Anal. 48 (2008) 1219–1224.

[38] A.A.C.M. Heestermans, J.W.V. Werkum, D. Taubert, T.H. Seesing, N.V. Beckerath, C.M. Hackeng, E. Schömig, F.W.A. Verheugt, J.M.T. Berg, Thromb. Res. 122 (2008) 776–781.

[39] Y. Tomoko, N. Mihoko, W. Mitsuhiro, N. Ken'ichiro, Chromatographia 27 (2006) 71–72.

[40] H. Ksycinska, P. Rudzki, M. Bukowska-Kiliszek, J. Pharm. Biomed. Anal. 41 (2006) 533–539.

[41] S.S. Singh, K. Sharma, D. Barot, P.R. Mohan, V.B. Lohray, J. Chromatogr. B 821 (2005) 173–180.

[42] E. Souri, H. Jalalizadeh, A. Kebriaee-Zadeh, M. Shekarchi, A. Dalvandi, Biomed. Chromatogr. 20 (2006) 1309–1314.

[43] G. Bahrami, B. Mohammadi, S. Sisakhtnezhad, J. Chromatogr. B 864 (2008) 168–172.

[44] J.-M. Pereillo, M. Maftouh, A. Andrieu, M.-F. Uzabiaga, O. Fedeli, P. Savi, M. Pascal, J.-M. Herbert, J.P. Maffrand, C. Picard, Drug Metab. Dispos. 30 (2002) 1288–1295.

[45] M. Takahashi, H. Pang, A. Kikuchi, K. Kawabata, A. Kurihara, N.A. Farid, T. Ikeda, ASMS Conference on Mass Spectroscopy and Allied Topics, 2006.

[46] D. Taubert, A. Kastrati, S. Harlfinger, O. Gorchakova, A. Lazar, N.V. Beckerath, A. Schömig, E. Schömig, Thromb. Haemost. 92 (2004) 311–316.

[47] A. Lainesse, Y. Ozalp, H. Wong, R.S. Alpan, Arzneimittelforschung 54 (2004) 600.

[48] P. Savi, E. Heilmann, P. Nurden, M.-C. Laplace, C. Bihour, G. Kieffer, A.T. Nurden, J.-M. Herbert, Clin. Appl. Thromb. Hemost. 2 (1996) 35–42.

[49] B.S. Shin, S.D. Yoo, Biomed. Chromatogr. 21 (2007) 883–889.

[50] A. Robinson, J. Hillis, C. Neal, A.C. Leary, J. Chromatogr. B 848 (2007) 344–354.

[51] R.V.S. Nirogi, V.N. Kandikere, M. Shukla, K. Mudigonda, S. Maurya, R. Boosi, Rapid Commun. Mass Spectrom. 20 (2006) 1695–1700.

[52] A. Mitakos, I. Panderi, Anal. Chim. Acta 505 (2004) 107–114.

[53] P. Lagorce, Y. Perez, J. Ortiz, J. Necciari, F. Bressole, J. Chromatogr. B 720 (1998) 107–117.

[54] H.-J. Kim, K.-J. Kim, J. Pharm. Sci. 97 (2008) 4473–4484.

[55] D. Antić, S. Filipić, D. Agbaba, Acta Chromatogr. 18 (2007) 199–206.

[56] N.V. Beckerath, D. Taubert, G. Pogatsa-Murray, E. Schomig, A. Kastrati, A. Schomig, Circulation 112 (2005) 2946–2950.

[57] S.A. Adcock, O.E. Espinosa, M.P. Sullivan, C.J.L. Bugge, J.G. Stark, A. Terry, AAPS Annual Meeting and Exposition, 2005.

[58] H. Mani, S. Toennes, B. Linnemann, D.A. Urbanek, J. Schwonberg, G.F. Kauert, E. Lindhoff-Last, Ther. Drug Monit. 30 (2008) 84–89.

[59] Examples of complex chiral separations on CHIRAL-AGP and CHIRAL-CBH, http://www.cromtech.co.uk (May 25, 2009).

Donepezil

Yousif A. Asiri* and Gamal A.E. Mostafa[†]

* Department of Clinical Pharmacy, College of Pharmacy, King Saud University, Riyadh, Saudi Arabia
† Department of Pharmaceutical Chemistry, College of Pharmacy, King Saud University, Riyadh, Saudi Arabia

Profiles of Drug Substances, Excipients, and Related Methodology, Volume 35
ISSN 1871-5125, DOI: 10.1016/S1871-5125(10)35003-5

1. DESCRIPTION

1.1. Nomenclature

1.1.1. Systematical chemical name

- 2,3-Dihydro-5,6-dimethyoxy-2-[[1-phenylmethyl)-4-piperidinyl] methyl]1*H*-inden-1-one.
- 5,6-Dimethoxy-2-[[1-(phenylmethyl)4-piperidinyl-methyl]-2,3-dihydro-1*H*-inden-1-one.
- 1-Benzyl-4-[(5,6-dimethoxy-1-indanon 2-yl)methyl]piperidine [1].

1.1.2. Nonproprietary names

Donepezil, donepezil hydrochloride (E2020).

1.1.3. Proprietary names

Aricept, Memac.

1.2. Formulae

1.2.1. Emperical formula, molecular weight, and CAS number

Donepezil	$C_{24}H_{29}NO_3$	379.5	[120014-06-4]
Donepezil hydrochloride	$C_{24}H_{29}NO_3 \cdot HCl$	415.95	[120011-70-3]

1.2.2. Structral formula

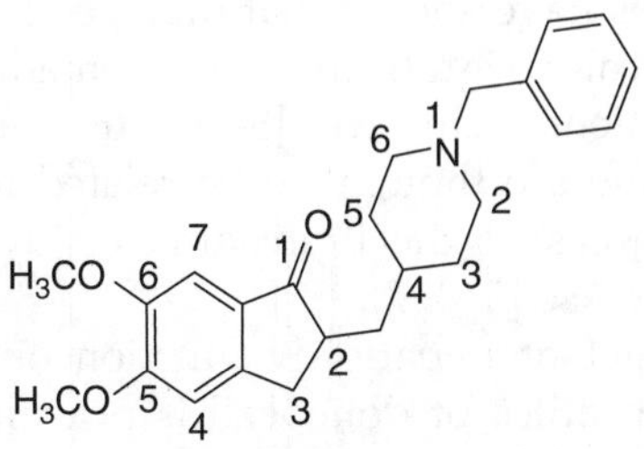

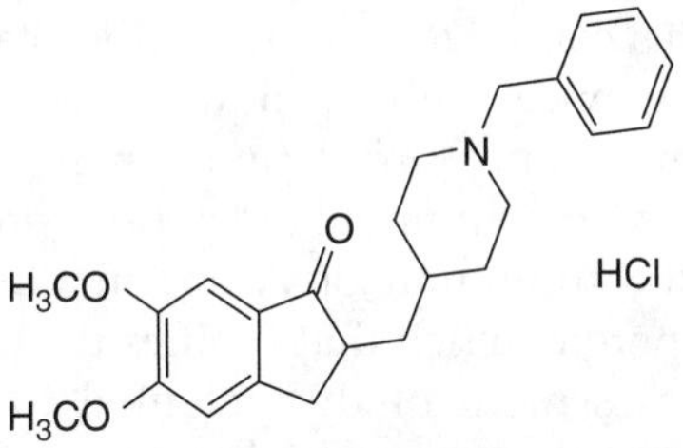

Donepezil

Donepezil hydrochloride

1.3. Elemental analysis

- Base: C 75.96%, H 7.70%, N 3.69%, O 12.6% [1].
- HCl salt: C 69.30%, H 7.27%, N 3.37%, O 11.54%, Cl 8.52 [1].

1.4. Physical properties

1.4.1. Appearance
A white to off-white solid [2].

1.4.2. Solubility
Donepezil hydrochloride is a white powder and is freely soluble in water, soluble in chloroform, sparingly soluble in glacial acetic acid and in ethanol, slightly soluble in acetonitrile, very slightly soluble in ethyl acetate, and insoluble in n-hexane [2].

1.4.3. Dissociation constant
pK_a 8.90 [2].

1.5. Uses and applications

Donepezil hydrochloride (E2020) is the second drug approved by the US FDA for the treatment of mild to moderate Alzheimer's diseases (AD). It is a new class of acetylcholinesterase (AChE) inhibitor having an

N-benzylpiperidine and an indanone moiety which shows longer and more selective action. It is now marketed in the United States and in some European and Asian countries under the trade name of Aricept®. In Japan, Aricept® is now under application to the Ministry of Welfare [3].

Donepezil HCl, a piperidine, is a highly selective inhibitor of the enzyme AChE [3, 4] that is chemically unique from other AChE inhibitors [5, 6]. *In vitro* and preclinical studies have demonstrated that donepezil is approximately 1200 times more selective for AChE in the brain than for butyrylcholinesterase (BuChE) in the periphery [3, 4, 7]. Phase II and III studies conducted in the United States have shown that donepezil (5 or 10 mg once daily) produces statistically significant improvements in cognition and global function in patients with AD [8–10]. Its clinical efficacy and minimal side-effect profile are thought to be related to its specific inhibition of AChE in the areas of the brain affected by the cholinergic deficit that typifies this disease [3, 4, 7].

Assessment of the potential impact of hepatic dysfunction on the pharmacokinetic and adverse event profiles of donepezil is of primary importance, as donepezil is orally administered and subject to extensive first-pass metabolism. In addition, both preclinical and clinical studies have demonstrated that donepezil is metabolized primarily in the liver [8–11]. *In vitro* studies using human hepatic microsomes have shown that the cytochrome P-450 isoenzyme (CYP-3A4) is mostly responsible for the metabolism of donepezil, with CYP-2D6 playing a minor role. With this in mind, this study was designed to assess the effects of compromised hepatic function on the pharmacokinetics of donepezil HCl [8–11].

2. METHOD OF PREPARATION [12–18]

1. Donepezil was prepared through three steps; the first one is coupling reaction in which compound of formula **1** is reacted with a compound of formula **2**. The reaction can be carried out at 30–70 °C in the presence of suitable base such as potassium carbonate in the presence of dimethylformamide (DMF). The second step is protecting group removal. The third step is substitution of nitrogen atom, and the last step is hydrolysis and decrboxylation [12, 13] as in Scheme 3.1.

2. Donepezil was prepared by reacting of 5,6-dimethy-1-indanone **1** with 1-benzyl-4-formylpiperidine **2** in the presence of strong base such as lithium diispopylamide to give 1-benzyl-4-[(5,6-dimethoxy-1-indanon)-2-ylidenyl]methyl piperidine hydrochloride **3** [12]. The next step is catalytic hydrogenation with 10% palladium on charcoal in THF. The product was purified by column chromatography. The process is illustrated in Scheme 3.2.

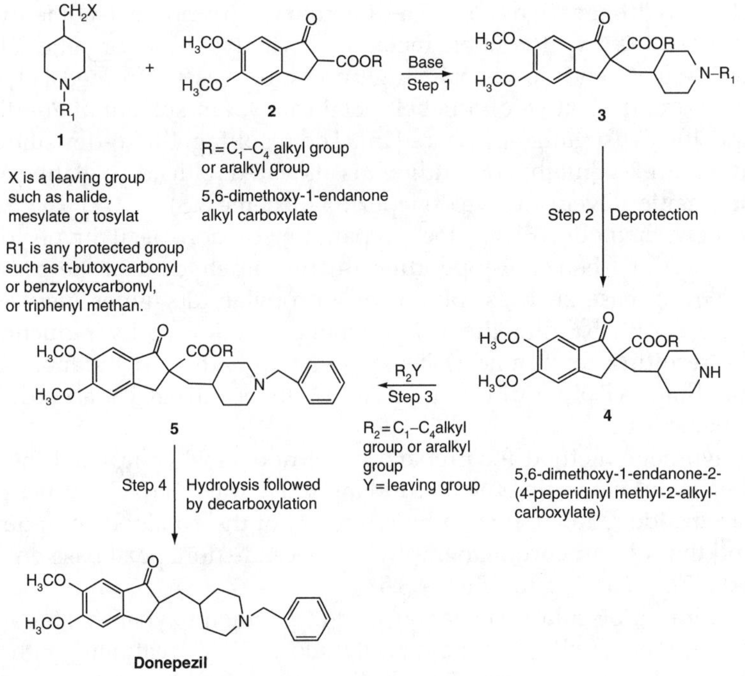
CH2X
R1
1
X is a leaving group
such as halide,
mesylate or tosylat
R1 is any protected group
such as t-butoxycarbonyl
or benzyloxycarbonyl,
or triphenyl methan.
H3CO
H3CO
COOR
2
R = C1–C4 alkyl group
or aralkyl group
5,6-dimethoxy-1-endanone
alkyl carboxylate
Base
Step 1
3
N–R1
Step 2
Deprotection
NH
4
5,6-dimethoxy-1-endanone-2-
(4-peperidinyl methyl-2-alkyl-
carboxylate)
R2Y
Step 3
R2 = C1–C4alkyl
group or aralkyl
group
Y = leaving group
5
Step 4
Hydrolysis followed
by decarboxylation
Donepezil

SCHEME 3.1

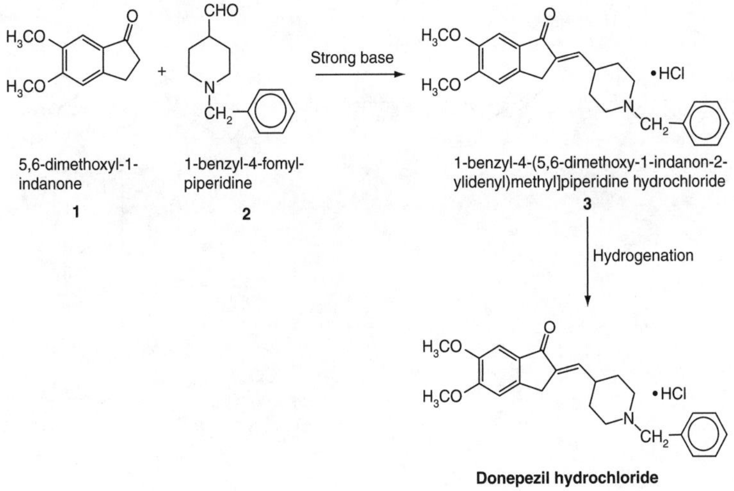
H3CO
H3CO
CHO
CH2
5,6-dimethoxyl-1-
indanone
1
1-benzyl-4-fomyl-
piperidine
2
Strong base
•HCl
1-benzyl-4-(5,6-dimethoxy-1-indanon-2-
ylidenyl)methyl]piperidine hydrochloride
3
Hydrogenation
Donepezil hydrochloride

SCHEME 3.2

3. The hydrogenation of 5,6-dimethoxy-2-(pyridine-4-yl)methylene indan-1-one **1** with a noble metal oxide catalyst in an organic solvent at 20–50 °C and 10–45 psi gauge pressure was performed [14] or a nonoxide derivative of a nobel metal catalyst in solvent at 20–100 °C and 10–90 psi gauge pressure [15]. The resulting 4-[(5,6-dimethoxy-1-indanon)-2-yl]methyl piperidine **2** is alkylated with an alkylating agent in organic solvent to give donepezil **3** (Scheme 3.3).

4. A new method [16] for the preparation of donepezil involved the reaction of 1-benzyl-4-piperidine-1-carboxaldehyde **1** in the presence of strong base such as lithium diisopropylamide under inert atmosphere with 5,6-dimethoxy-1-endanone **2** followed by reduction of the resulting compound (1-benzyl-4-[5,6-dimethoxy-1-indanon)-2-ylidinyl]methyl piperidine **3** to give the donepezil HCl **4** according to Scheme 3.4.

5. A synthetic method for preparation of donepezil comprised the condensation of 5,6-dimethoxy-1-indanone **1** with 1-benzyl-4-piperidinecarboxaldehyde **2** followed by reduction of the obtained compound **3** and the column chromatography of the crude donepezil base on silica gel [17], according to Scheme 3.5.

6. The intramolecular cyclization of 2-(3,4-dimethoxybenzyl)-3-(N-benzyl-4-piperidinyl)propionic acid **1** followed by treatment with HCl [18] as shown in Scheme 3.6. Cyclization of compound **1** was carried out under Friedel–Crafts reaction conditions, optionally with previous derivatization of the carboxylic group to a halocarbonyl group. Preferably, the cyclization process was carried out in the presence of protic acids or Lewis acids or a mixture of protic and Lewis acids gives donepezil **2** which is converted to the hydrochloride salt **3**.

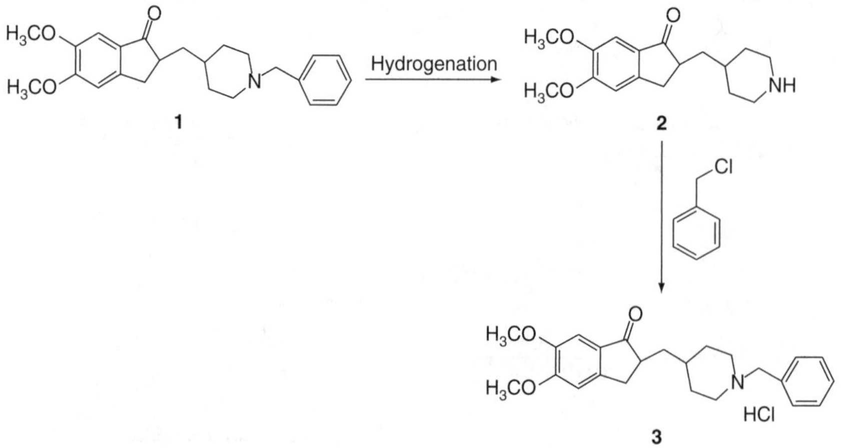

SCHEME 3.3

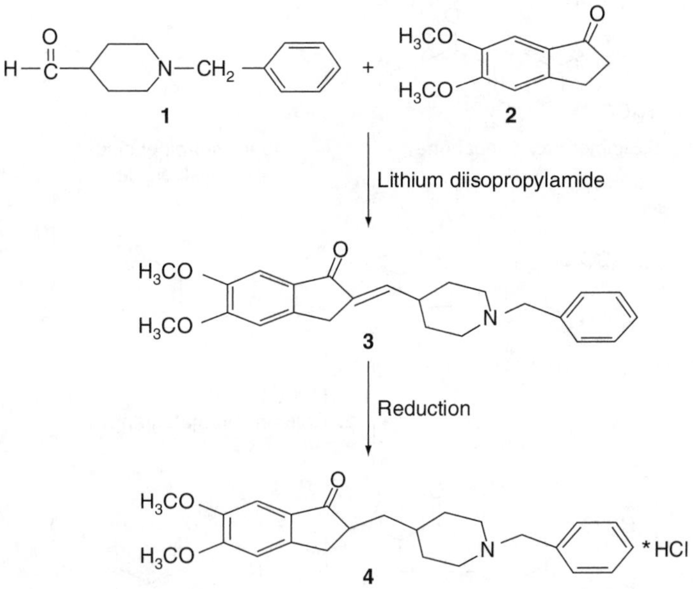

SCHEME 3.4

3. PHYSICAL PROPERTIES

3.1. Spectroscopy [2]

3.1.1. Ultraviolet spectroscopy

The ultraviolet (UV) absorption spectrum of donepezil in methanol was scanned from 200 to 400 nm, using UV/VIS spectrometer (Shimadzu ultraviolet–visible spectrophotometer 1601 PC). The compound exhibited four maxima at 313, 269, 230, and 209 (Fig. 3.1). Clarke [2] reported the following: aqueous acid 230, 271, and 316 nm [2].

3.1.2. Vibrational spectroscopy

The infrared absorption spectrum of donepezil was obtained in KBr pellet using a Perkin-Elmer infrared spectrophotometer. The IR spectrum is shown in Fig. 3.2. The principal peaks were observed at 3008, 2924, 2849, 1589, 1690, 1453, 1376, 1312, 1071, and 702 cm^{-1}. Assignments for the major infrared absorption band are provided in Table 3.1. Clarke [2] reported principal peaks at 1605, 1589, 1500, 759, and 702 cm^{-1} [2].

3.1.3. Nuclear magnetic resonance spectrometry

^{1}H and ^{13}C nuclear magnetic resonance (NMR) spectra of donepezil were registered with a Varian Gemini 200 spectrometer (200 MHz). Chemical shifts were expressed in parts per million (ppm) with respect to the tetramethysilane signal for ^{1}H and ^{13}C NMR (Figs. 3.3 and 3.4, respectively).

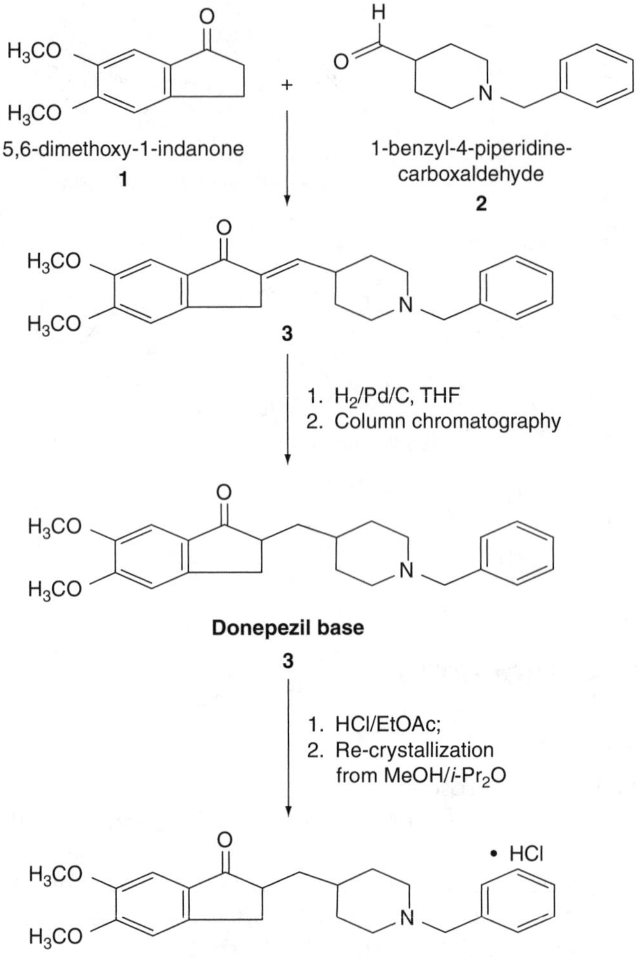

SCHEME 3.5

3.1.3.1. ^1H NMR spectrum The one-dimensional proton ^1H NMR spectrum of donepezil base dissolved in CDCl$_3$ is shown in Fig. 3.3. The corresponding spectral assignments ^1H NMR for donepezil are provided in Table 3.2.

3.1.3.2. ^{13}C NMR spectrum The one-dimensional ^{13}C NMR spectrum of donepezil dissolved in CDCl$_3$, which was recorded at 24 °C, is shown in Fig. 3.4. The assignments for the observed resonance bands associated with the various carbons are listed in Table 3.3. Figures 3.5–3.9 show the COSY, HSQC, HMBC, DEPT 90, and DEPT 135 NMR spectra, respectively.

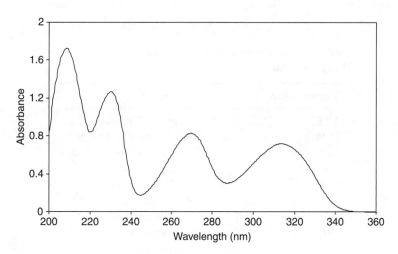

SCHEME 3.6

FIGURE 3.1   UV spectrum of donepezil in ethanol.

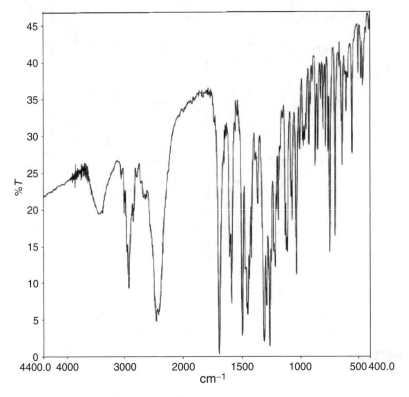

FIGURE 3.2   Infrared spectrum of donepezil (KBr disc).

TABLE 3.1   Vibrational assignments for donepezil infrared
absorption bands

Frequency (cm^{-1})	Assignment
3008	Aromatic CH stretch
2924	Aliphatic CH_2 stretch
2849	Aliphatic CH stretch
1589	C=C aromatic
1690	C=O carbonyl
1453	–CH_2– bending
1376	C–H bending (aliphatic)
1312	C–N stretch
1071	C–C stretch
702	C–H bending (aromatic)

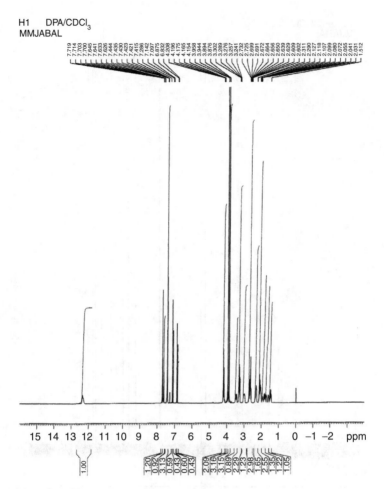

FIGURE 3.3 ¹H NMR spectrum of donepezil in CDCl₃.

3.2. Mass spectrum

Mass spectra of donepezil, carried out with electron impact method, were registered using Varian 320-GC/MS spectrometer. Figure 3.10 shows the detailed mass fragmentation pattern and Table 3.4 shows the mass fragmentation pattern of the drug substance. Clarke [2] reported the presence of the following principal peaks at $m/z = 288, 379, 91, 172, 189, 378, 191$ [2].

3.3. X-Ray powder diffraction pattern [19–22]

X-ray powder diffraction pattern has been obtained on D8-Advanced (Fig. 3.11), Bruker AXE Germany, diffractometer equipped with scintillation detector using copper Kα (=1.5406 Å) radiation with scanning range

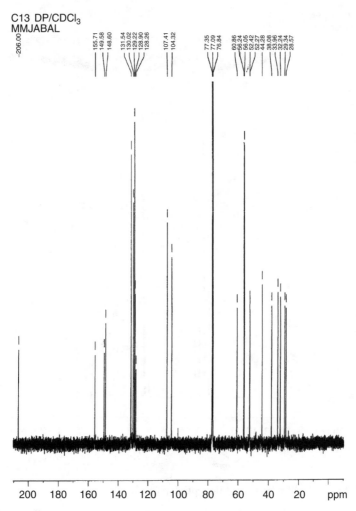

FIGURE 3.4 ^{13}C NMR spectrum of donepezil in CDCl$_3$.

between 2θ and 50θ (theta) at scanning speed of $2°\mathrm{min}^{-1}$. Detail process for preparing the donepezil hydrochloride amorphous and its polymorph form was described [20]. A full data summary is compiled in Table 3.5.

3.4. Thermal method of analysis

3.4.1. Melting point
M.P. 224 °C.

TABLE 3.2 Assignment of the resonance bands in the ^{1}H NMR spectrum of donepezil

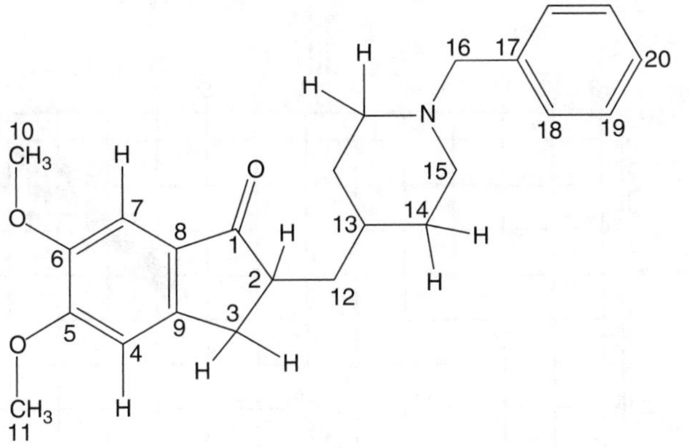

Chemical shift (ppm) relative to TMS	Number of protons	Multiplicity	Assignment (proton at carbon atom)
3.43–3.51	1	m	2
3.25–3.30	2	m	3
7.11	1	s	4
6.84	1	s	7
3.86	3	s	10
3.95	3	s	11
1.48–1.54	2	m	12
1.92–1.96	1	m	13
1.77–1.87	4	m	14
2.04–2.14	4	m	15
4.15	2	s	16
7.48	2	d	18
7.43–7.47	2	m	19
7.63–7.65	1	m	20

s, singlet; d, doublet; m, multiplet.

3.4.2. Differential scanning calorimetry

The donepezil hydrochloride has been charchaterized by differential scanning calorimetry DSC, which exhibits a significant endo peak around 229.85 °C [19]. The DSC thermogram of donepezil hydrochloride is substantially depicted as in Fig. 3.12.

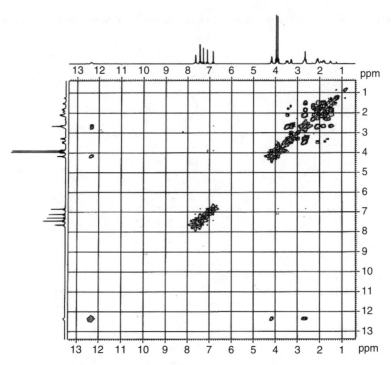

FIGURE 3.5   COSY ^{1}H NMR spectrum of donepezil in CDCl$_3$.

4. METHODS OF ANALYSIS

4.1. Spectrophotometry

Stability indicating assay methods for determination of donepezil hydro-chloride in the presence of its oxidative degradate were developed and validated. The first three methods are spectrophotometric depending on using zero order (D(0)), first order (D(1)), and second order (D(2)) spectra [23]. The absorbance was measured at 315 nm for D(0) while the ampli-tude was measured at 332.1 nm for D(1) and 340 nm for D(2) using deionized water as a solvent. Donepezil hydrochloride can be determined in the presence of up to 70% of its oxidative degradate (II) using D(0), 80% using D(1), and 90% using D(2). The linearity range was found to be 8–56 μg/ml for D(0), D(1), and D(2). These methods were applied for the analysis of donepezil HCl in both powder and tablet form. Also, a spectrofluorimetric method, depending on measuring the native fluores-cence of donepezil hydrochloride in deionized water using lambda exci-tation 226 nm and lambda emission 391 nm, is suggested. The linearity range was found to be 0.32–3.20 μg/ml using this method, donepezil hydrochloride was determined in the presence of up to 90% of oxidative

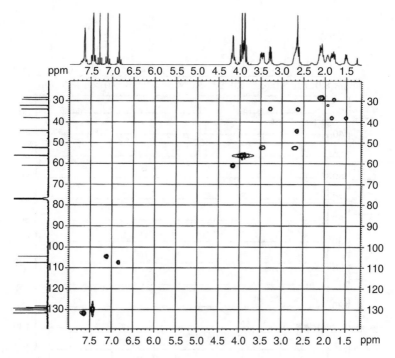

FIGURE 3.6 The HSQC NMR spectrum of donepezil in CDCl₃.

product. The proposed method was applied for the analysis of donepezil HCl in tablet form as well as in human plasma. The last method depends on using TLC separation of donepezil hydrochloride from its oxidative degradate, and donepezil HCl was then determined spectrodensitometrically. The mobile phase was methanol:chloroform:25% ammonia (16:64:0.1, v/v/v). The linearity range was found to be 2–15 μg/spot. This method was applied to the analysis of donepezil HCl in both powder and tablet form using acetonitrile as a solvent.

4.2. Potentiometry

The construction and electrochemical response characteristics of poly (vinyl chloride) membrane sensors for donepezil HCl are described. The sensing membranes incorporate ion association complexes of donepezil HCl cation and sodium tetraphenyl borate (sensor 1), or phosphomolybdic acid (PMA) (sensor 2), or phosphotungstic acid (sensor 3) as electroactive materials. The sensors display a fast, stable, and near-Nernstian response over a relative wide donepezil HCl concentration

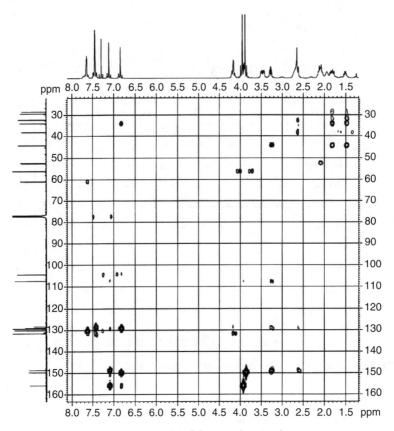

FIGURE 3.7 The HMBC NMR spectrum of donepezil in CDCl₃.

range (1×10^{-2} to $\sim 1 \times 10^{-6}$ M), with cationic slopes of 53.0, 54.0, and 51.0 mV per concentration decade over a pH range of 4.0 and 8.0. The sensors show good discrimination of donepezil HCl from several inorganic and organic compounds. The direct determination of 2.5–4000.0 μg/ml of donepezil HCl shows an average recovery of 99.0%, 99.5%, and 98.5% and a mean relative standard deviation (RSD) of 1.6%, 1.5%, and 1.7% at 100.0 μg/ml for sensors 1, 2, and 3, respectively. The proposed sensors have been applied for direct determination of donepezil HCl in some pharmaceutical preparations. The results obtained by determination of donepezil HCl in tablets using the proposed sensors are comparable favorably with those obtained using the high-performance liquid chromatography (HPLC) method. The sensors have been used as indicator electrodes for potentiometric titration of donepezil [24].

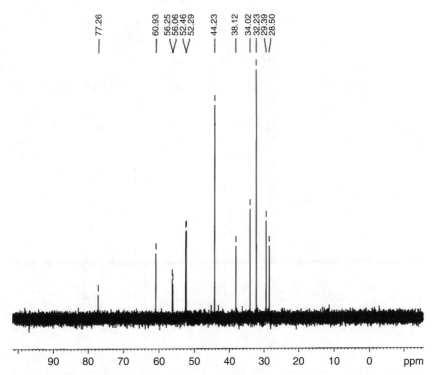

FIGURE 3.8 The DEPT 90 ^{13}C NMR of donepezil in CDCl$_3$.

4.3. Voltammetry

The voltammetric behavior of donepezil was studied at a glassy carbon electrode using cyclic, linear sweep, differential pulse voltammetry (DPV), and Osteryoung square-wave (OSWV) voltammetric techniques [25]. Donepezil exhibited irreversible anodic waves within the pH range 1.80–9.00 in different supporting electrolytes. The peak was characterized as being irreversible and diffusion-controlled. The possible mechanism of the oxidation process is discussed. The current–concentration plot was rectilinear over the range from 1×10^{-6} to 1×10^{-4} M in Britton–Robinson buffer at pH 7.0 with a correlation coefficient between 0.997 and 0.999 in supporting electrolyte and human serum samples using the DPV and OSWV techniques. The repeatability and reproducibility of the methods for both media (supporting electrolyte and serum sample) were determined. Precision and accuracy of the developed methods were demonstrated by recovery studies. The standard addition method was used for the recovery studies. No electroactive interferences were found in biological fluids from endogenous substances or additives present in

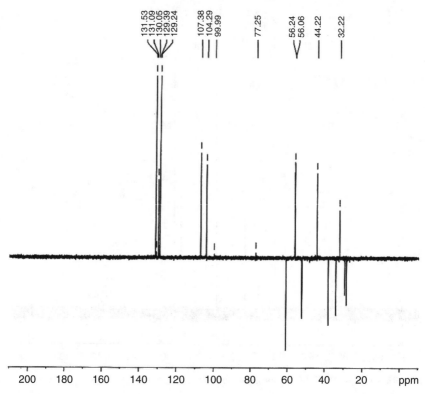

FIGURE 3.9 The DEPT 135 ^{13}C NMR of donepezil in CDCl$_3$.

tablets. The methods developed were successfully applied to the determination of donepezil in tablets and in spiked human serum.

4.4. Chromatographic method

4.4.1. High-performance liquid chromatography/ultraviolet

A new gradient HPLC method has been developed and validated for the determination of both assay and related substances of donepezil hydrochloride in oral pharmaceutical formulations [26]. Different kinds of columns and gradient elution programs were tested to achieve satisfactory separation between the active substance, four impurities, and an interfering excipients used in the formulation. The best results were obtained using an Uptisphere ODB C$_{18}$ column 250 mm × 4.6 mm, 5 μm, UV detection at 270 nm and a gradient elution using phosphate buffer (0.005 M, pH 3.67), and methanol as the mobile phase. The method was validated with respect to linearity, precision, accuracy, specificity, and robustness. It was also found to be stability indicating, and therefore

TABLE 3.3 Assignment for the resonance bands in the ^{13}C NMR spectrum of donepezil

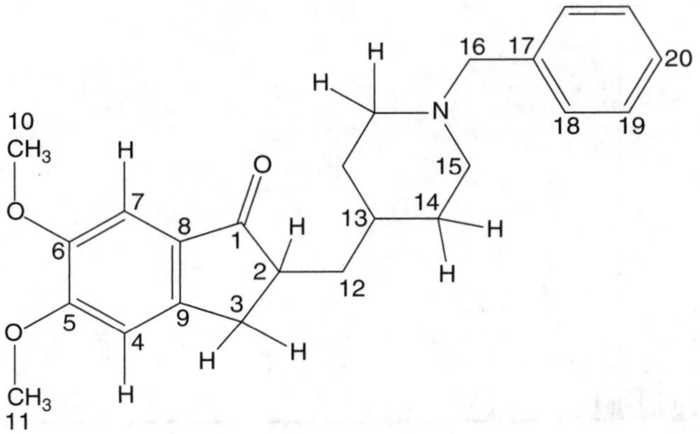

Chemical shift (ppm)	Carbon number
206.77	1
44.24	2
34.24	3
104.36	4
155.74	5
149.62	6
107.41	7
130.04	8
148.58	9
56.05	10
56.22	11
38.11	12
32.24	13
28.49	14
52.49	15
60.93	16
131.53	17
129.22	18
128.92	19
128.21	20

suitable for the routine analysis of donepezil hydrochloride and related substances in the pharmaceutical formulations

A simple and sensitive HPLC method with UV absorbance detection is described for the quantification of donepezil, in human plasma [27].

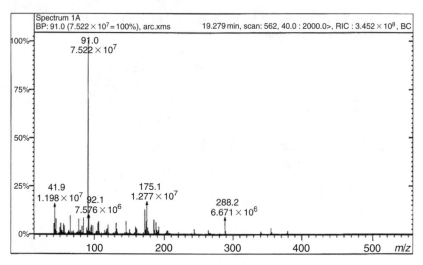

FIGURE 3.10   Mass spectrum of donepezil.

After sample alkalinization with 0.5 ml of NaOH (0.1 M), the test compound was extracted from 1 ml of plasma using isopropanol:hexane (3:97, v/v). The organic phase was back-extracted with 75 μl of HCl (0.1 M) and 50 μl of the acid solution was injected into a C_{18} STR ODS-II analytical column (5 μm, 150 $\times$ 4.6 mm i.d.). The mobile phase consisted of phosphate buffer (0.02 M, pH 4.6), perchloric acid (6 M), and acetonitrile (59.5:0.5:40, v/v/v) and was delivered at a flow rate of 1.0 ml/min at 40 °C. The peak was detected using a UV detector set at 315 nm, and the total time for a chromatographic separation was ~8 min. The method was validated for the concentration range 3–90 ng/ml. Mean recoveries were 89–98%. Intra- and inter-day RSDs were less than 7.3% and 7.6%, respectively, at the concentrations ranging from 3 to 90 ng/ml. The method shows good specificity with respect to commonly prescribed psychotropic drugs, and it could be successfully applied for pharmacokinetic studies and therapeutic drug monitoring.

4.4.2. High-performance liquid chromatography/fluorescence

A simple and sensitive HPLC method with fluorescence detection for determination of donepezil in plasma and microdialysate samples was developed [28]. A rapid isocratic separation of donepezil could be achieved by a short C_{30} column using mobile phases of 25 mM citric acid/50 mM Na_2HPO_4 (pH 6.0):CH_3CN (73:27%, v/v) containing 3.5 mM sodium 1-octanesulfonate for plasma and H_2O:CH_3CN:CH_3OH (80:17:3%, v/v/v) containing 0.01% acetic acid for microdialysate. The flow rate of eluent was set at 1 ml/min, the eluate was monitored at 390 nm with an excitation at 325 nm. The detection limits ($S/N = 3$) of

TABLE 3.4   Mass spectral fragmentation patter of donepezil

m/z	Relative intensity (%)	Formula	Structure
379.2	1	$C_{23}H_{27}NO_3$	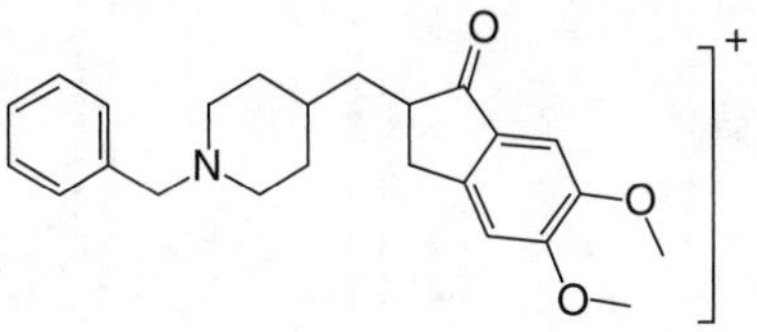
288.2	7.5	$C_{17}H_{23}NO_3$	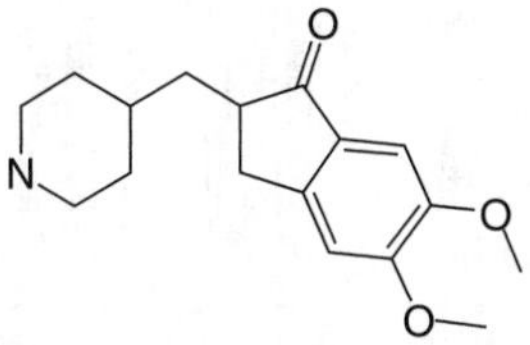
243.1	2.5	$C_{15}H_{17}NO_2$	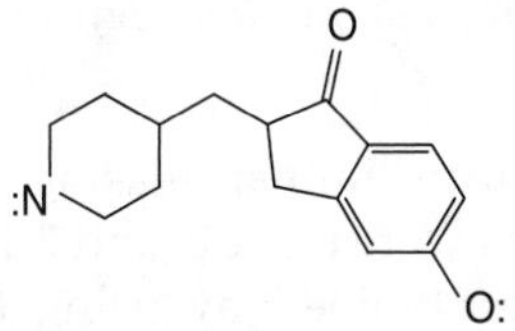
192.1	1	$C_{11}H_{12}O_3$	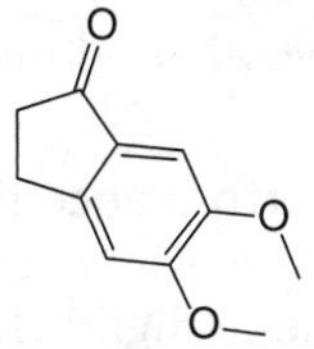
175.1	2	$C_{12}H_{17}N$	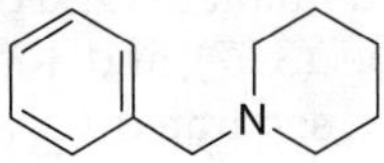
91.1	100	C_7H_7	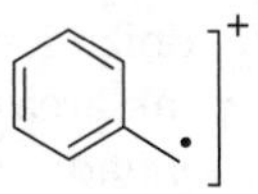
92.1	2.5	C_7H_8	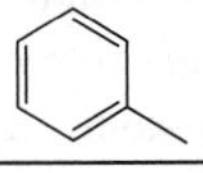

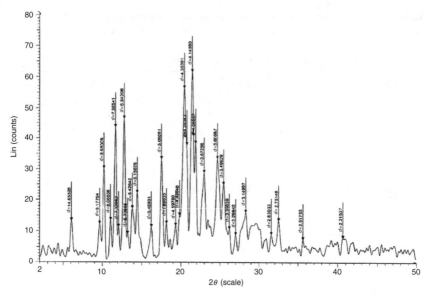

FIGURE 3.11 X-Ray powder diffraction pattern of donepezil hydrochloride [20].

donepezil for human plasma, rat plasma, and rat brain or blood micro-dialysates were 0.2, 1.0, and 2.1 ng/ml, respectively. Reproducible results could be obtained by using (±)-2-[(1-benzyl-piperidine-4-yl)ethyl]-5,6-dimethoxyindan-1-one hydrochloride as an internal standard. The method was successfully applied for monitoring of donepezil levels in rat plasma, blood and brain microdialysates, and patient plasma.

4.4.3. High-performance liquid chromatography/mass spectrometry

To establish a sensitive and specific liquid chromatography–mass spectrometry (time-of-flight) [LC–MS (TOF)] method for the determination of donepezil in human plasma after an oral administration of 5 mg donepezil hydrochloride tablet [29]. Alkalized plasma was extracted with isopropanol–*n*-hexane (3:97) and loratadine was used as internal standard (IS). Solutes were separated on a C_{18} column with a mobile phase of methanol:acetate buffer (pH 4.0) (80:20). Detection was performed on a TOF mass spectrometry equipped with an electrospray ionization interface and operated in positive-ionization mode. Donepezil quantitation was realized by computing the peak area ratio (donepezil–loratadine) (donepezil m/z 380 $[M + H]^+$ and loratadine m/z 383$[M + H]^+$) and comparing them with calibration curve ($r = 0.9998$). The linear calibration curve was obtained in the concentration range of 0.1–15 μg/l. The detection limit of donepezil was 0.1 μg/l. The average recovery was more than 90%. The intra- and inter-run precision was measured to be below 15% of RSD

TABLE 3.5 Data deduced from X-ray diffraction pattern of donepezil HCl [20]

S. No.	Diffraction angel (2θ)	Intensity, I/I_0 (%)
1	6.026	21.2
2	9.630	19.6
3	10.183	48.8
4	11.043	21.6
5	11.657	70.5
6	12.065	18.0
7	12.741	75.1
8	13.186	14.4
9	13.769	27.7
10	14.769	35.9
11	16.194	18.1
12	17.510	53.7
13	18.140	19.8
14	19.289	18.7
15	19.799	24.0
16	20.381	91.4
17	20.720	61.3
18	21.400	100.0
19	21.841	62.1
20	22.944	46.6
21	24.944	54.0
22	24.649	40.4
23	25.433	17.1
24	26.203	14.0
25	27.011	25.6
26	28.309	14.2
27	31.586	21.4
28	32.516	11.6
29	35.633	12.5

A sensitive and selective liquid chromatography–tandem mass spectrometry (LC–MS–MS) assay for the simultaneous determination of donepezil (D) and its pharmacologically active metabolite, 6-O-desmethyl donepezil (6-ODD) in human plasma is developed using galantamine as IS [30]. The analytes and IS were extracted from 500 μl aliquots of human plasma via solid-phase extraction (SPE) on Waters Oasis HLB cartridges. Chromatographic separation was achieved in a run time of 6.0 min on a Waters Novapak C_{18} (150 mm × 3.9 mm, 4 μm) column under isocratic conditions. Detection of analytes and IS was done by tandem mass

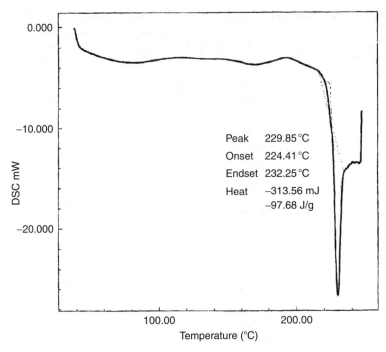

Peak   229.85 °C
Onset   224.41 °C
Endset   232.25 °C
Heat   −313.56 mJ
          −97.68 J/g

FIGURE 3.12 Differential scanning calorimetry thermogram of donepezil HCl [19].

spectrometry, operating in positive ion and multiple reaction monitoring acquisition mode. The protonated precursor to product ion transitions monitored for donepezil, 6-ODD, and IS were at m/z 380.1 → 91.2, 366.3 → 91.3, and 288.2 → 213.2, respectively. The method was fully validated for its selectivity, interference check, sensitivity, linearity, precision and accuracy, recovery, matrix effect, ion suppression/enhancement, cross-specificity, stability, and dilution integrity. A linear dynamic range of 0.10–50.0 ng/ml for donepezil and 0.02–10.0 ng/ml for 6-ODD was evaluated with mean correlation coefficient (r) of 0.9975 and 0.9985, respectively. The intra- and inter-batch precision (%CV, coefficient of variation) across five quality control levels was less than 7.5% for both the analytes. The method was successfully applied to a bioequivalence study of 10 mg donepezil tablet formulation in 24 healthy Indian male subjects under fasting condition

An automated high-throughput liquid chromatography–tandem mass spectrometry (LC–MS–MS) method was developed for quantitative determination of donepezil in human plasma [31]. Samples of 150 μl of plasma were placed in 2.2 ml 96-deepwell plates and both donepezil and loratadine (IS) were extracted from human plasma by liquid–liquid extraction, using hexane as the organic solvent. Robotic liquid handling work stations were employed for all liquid transfer and solution preparation steps and resulted in a short sample preparation time. After vortexing, centrifugation and freezing, the supernatant organic solvent was evaporated and reconstituted in a small volume of reconstitution solution. The method developed, includes a sample analysis performed by reversed phase LC–MS–MS, with positive ion electrospray ionization, using multiple reaction monitoring. The chromatographic run time was set for 2.0 min with a flow rate of 0.7 ml/min in a C_{18} analytical column. The method was significantly sensitive, specific, accurate, and precise for the determination of donepezil in human plasma and had the shortest run time. The curve was proved to be linear for the concentration range of 0.1–100 ng/ml. After validation, the method was applied to the rapid and reliable quantitative determination of donepezil in a bioequivalence study after administration of a 5 mg donepezil tablet.

A selective, sensitive, and rapid hydrophilic interaction liquid chromatography with electrospray ionization tandem mass spectrometry was developed for the determination of donepezil in human plasma [32]. Donepezil was twice extracted from human plasma using methyl-*tert*-butyl ether at basic pH. The analytes were separated on an Atlantis HILIC Silica column with the mobile phase of acetonitrile:ammonium formate (50 mM, pH 4.0) (85:15, v/v) and detected by tandem mass spectrometry in the selective reaction monitoring mode. The calibration curve was linear ($r = 0.9994$) over the concentration range of 0.10–50.0 ng/ml and the lower limit of quantification was 0.1 ng/ml using 200 μl plasma sample. The CV and relative error for intra- and inter-assay at four quality control levels were 2.7% to 10.5% and -10.0% to 0.0%, respectively. There was no matrix effect for donepezil and cisapride. The present method was successfully applied to the pharmacokinetic study of donepezil after oral dose of donepezil hydrochloride (10 mg tablet) to male healthy volunteers.

A liquid chromatography/tandem mass spectrometry (LC/MS/MS) method was developed [33] and validated for the determination of donepezil in human plasma samples. Diphenhydramine was used as the IS. The collision-induced transition m/z 380 → 91 was used to analyze donepezil in selected reaction monitoring mode. The signal intensity of the m/z 380 → 91 transition was found to relate linearly with donepezil concentrations in plasma from 0.1 to 20.0 ng/ml. The lower limit of quantification of the LC/MS/MS method was 0.1 ng/ml. The intra- and inter-day precisions were below 10.2% and the accuracy was between -2.3% and $+2.8\%$. The validated LC/MS/MS method was applied to a pharmacokinetic study in which healthy Chinese volunteers each received a single oral dose of 5 mg donepezil hydrochloride. The non-compartmental pharmacokinetic model was used to fit the donepezil plasma concentration–time curve. Maximum plasma concentration was

12.3 ± 2.73 ng/ml which occurred at 3.50 ± 1.61 h postdosing. The apparent elimination half-life and the area under the curve were, respectively, 60.86 ± 12.05 h and 609.3 ± 122.2 ng h/ml. LC/MS/MS is a rapid, sensitive, and specific method for determining donepezil in human plasma samples.

A sensitive, simple, and specific liquid chromatographic method coupled with electrospray ionization-mass spectrometry for the determination of donepezil in plasma was developed, and its pharmacokinetics in healthy, male, Chinese was studied [34]. Using loratadine as the IS, after extraction of the alkalized plasma by isopropyl alcohol–*n*-hexane (3:97, v/v), solutes are separated on a C_{18} column with a mobile phase of methanol–acetate buffer (pH 4.0) (80:20, v/v). Detection is performed with a TOF mass spectrometer equipped with an electrospray ionization source operated in the positive-ionization mode. Quantitation of donepezil is accomplished by computing the peak area ratio (donepezil [M + H](+) m/z 380-loratadine [M + H](+) m/z 383) and comparing them with the calibration curve ($r = 0.9998$). The linear calibration curve is obtained in the concentration range 0.1–15 ng/ml. The limit of quantitation is 0.1 ng/ml. The mean recovery of donepezil from human plasma is 99.4 ± 6.3% (range 93.4–102.6%). The inter- and intra-day RSD is less than 15%. After an oral administration of 5 mg donepezil to 20 healthy Chinese volunteers, the main pharmacokinetic parameters of donepezil are as follow: T(max), 3.10 ± 0.55 h; $t_{1/2}$, 65.7 ± 12.8 h; C(max), 10.1 ± 2.02 ng/ml; MRT, 89.4 ± 13.4 h; and CL/F, 9.9 ± 4.3 l/h.

4.4.4. Chiral analysis

A new precise, sensitive, and accurate stereoselective HPLC method for the simultaneous determination of donepezil enantiomers in tablets and plasma with enough sensitivity to follow its pharmacokinetics in rats up to 12 h after single oral dosing is investigated [35]. Enantiomeric resolution was achieved on a cellulose tris(3,5-dimethylphenyl carbamate) column known as Chiral OD, with UV detection at 268 nm, and the mobile phase consisted of *n*-hexane, isopropanol, and triethylamine (87:12.9:0.1). Donepezil enantiomers were well resolved with mean retention times of 12.8 and 16.3 min, respectively. Linear response ($r > 0.994$) was observed over the range of 0.05–2 μg/ml of donepezil enantiomers, with detection limit of 20 ng/ml. The mean RSD% of the results of within-day precision and accuracy of the drug were ≤10%. There was no significant difference ($p > 0.05$) between inter- and intra-day studies for each enantiomer which confirmed the reproducibility of the assay method. The mean extraction efficiency was 92.6–93.2% of the enantiomers. The method was found to be suitable and accurate for the quantitative determination of donepezil enantiomers in tablets. The assay method also shows good specificity to donepezil enantiomers, and it could be applied to its pharmacokinetic studies and to therapeutic drug monitoring

Two HPLC methods for the determination of enantiomers of donepezil HCl in rat plasma have been developed [36].The first method involves chiral separation of donepezil HCl on an ovomucoid-bonded column, and native fluorescence detection of donepezil HCl with excitation at 318 nm and emission at 390 nm. The fluorometric detection is without interference from background components and is about five times more sensitive than UV detection at 271 nm. The method was applied to monitoring the racemization of each enantiomer of donepezil HCl in buffer solutions and in rat plasma. The second method involves separation of donepezil HCl from background components of rat plasma on an a chiral column, collection of the donepezil HCl fraction into a sample loop, concentration to a trap column, transfer of donepezil HCl to a chiral column, resolution of the enantiomers of donepezil HCl on the chiral column, and fluorometric detection of the enantiomers of donepezil HCl with excitation at 318 nm and emission at 390 nm. The detection limits of donepezil HCl and each enantiomer of donepezil HCl were 1 ng/ml, respectively, with a 200 μl injection of deproteinized plasma samples.

A rapid, sensitive, and enantioselective LC–MS–MS method using deuterium-labeled IS was developed and evaluated for the simultaneous quantitative determination of donepezil enantiomers in human plasma without interconversion during clean-up process and measurement [37]. The use of an avidin column allowed the separation of donepezil enantiomers, which were specifically detected by MS–MS without interference from its metabolites and plasma constituents. Evaluation of this assay method shows that samples can be assayed with acceptable accuracy and precision within the range from 0.0206 to 51.6 ng/ml for both R-donepezil and S-donepezil. This analytical method was applied to the simultaneous quantitation of donepezil enantiomers in human plasma.

HPLC with column switching and mass spectrometry was applied to the online determination and resolution of the enantiomers of donepezil HCl in plasma [38]. This system employs two avidin columns and fast atom bombardment-mass spectrometry (FAB-MS). A plasma sample was injected directly into an avidin trapping column (10 mm × 4.0 mm i.d.). The plasma protein was washed out from the trapping column immediately while donepezil HCl was retained. After the column-switching procedure, donepezil HCl was separated enantioselectivity in an avidin analytical column. The separated donepezil HCl enantiomers were specifically detected by FAB-MS without interference from metabolites of donepezil HCl and plasma constituents. The limit of quantification for each enantiomer of donepezil HCl in plasma was 1.0 ng/ml and the intra- and inter-assay RSDs for the method were less than 5.2%. The assay was validated for enantioselective pharmacokinetic studies in the dog.

To establish chiral separation method for donepezil hydrochloride enantiomers by capillary electrophoresis (CE) and to determine the two enantiomers in plasma [39], alkalized plasma was extracted by isopropanol–n-hexane (3:97) and L-butefeina was used as the IS. Enantioresolution was achieved using 2.5% sulfated-beta-cyclodextrin as chiral selector in 25 mmol/l triethylammonium phosphate solution (pH 2.5) on the uncoated fused-silica capillary column (70 cm × 50 μm i.d.). The feasibility of the method to be used as quantitation of donepezil HCl enantiomers in rabbit plasma was also investigated. Donepezil HCl enantiomers were separated at a baseline level under the above condition. The linearity of the response was evaluated in the concentration range from 0.1 to 5 mg/l. The linear regression analysis obtained by plotting the peak area ratio ($A(s)/A(i)$) of the analyte to the IS versus the concentration (C) showed excellent correlation coefficient The low limit of detection was 0.05 mg/l. The inter- and intra-day precisions (RSD) were all less than 20%. Compared with chiral stationary phase by HPLC, the CE method is simple, reliable, inexpensive, and suitable for studying the stereoseletive pharmacokinetics in rabbit.

4.4.5. Capillary electrophoresis
Field-amplified sample stacking (FASS) in CE was used to determine the concentration of donepezil in human plasma [40]. A sample pretreatment by liquid–liquid extraction with isopropanol/n-hexane (3:97, v/v) and subsequent quantification by FASS-CE was used. Before sample loading, a water plug (0.5 psi, 6 s) was injected to permit FASS. Electrokinetic injection (7 kV, 90 s) was used to introduce sample cations. The separation condition for donepezil was performed in electrolyte solutions containing Tris buffer (60 mM, pH 4.0) with sodium octanesulfonate 40 mM and 0.01% polyvinyl alcohol as a dynamic coating to reduce analytes' interaction with capillary wall. The separation was performed at 28 kV and detected at 200 nm. Using atenolol as an IS, the linear ranges of the method for the determination of donepezil in human plasma were over a range of 1–50 ng/ml. The limit of detection was 0.1 ng/ml ($S/N = 3$, sampling 90 s at 7 kV). One female volunteer (54 years old) was orally administered a single dose of 10 mg donepezil (Aricept, Eisai), and blood samples were drawn over a 60-h period for pharmacokinetic study. The method was also applied to monitor donepezil in 16 AD patients' plasmas.

5. PHARMACOLOGY

It has been demonstrated that AD is associated with a relative decrease in the activity of the cholinergic system in the cerebral cortex and other areas of the brain. Studies suggest that donepezil hydrochloride exerts its therapeutic effect by enhancing cholinergic function in the central nervous system. This is accomplished by increasing the concentration of acetylcholine through reversible inhibition of AChE [41–45].

5.1. Pharmacokinetics

5.1.1. Absorption
Donepezil is well absorbed with a relative oral bioavailability of 100% and reaches peak plasma concentrations in 3–4 h. Oral administration of Aricept produces highly predictable plasma concentrations with plasma concentrations and area under the curve rise in proportion to the dose. The terminal disposition half-life is approximately 70 h, thus administration of multiple single-daily doses results in gradual approach to steady state. Approximate steady state is achieved within 3 weeks after the initiation of therapy. Once at steady state, plasma donepezil hydrochloride concentrations and the related pharmacodynamic activity show little variability over the course of the day. Neither food nor time of administration (morning versus evening dose) affect the absorption of donepezil hydrochloride [46–51].

5.1.2. Distribution
The steady-state volume of distribution is 12 l/kg. Donepezil hydrochloride is approximately 96% bound to human plasma proteins. The distribution of donepezil hydrochloride in various body tissues has not been definitively studied. However, in a mass balance study conducted in healthy male volunteers, 240 h after the administration of a single 5 mg dose of ^{14}C-labeled donepezil hydrochloride, approximately 28% of the label remained unrecovered. This suggests that donepezil and/or its metabolites may persist in the body for more than 10 days.

The average CSF:plasma ratio for both doses, expressed as a percent of the concentration in plasma, was 15.7% [46–51].

5.1.3. Metabolism and excretion
Donepezil is both excreted in the urine intact and extensively metabolized to four major metabolites, two of which are known to be active, and a number of minor metabolites, not all of which have been identified. Three of the human metabolites of donepezil have not undergone extensive safety tests in animals. These comprise two O-demethylated derivatives and an N-oxidation product. Donepezil is metabolized by CYP 450 isoenzymes 2D6 and 3A4 and undergoes glucuronidation. The rate of metabolism of donepezil is slow and does not appear to be saturable. These findings are consistent with the results from formal pharmacokinetic studies which showed that donepezil and/or its metabolites do not inhibit the metabolism of theophylline, warfarin, cimetidine, or digoxin.

in humans. Pharmacokinetic studies also demonstrated that the metabolism of donepezil is not affected by concurrent administration of digoxin or cimetidine. Following administration of ^{14}C-labeled donepezil, plasma radioactivity, expressed as a percent of the administered dose, was present primarily as intact donepezil (53%) and as 6-O-desmethyl donepezil (11%), which has been reported to inhibit AChE to the same extent as donepezil *in vitro* and was found in the plasma at concentrations equal to about 20% of donepezil. Approximately 57% and 15% of the total radioactivity was recovered in urine and feces, respectively, over a period of 10 days, while 28% remained unrecovered, with about 17% of the donepezil dose recovered in the urine as unchanged drug. There is no evidence to suggest enterohepatic recirculation of donepezil and/or any of its metabolites. Plasma donepezil concentrations decline with a half-life of approximately 70 h. Sex, race, and smoking history have no clinically significant influence on plasma concentrations of donepezil [46–51].

5.2. Pharmacodynamics

Donepezil hydrochloride is a specific and reversible inhibitor of AChE the predominant cholinesterase in the brain. Donepezil hydrochloride was found *in vitro* to be over 1000 times more potent an inhibitor of this enzyme than that of BuChE, an enzyme which is present mainly outside the central nervous system. In patients with Alzheimer's dementia participating in clinical trials, administration of single daily doses of 5 or 10 mg Aricept produced steady-state inhibition of AChE activity (measured in erythrocyte membranes) of 63.6% and 77.3%, respectively, when measured post dose. The inhibition of AChE in red blood cells by donepezil hydrochloride has been shown to correspond closely to the effects in the cerebral cortex. In addition, significant correlation was demonstrated between plasma levels of donepezil hydrochloride, AChE inhibition and change in Alzeimer's Disease Assessment Scale-cognitive subscale (ADAS-cog), a sensitive and well-validated scale which examines cognitive performance including memory, orientation, attention, reason, language, and praxis [46–51].

5.3. Drug metabolism

Three metabolic pathways were identified: (i) O-dealkylation to metabolites M1 and M2, with subsequent glucuronidation to metabolites M3 and M5; (ii) hydrolysis to metabolite M4; and (iii) N-oxidation to metabolite M6 (Fig. 3.13). Additional metabolic pathways may be operative, as represented by the number of unknown compounds observed. However, as each unknown metabolite represented an average of <2% of the dose, these pathways are considered minor contributors to the metabolic process [52].

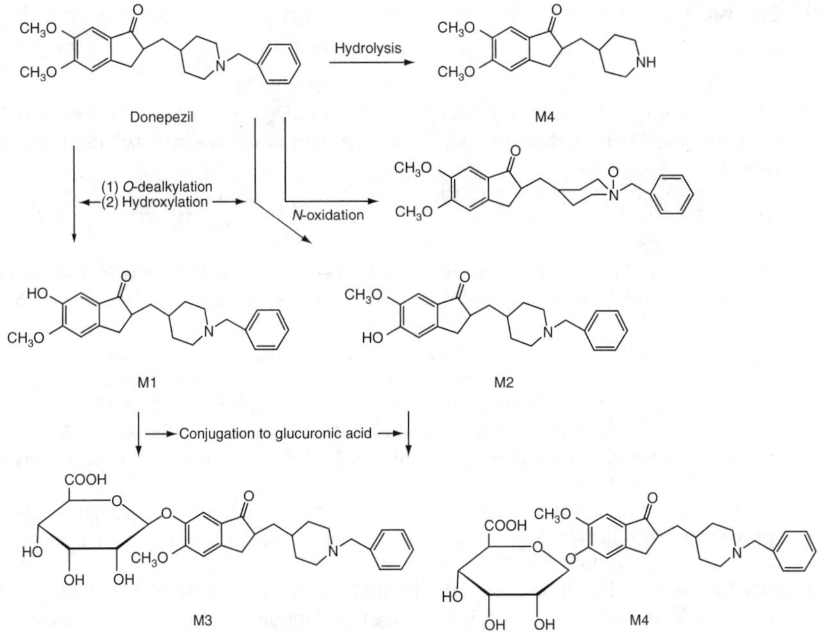

FIGURE 3.13 Proposed metabolic pathways for donepezil [52].

In plasma, the parent compound accounted for about 25% of the dose recovered during each sampling period, as well as of the cumulative dose recovered. The recovered residue showed higher levels of the hydroxylated metabolites M1 and M2 than of their glucuronide conjugates M3 and M5, respectively [52].

In urine, the parent compound accounted for 17%, on average, of the dose recovered from each pooled sample, as well as of the total recovered dose. The major metabolite was the hydrolysis product M4, followed by the glucuronidated conjugates M3 and M5. In feces, the parent compound also predominated, although it accounted for only 1% of the recovered dose. A large percentage of the radioactivity in feces consisted of unidentified very polar metabolites, which were retained at the TLC origin. Of the extracted metabolites, the hydroxylation products M1 and M2 were the most abundant, followed by the hydrolysis product M4 and the N-oxidation product M6 [52].

ACKNOWLEDGMENTS

The authors thank Prof. Abdullah A. Al-Badr, Prof. of Medicinal Chemistry, Department of Pharmaceutical Chemistry, College of Pharmacy King Saudi University, and Mr. Tanvir A. Butt for their help and support to this work.

REFERENCES

[1] The Merck Index, an Encyclopedia of Chemicals, Drugs, and Biologicals, 14th ed., Merck & Co., Inc., Whitehouse Station, NJ, USA, 2006, p. 3419.

[2] A.C. Maffat, M.D. Osselton, B. Widdop, Clarck's Analysis of Drug and Poisons, vol. 2, third ed., The Pharmaceutical Press, Royal Pharmaceutical of Great Britain, London, 2004, p. 954.

[3] H. Sugimoto, Structure activity relationships of acetylcholinesterase inhibitors: donepezil hydrochloride for the treatment of Alzheimer's Disease, Pure Appl. Chem. 71 (1999) 2031–2037.

[4] S.L. Rogers, Y. Yamanishi, K. Yamatsu, E2020—the pharmacology of a piperidine cholinesterase inhibitor, in: R. Becker, E. Giacobini (Eds.), Cholinergic Basis for Alzheimer Therapy, Birkhäuser, Boston, 1991, pp. 314–320.

[5] H. Sugimoto, Y. Iimura, Y. Yamanishi, K. Yamatsu, Synthesis and antiacetylcholinesterase activity of 1-benzyl-4-[(5,6-dimethoxy-1-indanon-2-yl)methyl]piperidine hydrochloride (E2020) and related compounds, Bioorg. Med. Chem. Lett. 2 (1992) 871–876.

[6] Y. Iimura, M. Mishima, H. Sugimoto, Synthesis of 1-benzyl-4-[(5,6-dimethoxy [2-^{14}C]-1-indanon)-2-yl]-methylpiperidine hydrochloride (E2020-^{14}C), J. Label. Comp. Radiopharm. XXVI (1989) 835–839.

[7] K. Sherman, Pharmacodynamics of oral E2020 and tacrine in humans: novel approaches, in: R. Becker, E. Giacobini (Eds.), Cholinergic Basis for Alzheimer Therapy, Birkhäuser, Boston, 1991, pp. 321–328.

[8] S.L. Rogers, L.T. Friedhoff, The efficacy and safety of donepezil in patients with Alzheimer's disease: results of a US multicenter, randomized, double-blind, placebo-controlled trial, Dementia 7 (1996) 293–303.

[9] S.L. Rogers, L.T. Friedhoff, The Donepezil Study Group, The efficacy and safety of Donepezil in patients with Alzheimer's disease: results of a US multicentre, randomized, double-blind, placebo trial, Dementia 7 (1996) 293–303.

[10] S.L. Rogers, M.R. Farlow, R.S. Doody, R. Mohs, L.T. Friedhoff, The Donepezil Study Group, A 24-week, double-blind, placebo-controlled trial of donepezil in patients with Alzheimer's disease, Neurology 50 (1998) 136–145.

[11] S.L. Rogers, R.S. Doody, R. Mohs, L.T. Friedhoff, The Donepezil Study Group, Donepezil improves cognition and global function in Alzheimer's disease: a 15-week, double-blind, placebo-controlled study, Arch. Intern. Med. 158 (1998) 1021–1031.

[12] O.L. Lerman, J. Kaspi, O. Arad, M. Alnabari, Y. Sery. Process for the preparation of donepzil. U.S. Patent No., US 2004/0048893 A1.

[13] L. Zelikovitch, O. Arad, M. Alnabari, Y. Sery, O. Kurlat, M. Bentolila et al., Process for alkylating secondary amines and the use in donepezil preparation thereof, U.S. Patent No., US 2006/0122227 A1.

[14] J.S. Vidyadhar, N.A. Venkatraman, S.R. Pandurang, Process for the preparation of 1-benzyl-4(5,6-dimethoxy-1-indanon)-2-yl) methyl piperidine hydrochloride(donepzil HCl), U.S. Patent No., US, 2003, 6,649,765 B1.

[15] T.V. Radhakrishnan, S.D. Govind, N.A. Venkatraman, Process for the preparation of 1-benzyl-4(5,6-dimethoxy-1-indanon)-2-yl) methyl piperidine hydrochloride (donepzil HCl), U.S. Patent No., US 2004, 0158070 A1.

[16] N. Mahesh, G.A. Kumar, D. Ramesh, M. Sivakumaran, Process for the preparation of donepezil hydrochloride, United State Patent Application, US 2007/0191610 A1.

[17] H. Sugimoto, T. Yutaka, H. Kunizou, K. Norio, L. Youichi, S. Atsushi, et al., J. Med. Chem. 38 (1995) 481. U.S. Patent No. 5,100,901.

[18] A.L. Gutman, B. Tishin, A. Vilensky, P. Potyabin, G.A. Nisnevich, Process for production of highly pure donepezil hydrochloride, U.S. Patent No., US 2004 /0192919 A1.

[19] M.S. Reddy, S. Eswaraiah, M.V. Thippannachar, E.R. Chandrashekar, P.A. Kumar, United States Patent Application Publication, Novel crystalline form-VI of donepezil hydrochloride and process for the preparation thereof, United State of Patents, US2004/0229914A1.

[20] U.P. Aher, V.R. Tarur, D.G. Sathe, A.V. Naidu, K.D. Sawant, Donepezil hydrochloride form VI, U.S. Patent No., US 2007/0123565 A1.

[21] U.P. Aher, V.R. Tarur, D.G. Sathe, A.V. Naidu, K.D. Sawant, Polymorph of (1-benzyl-4-[(5,6-dimeththoxy-1-indanone)-2-yl] methyl piperidine hydrochloride(donepezil hydrochloride) and a process for producing thereof, U.S. Patent No., US 2005/0272775 A1.

[22] I. Adin, C. Iustain, O. Arad, J. Kaspe, United States Patent Application, Crystalline forms of donepezil base, U.S. Patent No., US 2006/0122226 A1.

[23] S.S. Abbas, Y.M. Fayez, L.-S. Abdel Fattah, Stability indicating methods for determination of donepezil hydrochloride according to ICH guidelines, Chem. Pharm. Bull. 54 (2006) 1447–1450.

[24] G.A.E. Mostafa, M.M. Hefnawy, A. Al-Majed, Membrane sensors for the selective determination of donepzil hydrochloride, J. AOAC Int. (2009) in press.

[25] A. Golcu, S.A. Ozkan, Electroanalytical determination of donepezil HCl in tablets and human serum by differential pulse and Osteryoung square wave voltammetry at a glassy carbon electrode, Pharmazie 61 (2006) 760–765.

[26] K. Stella, M. Stella, A. Pandora, A. Morfis, B. Antonios, K. Maria, New gradient high-performance liquid chromatography method for determination of donepezil hydrochloride assay and impurities content in oral pharmaceutical formulation, J. Chromatogr. A 1189 (2008) 392–397.

[27] Y.-F. Norio, F. Rie, T. Takenori, T. Tomonori, Determination of donepezil, an acetylcholinesterase inhibitor, in human plasma by high-performance liquid chromatography with ultraviolet absorbance detection, J. Chromatogr. B 768 (2002) 261–265.

[28] K. Nakashima, K. Itoh, M. Kono, M.N. Nakahima, M. Wada, Determination of donepezil hydrochloride in human and rat plasma, blood and brain microdialysates by HPLC with a short C30 column, J. Pharm. Biomed. Anal. 41 (2006) 201–206.

[29] Y.H. Lu, H.M. Wen, W. Li, Y.M. Chi, Z.X. Zhang, Determination of donepezil in human plasma by HPLC–MS, Yao Xue Xue Bao 38 (2003) 203–206.

[30] N.P. Bhavin, S. Naveen, S. Mallika, S.S. Pranav, Quantitation of donepezil and its active metabolite 6-O-desmethyl donepezil in human plasma by a selective and sensitive liquid chromatography–tandem mass spectrometric method, Anal. Chim. Acta 629 (2008) 145–157.

[31] A. Constantinos, D. Yannis, K. Constantinos, L.L. Yannis, Quantitative determination of donepezil in human plasma by liquid chromatography/tandem mass spectrometry employing an automated liquid–liquid extraction based on 96-well format plates: application to a bioequivalence study, J. Chromatogr. B 848 (2007) 239–244.

[32] E.J. Park, H.W. Lee, H.Y. Ji, H.Y. Kim, M.H. Lee, E.S. Park, et al., Hydrophilic interaction chromatography–tandem mass spectrometry of donepezil in human plasma: application to a pharmacokinetic study of donepezil in volunteers, Arch. Pharm. Res. 31 (2008) 1205–1211.

[33] Z. Xie, Q. Liao, X. Xu, M. Yao, J. Wan, D. Liu, Rapid and sensitive determination of donepezil in human plasma by liquid chromatography/tandem mass spectrometry: application to a pharmacokinetic study, Rapid Commun. Mass Spectrom. 20 (2006) 3193–3198.

[34] Y. Lu, H. Wen, W. Li, Y. Chi, Z. Zhang, Determination of donepezil hydrochloride (E2020) in plasma by liquid chromatography-mass spectrometry and its application to pharmacokinetic studies in healthy, young, Chinese subjects, J. Chromatogr. Sci. 42 (2004) 234–237.

[35] A.R. Mahasen, H.A. Heba, T.A. Bushra, Y.A.-E. Hassan, N. Kenichiro, Stereoselective HPLC assay of donepezil enantiomers with UV detection and its application to pharmacokinetics in rats, J. Chromatogr. B. 830 (2006) 114–119.

[36] J. Haginaka, C. Seyama, Determination of enantiomers of 1-benzyl-4-[(5,6-dimethoxy-1-indanon)-2-yl]methylpiperidine hydrochloride (E2020), a centrally acting acetylcholine esterase inhibitor, in plasma by liquid chromatography with fluorometric detection, J. Chromatogr. 577 (1992) 95–102.

[37] M. Kenji, O. Yoshiya, N. Hiroshi, Y. Tsutomu, Simultaneous determination of donepezil (Aricept®) enantiomers in human plasma by liquid chromatography–electrospray tandem mass spectrometry, J. Chromatogr. B. 729 (1999) 147–155.

[38] Y. Odd, H. Ohe, S. Tanaka, N. Asakawa, Direct determination of E2020 enantiomers in plasma by liquid chromatography–mass spectrometry and column-switching techniques, J. Chromatogr. A. 694 (1995) 209–218.

[39] Y.H. Lu, M. Zhang, Q. Meng, Z.X. Zhang, Separation and determination of donepezil hydrochloride enantiomers in plasma by capillary electrophoresis, Yao Xue Hue Bao 41 (2006) 471–475.

[40] H.H. Yeh, Y.H. Yang, J.Y. Ko, S.H. Chen, Sensitive analysis of donepezil in plasma by capillary electrophoresis combining on-column field-amplified sample stacking and its application in Alzheimer's disease, Electrophoresis 29 (2008) 3649–3657.

[41] P.J. Tiseo, K. Foley, L.T. Friedhoff, The effect of multiple doses of donepezil HCl on the pharmacokeinetic and pharmacodynamic profile of warfarin, Br. J. Clin. Pharmacol. 46 (1998) 45–50.

[42] P.J. Tiseo, K. Foley, L.T. Friedhoff, Pharmacodynamic and pharmacokinetic of donepezil HCl profile following evening administration, Br. J. Clin. Pharmacol. 46 (1998) 13–18.

[43] J.F. Reyes, R. Vargas, D. Kumar, E.I. Cullen, C.A. Perdomo, R.D. Pratt, Steady-state pharmacokinetics, pharmacodynamics and tolerability of donepezil hydrochloride in hepatically impaired patients, Br. J. Clin. Pharmacol. 58 (2004) 9–17.

[44] S.L. Rogers, N.M. Coope, R. Snkoventy, J.E. Pederson, J.N. Lee, L.T. Friedhoff, Pharmacodynamic and pharmacokinetic of donepezil HCl profile following multiple oral doses, Br. J. Clinc. Pharmacol. 46 (1998) 7–13.

[45] C.F. Nagy, D. Kumar, E.I. Cullen, W.K. Bolton, T.C. Marbury, M.J. Gutierrez, H.W. Hutman, R.D. Pratt, Steady-state pharmacokinetics and safety of donepezil HCl in subjects with moderately impaired renal function, Br. J. Clin. Pharmacol. 58 (2004) 18–24.

[46] P.J. Tiseo, C.A. Perdomo, L.T. Friedhoff, Concurrent administration of donepezil HCl and cimetidine: assessment of pharmacokinetic changes following single and multiple doses, Br. J. Clinc. Pharmacol. 46 (1998) 25–29.

[47] M. Mihara, A. Ohnishi, Y. Tomono, Pharmacokinetics of E2020, a new compound for Alzheimer's disease, in healthy male volunteers, Int. J. Clin. Pharmacol. Ther. Toxicol. 31 (1993) 223–229.

[48] A. Ohnishi, M. Mihara, H. Kamakura, Comparison of the pharmacokinetics of E2020, a new compound for Alzheimer's disease, in healthy young and elderly subjects, J. Clin. Pharmacol. 33 (1993) 1086–1091.

[49] S.L. Rogers, L.T. Friedhoff, Long-term efficacy and safety of donepezil in the treatment of Alzheimer's disease: an interim analysis of the results of a US multicentre open label extension study, Eur. Neuropsychopharmacol. 8 (1998) 67–75.

[50] A.J. Wagstaff, D.T. McTavish, A review of its pharmacodynamic and pharmacokinetic properties, and therapeutic efficacy in Alzheimer's disease, Drugs Aging 4 (1994) 510–540.

[51] M.L. Crimson, First drug approved for Alzheimer's disease, Ann. Pharmacother. 28 (1994) 744–751.

[52] P.J. Tiseo, C.A. Perdomo, L.T. Friedhoff, Metabolism and elimination of 14C-donepezil in healthy volunteers: a single-dose study, Br. J. Clin. Pharmacol. 46 (1998) 19–24.

CHAPTER 4

Omeprazole

Abdullah A. Al-Badr

Contents

Department of Pharmaceutical Chemistry, College of Pharmacy, King Saud University, Riyadh, Kingdom of Saudi Arabia

Profiles of Drug Substances, Excipients, and Related Methodology, Volume 35
ISSN 1871-5125, DOI: 10.1016/S1871-5125(10)35004-7

1. DESCRIPTION

1.1. Nomenclature

1.1.1. Systematic chemical names

- (*RS*)-5-Methoxy-2-(4-methoxy-3,5-dimethyl-2-pyridinyl-methyl-sulfinyl)-benzimidazole
- 5-Methoxy-2-[[(4-methoxy-3,5-dimetyl-2-pyridinyl)methyl]sulfinyl]-1*H*-benzimidazole
- 5-Methoxy-2-[[(4-methoxy-3,5-dimethyl-2-pyridinyl)methyl]sulfinyl]-1*H*-benzimidazole
- 2-[[(3,5-Dimethyl-4-methoxy-2-pyridyl)methyl]sulfinyl-5-methoxy]-1*H*-benzimidazole [1–4]

1.1.2. Nonproprietary names
Omeprazole, H-168/68, Omeprazol, Omeprazolas, Omeprazolum, Omepratosoli [1, 2].

1.1.3. Proprietary names
Antra, Gastroloc, Gastroguard, Logastric, Losec, Mepral, Mopral, Omapren, Omelich, Omelind, Omepral, Omeprazen, Omeprazole, Ompanyt, Osiren, Parizac, Pepticum, Prilosec, Prilosico, Zegerid, Zoltum [1, 3].

1.2. Formulae

1.2.1. Empirical, molecular weight, and CAS number

Omeprazole	$C_{17}H_{19}N_3O_3S$	345.42	[73590-58-6]
Omeprazole sodium	$C_{17}H_{18}N_3O_3S,Na$	367.42	[95510-70-6]
Omeprazole magnesium	$C_{34}H_{36}MgN_6O_6S_2$	713.1	[95382-33-5]

1.2.2. Structural formula

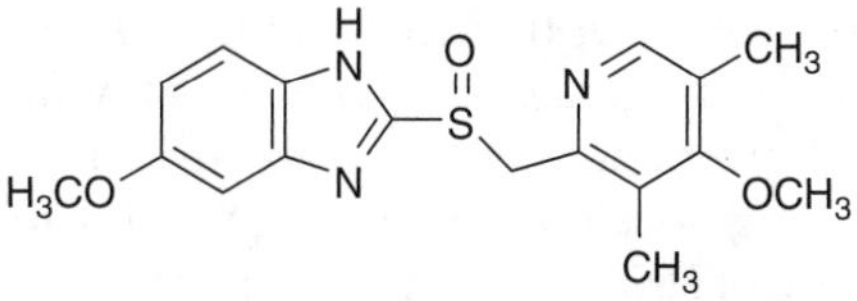

1.3. Elemental analysis

C 59.11%, H 5.54%, N 12.16%, O 13.90%, S 9.28%.

1.4. Appearance

Omeprazole: A white or almost white powder [3].
Omeprazole sodium: A white or almost white powder, hygroscopic [3].

1.5. Uses and applications

Omeprazole is a proton pump inhibitor which inhibits secretion of gastric acid by irreversibly blocking the enzyme system of hydrogen/potassium adenosine triphosphatase (H^+/K^+-ATPase), the "proton pump" of the gastric parietal cell. The drug is used in conditions where the inhibition of gastric acid secretion may be beneficial, including aspiration syndromes [5], dyspepsia [6], gastro-oesophageal reflux disease [7], peptic ulcer disease [8], and the Zollinger–Ellison syndrome [9]. Esomeprazole which is an isomer of omeprazole is also used [10]. The dose of omeprazole may need to be reduced in patients with hepatic impairment [1, 11].

Omeprazole may be given by mouth as the base or magnesium salt or intravenously as the sodium salt. Doses are expressed in terms of the base. Omeprazole magnesium 10.32 mg and omeprazole sodium 10.64 mg are each equivalent to about 10 mg of omeprazole. For the relief of the acid-related dyspepsia, the drug is given in usual doses of 10 or 20 mg daily by mouth for 2–4 weeks. The usual dose for the treatment of gastro-oesophageal reflux disease is 20 mg by mouth once daily for 4 weeks, followed by a further 4–8 weeks if not fully healed. In refractory oesopha-gitis, a dose of 40 mg daily may be used. Maintenance therapy after healing of oesophagitis is 20 mg once daily, and for acid reflux is 10 mg daily. In children, over 1 year of age, licensed UK oral doses for treatment are 10 mg daily in those weighing 10–20 kg, and 20 mg daily in those weighing over 20 kg. These doses may be doubled if necessary. The British National Formulary for Children (BNFC) recommends a dose of 700 μg/kg daily in children 1 month to 2 years of age, increased if necessary up to 3 mg/kg daily, or 20 mg daily, whichever is less. Similar initial doses are suggested in neonates [1].

In the management of peptic ulcer a single daily dose of 20 mg by mouth, or 40 mg in severe cases, is given. Treatment is continued for 4 weeks for duodenal ulcer and 8 weeks for gastric ulcer. Where appropri-ate, a dose of 10–20 mg once daily may be given for maintenance [1].

For the eradication of *Helicobacter pylori* in peptic ulceration, omeprazole may be combined with antibacterials in dual or triple therapy. Effective triple therapy regimens include omeprazole 20 mg twice daily combined with: amoxycillin 500 mg and metronidazole 400 mg, both three times daily; clarithromycin 500 mg and metronidazole 40 mg (or tinidazole 500 mg) both twice daily; or with amoxycillin 1 g and clarithromycin 500 mg both twice daily. These regimens are given for 1 week. Dual therapy regimens, such as omeprazole 40 mg daily with either amoxycillin 750 mg to 1 g twice daily or clarithromycin 500 mg three times daily, are less effective and must be given for 2 weeks. Omeprazole alone may be continued for a further 4–8 weeks [1].

Doses of 20 mg are used in the treatment of Non-steroidal Anti-inflammatory Drugs (NSAID)-associated ulceration; a dose of 20 mg daily may also be used for prophylaxis in patients with a history of gastroduodenal lesions who require continued NSAID treatment [1].

The initial recommended dosage for patients with the Zollinger–Ellison syndrome is 60 mg by mouth once daily, adjusted as required. The majority of patients are effectively controlled by doses in the range 20–120 mg daily, but doses up to 120 mg three times daily have been used. Daily doses above 80 mg should be given as divided doses (usually 2) [1].

Omeprazole is also used for the prophylaxis of acid aspiration during general anesthesia, in dose of 40 mg the evening before surgery and a further 40 mg twice to 6 h before the procedure [1].

Patients who are unsuited to receive oral therapy, omeprazole sodium may be given on a short-term basis by intravenous infusion, in a usual dose equivalent to 40 mg of the base over a period of 20–30 min in 100 ml of sodium chloride 0.9% or glucose 5%. It may also be given by slow intravenous injection. Higher intravenous doses have been given to patients with Zollinger–Ellison syndrome [1].

2. METHODS OF PREPARATION

2.1. Brandstrom and Lamm [12] used the following method for the preparation of omeprazole:

$$\mathbf{1} \xrightarrow[\text{H}_2\text{O}_2]{\text{CH}_3\text{COOH}} \mathbf{2} \xrightarrow{\text{H}_2\text{SO}_4/\text{HNO}_3} \mathbf{3} \longrightarrow$$

$$\xrightarrow{\text{NaOH/CH}_3\text{OH}} \mathbf{4} \xrightarrow{(\text{CH}_3\text{CO})_2\text{O}} \mathbf{5} \xrightarrow{\text{NaOH}}$$

$$\longrightarrow \mathbf{6} \xrightarrow{\text{SOCl}_2} \mathbf{7} \xrightarrow{\mathbf{8}}$$

$$\longrightarrow \mathbf{9} \xrightarrow{\text{H}_2\text{O}_2} \mathbf{10}$$

Omeprazole

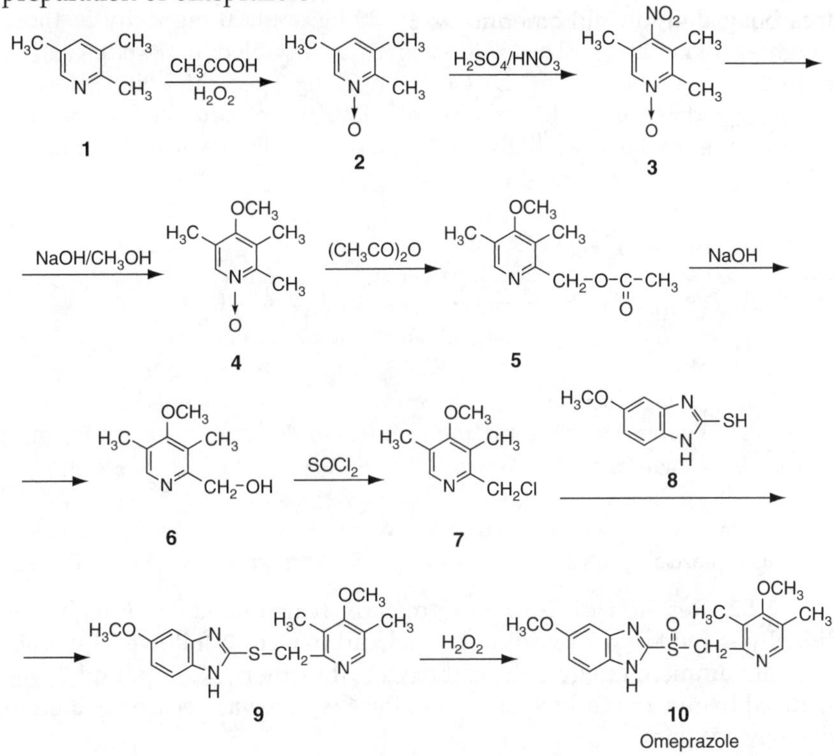

2,3,5-Trimethyl pyridine **1** was oxidized by hydrogen peroxide in acetic acid to give the N-oxide **2** and the latter was nitrated using a mixture of sulfuric acid and nitric acid to give the 4-nitro derivative **3**. The nitro group in **3** was displaced by hydroxymethylation to yield **4**. Treatment of compound **4** with acetic acid anhydride reduces the ring and forms an ester derivative **5**. The corresponding alcohol **6** was formed by the treatment with base, followed by displacement of the hydroxyl group with a chloride using thionyl chloride to give 2-chloromethyl-4-methoxy-2,3,5-trimethyl pyridine **7**. The benzimidazole portion **8** displaces the chloride giving 5-methoxy-2-[((4-mrthoxy-3,5-dimethyl-2-pyridinyl)methyl)thio]-1H-benzimidzole **9**. Omeprazole **10** is formed in a final step where the thioether group is oxidized by hydrogen peroxide to the corresponding sulfoxide.

2.2. Slemon and Macel [13] used several intermediates for the preparation of omeprazole. The drug was produced from the corresponding acetamide–sulfide compounds by a process of oxidation to form the amide sulfinyl compound, followed by alkaline hydrolysis to the sulfinyl carboxylate or salt, and decarboxylation. The methods are as follows:

2.2.1. Reaction of 5-methoxy-2-chloromercaptobenzimidazole **1** with 3,5-dimethyl-4-methoxy-2-methylamidopyridine **2** yielded the acetamide thioether **3**. Compound **3** was oxidized to give the acetamide sulfoxide **4**. The acetamide **4** was hydrolyzed to give the carboxylate **5** which was decarboxylated to yield omeprazole **6**.

6
Omeprazole

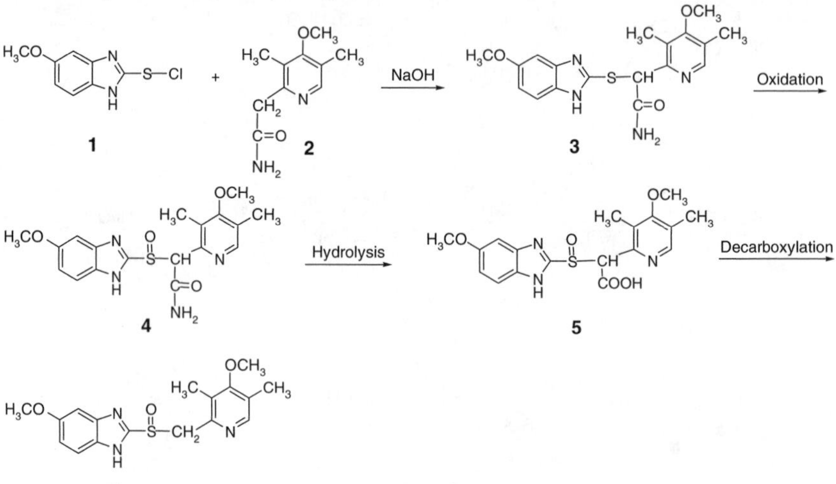

2.2.2. Reaction of 2-*S*-*S*-bis(5-methoxybenzimidazole) **1** with 3,5-dimethyl-4-methoxy-2-pyridine methyl carboxylate **2** followed by reaction with ammonia to give the carboxylate thioether **3**. Compound **3** was oxidized to give the carboxylate sulfoxide **4** which was decarboxylated to give omeprazole **5**.

5
Omeprazole

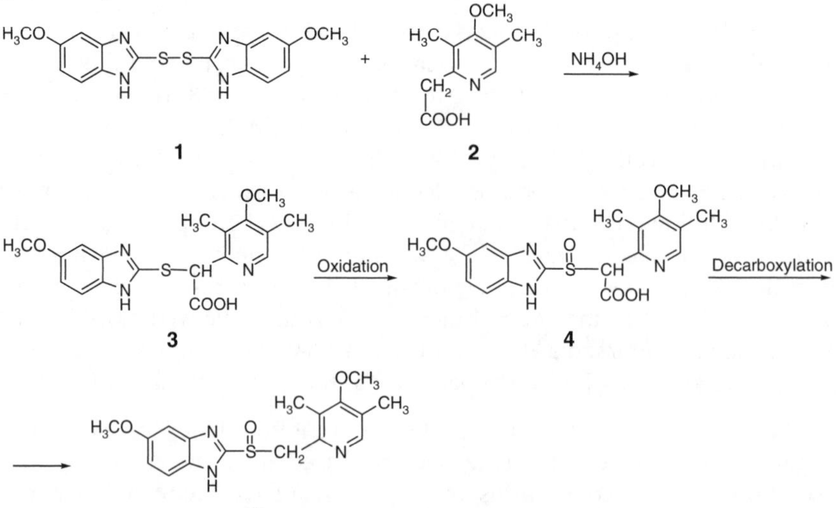

2.2.3. Reaction of 3,5-dimethyl-4-methoxy-2-chloropyridine **1** with 5-methoxy-2-(methylcarboxylate)-thiobenzimidazole **2** followed by treatment with ammonia to give the carboxylate thioether **3**. Compound **3** was oxidized to give the carboxylate sulfoxide **4** which is decarboxylated to give omeprazole **5**.

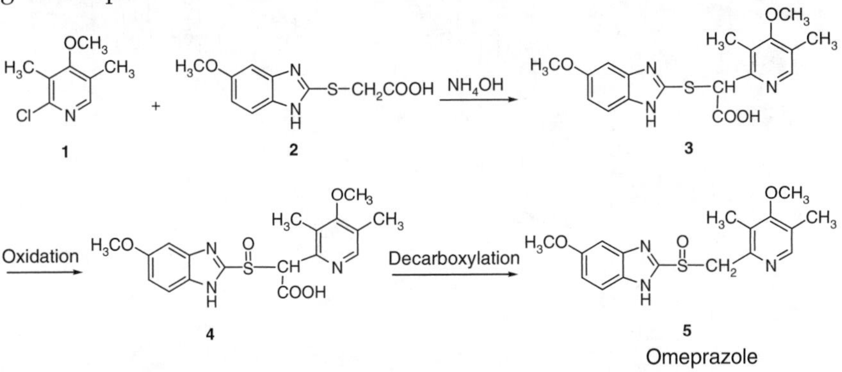

2.2.4. Reaction of 3,5-dimethyl-4-methoxy-2-chloromethylamido pyridine **1** with 5-methoxy-2-mercaptoimidazole **2** in a base to yield the acetamide thioether **3**. Compound **3** was oxidized to the acetamide sulfoxide **4**. Compound **4** was hydrolyzed to yield the carboxylate sulfoxide **5** which was then decarboxylated to give omeprazole **6**.

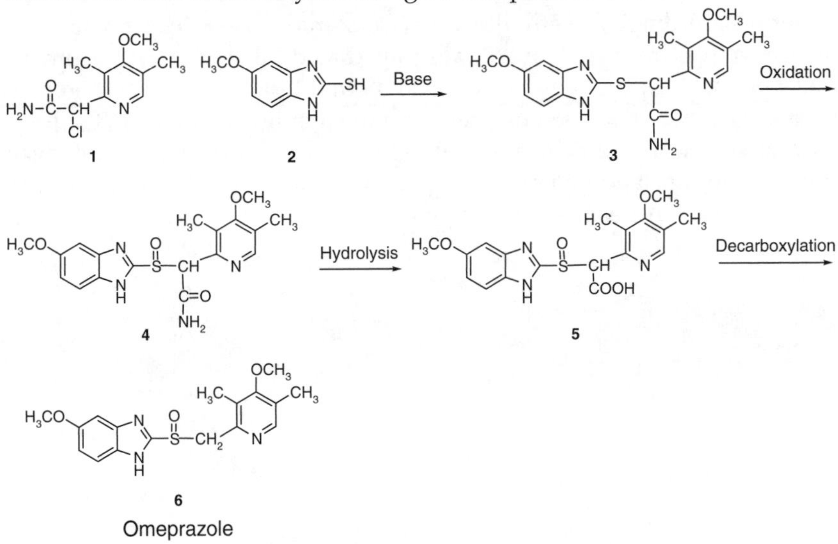

Omeprazole

2.2.5. Reaction of 3,5-dimethyl-4-methoxy-2-mercaptomethyl amido-pyridine **1** with 5-methoxy-2-chlorobenzimidazole **2** in a base to yield to the acetamide thioether **3**. Compound **3** was oxidized to give the

acetamide sulfoxide **4**. Compound **4** was hydrolyzed to yield the carboxylate sulfoxide **5** which was decarboxylated to give omeprazole **6**.

6
Omeprazole

2.3. Cotton *et al.* [14] described an asymmetric synthesis of esomeprazole.
Esomeprazole, the (S)-enantiomer of omeprazole, was synthesized via
asymmetric oxidation of prochiral sulfide 5-methoxy-2-[[(4-methoxy-3,5-
dimethyl pyridin-2-yl)methyl]thio]-1*H*-benzimidazole **1**. The asymmetric
oxidation was achieved by titanium-mediated oxidation with cumene
hydroperoxide in the presence of (S,S)-diethyl tartarate (DET). The enantioselectivity was provided by preparing the titanium complex in the presence of sulfide **1** at an elevated temperature and/or during a prolonged
preparation time and by performing the oxidation of sulfide **1** in the presence of amine. An enantioselectivity of $\sim$94% ee was obtained using this
method.

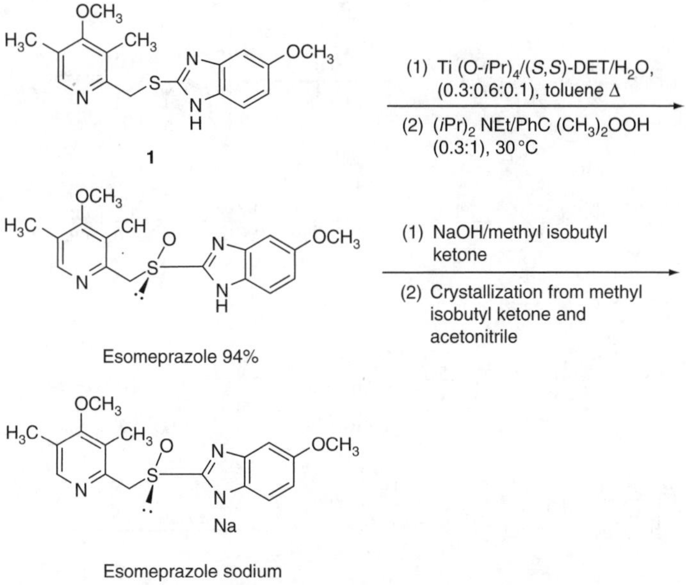

(1) Ti (O-iPr)₄/(S,S)-DET/H₂O,
    (0.3:0.6:0.1), toluene Δ

(2) (iPr)₂ NEt/PhC (CH₃)₂OOH
    (0.3:1), 30 °C

Esomeprazole 94%

(1) NaOH/methyl isobutyl
    ketone

(2) Crystallization from methyl
    isobutyl ketone and
    acetonitrile

Esomeprazole sodium

2.4. Omeprazole is obtained [15] by the reaction of acetyl ethyl propionate **1** with ammonia to give ethyl -3-amino-2,3-dimethyl acrylate **2**. Compound **2** was converted to to 2,4-dihydroxy-3,5,6-trimethyl pyridine **3** by treatment with methyl diethylmalonate. Treatment of compound **3** with phosphorous oxychloride produced 2,4-dichloro-3,5,6-trimethyl pyridine **4**. 4-Chloro-3,5,6-trimethyl pyridine **5** was obtained by treatment of compound **4** with hydrogen. On treatment of compound **5** with hydrogen peroxide and acetic acid, 4-chloro-3,5,6-trimethyl-pyridine-N-oxide **6** was produced. Treatment of compound **6** with acetic anhydride gave 4-chloro-2-hydroxymethyl-3,5-dimethyl pyridine **7** which was converted to 2-hydroxymethyl-3,5-dimethyl-4-methoxypyridine **8** by treatment with sodium methoxide. Compound **8** was treated with thionyl chloride to produce 2-chloromethyl-3,5-dimethyl-4-methoxypyridine **9**. Compound **9** interacts with 5-methoxy-2-mercaptobenzimidazole to give 5-methoxy 2-[((4-methoxy-3,5-dimethyl-2-pyridinyl)methyl)thio]-1H-benzimidazole **10** which is oxidized to omeprazole **11**.

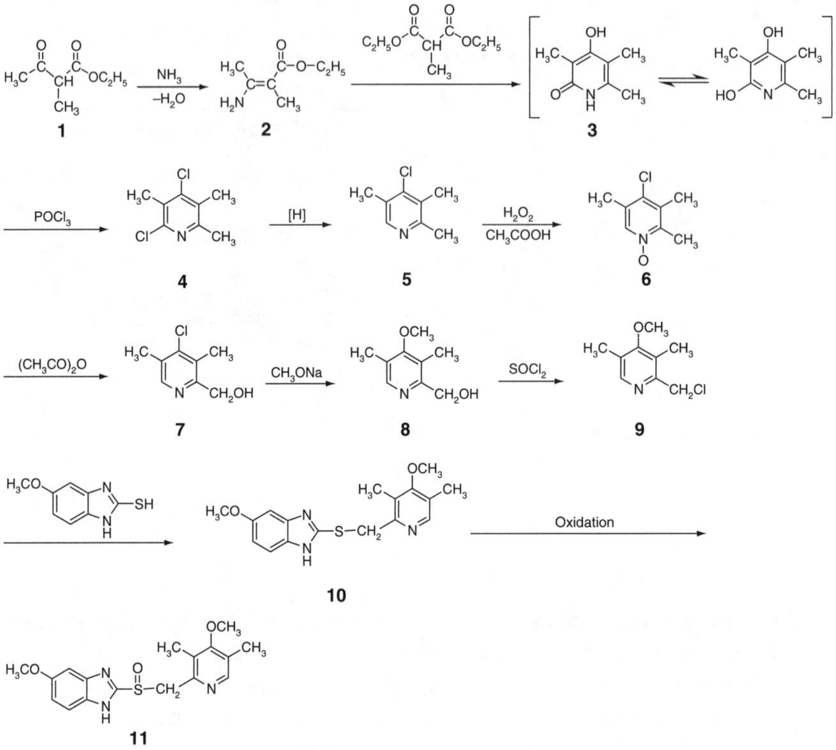

2.5. Baldwin *et al.* [16] used the following scheme for the preparation of omeprazole:

2-Methyl-1-penten-3-one-1-ol **1** and glacial acetic acid in benzene was added to pyrrolidine to give 2-methyl-1-penten-1-[*N*-pyrrolidinyl]-3-one **2**. Compound **2** when treated with oxalyl chloride and methanol was added, 3,5-dimethyl-2-methoxycarbonyl-4-pyrone **3** was produced. Treatment of compound **3** with sodium borohydride in methanol gives 3,5-dimethyl-2-hydroxymethyl-4-pyrone **4**. Compound **4** was converted to 3,5-dimethyl-2-hydroxymethyl-4-pyridone **5** by heating compound **4** with aqueous ammonia in a sealed flask. Compound **5** was converted to 4-chloro-2-chloromethyl-3,5-dimethyl pyridine **6** by treatment with phosphorous oxychloride. Treatment of compound **6** with 5-methoxy-2-mercaptobenzimidazole in tetrahydrofuran gave 2-[2-(4-chloro-3,5-dimethyl pyridinyl)methylthio]-5-methoxy benzimidazole **7**. When compound **7** was treated with potassium hydroxide in dimethyl sulfoxide containing methanol, 2-[2-(3,5-dimethyl-4-methoxypyridinyl)methylthio]-5-methoxy benzimidazole **8** was obtained. Compound **8** was dissolved in methanol and treated with hydrogen chloride to give the hydrochloride salt **9**. Compound **9** was then converted to omeprazole **10** by treatment with *m*-chloroperbenzoic acid (*m*-CPBA).

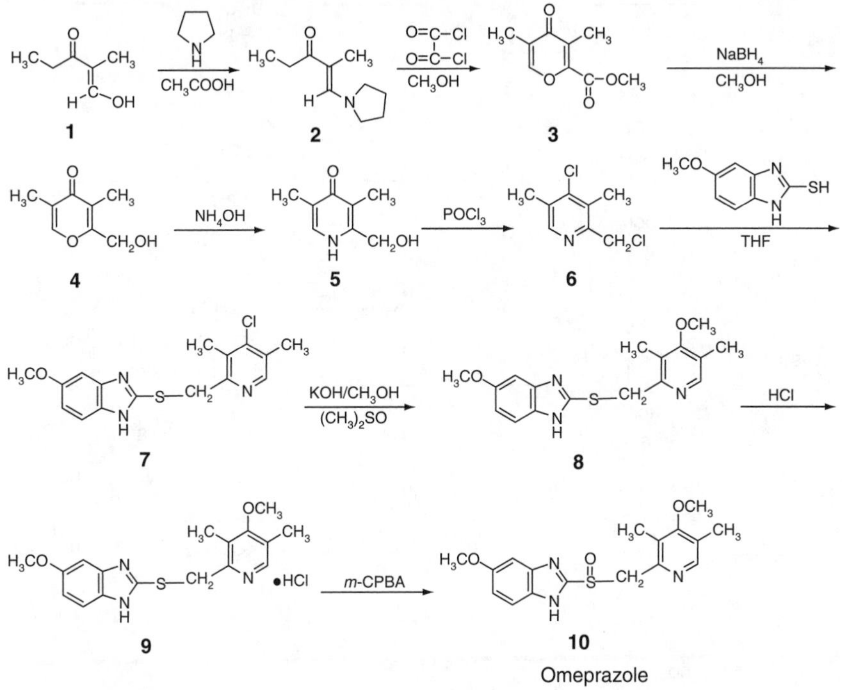

2.6. Singh *et al.* [17] used the following method for the preparation of omeprazole:

3,5-Dimethyl pyridine **1** was treated with hydrogen peroxide in acetic acid to give 3,5-dimethyl pyridine-*N*-oxide **2**. Compound **2** was converted to 3,5-dimethyl-4-nitropyridine-*N*-oxide **3** by treatment with a mixture of sulfuric and nitric acids. Compound **3** was treated with dimethyl sulfate to give 3,5-dimethyl-4-nitro-1-methoxypyridinium methyl sulfate **4**. Compound **4** was converted to 3,5-dimethyl-4-nitro-2-hydroxy methyl pyridine **5** by treatment with ammonium persulfate and methanol. Compound **5** was treated with thionyl chloride to give 3,5-dimethyl-4-nitro-2-chloromethyl pyridine hydrochloride **6**. When compound **6** was reacted with 5-methoxy-2-mercaptobenzimidazole **7** in sodium hydroxide and dichloromethane, 5-methoxy-2-[(3,5-dimethyl-4-nitro-2-pyridinyl)methylthio]-1*H*-benzimidazole **8** was obtained. Treatment of compound **8** with sodium hydroxide and methanol replaces the 4-nitro group in **8** by a 4-methoxy group to give **9**. Compound **9** was converted to the hydrochloride salt **10**. Compound **10** was oxidized by hydrogen peroxide in phthalic anhydride to give omeprazole **11**.

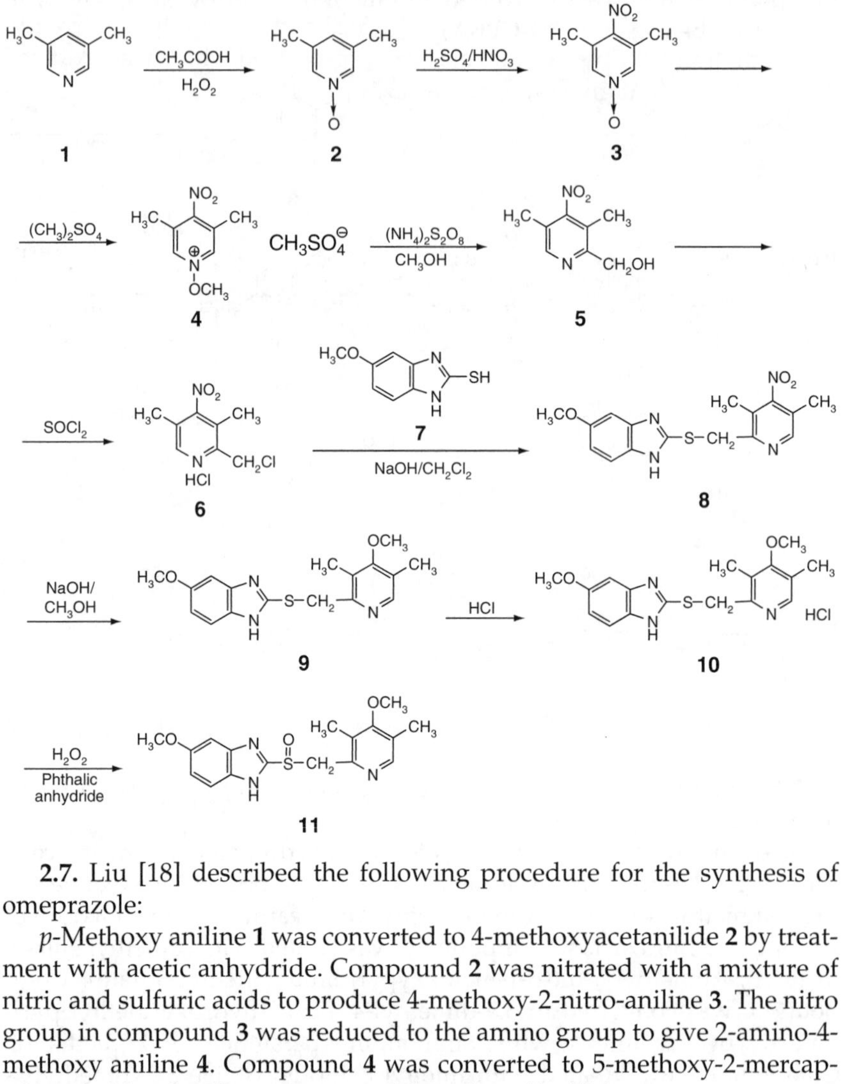

2.7. Liu [18] described the following procedure for the synthesis of omeprazole:

p-Methoxy aniline **1** was converted to 4-methoxyacetanilide **2** by treatment with acetic anhydride. Compound **2** was nitrated with a mixture of nitric and sulfuric acids to produce 4-methoxy-2-nitro-aniline **3**. The nitro group in compound **3** was reduced to the amino group to give 2-amino-4-methoxy aniline **4**. Compound **4** was converted to 5-methoxy-2-mercaptobenzimidazole **5** by treatment with carbon disulfide.

3,5-Dimethyl pyridine **6** was treated with hydrogen peroxide and 3,5-dimethyl pyridine-*N*-oxide **7** was produced. Compound **7** was nitrated with a mixture of nitric and sulfuric acids to give 3,5-dimethyl-4-nitropyridine-*N*-oxide **8**. Compound **8** was converted to 3,5-dimethyl-4-methoxypyridine-*N*-oxide **9** by reaction with methanol. Compound **9** was treated with a mixture of methanol and ammonium dithionite to give 3,5-dimethyl-

4-methoxy-2-hydroxymethyl-pyridine-*N*-oxide **10**. Compound **10** was converted to 3,5-dimethyl-4-methoxy-2-chloromethyl-pyridine **11** by treatment with thionyl chloride. Compound **11** was reacted with compound **5** in sodium hydroxide and 5-methoxy-2-[(3,5-dimethyl-4-methoxy-2-pyridinyl)methylthio]-1*H*-benzimidazole **12** was produced which was oxidized by *m*-CPBA acid to give omeprazole **13**.

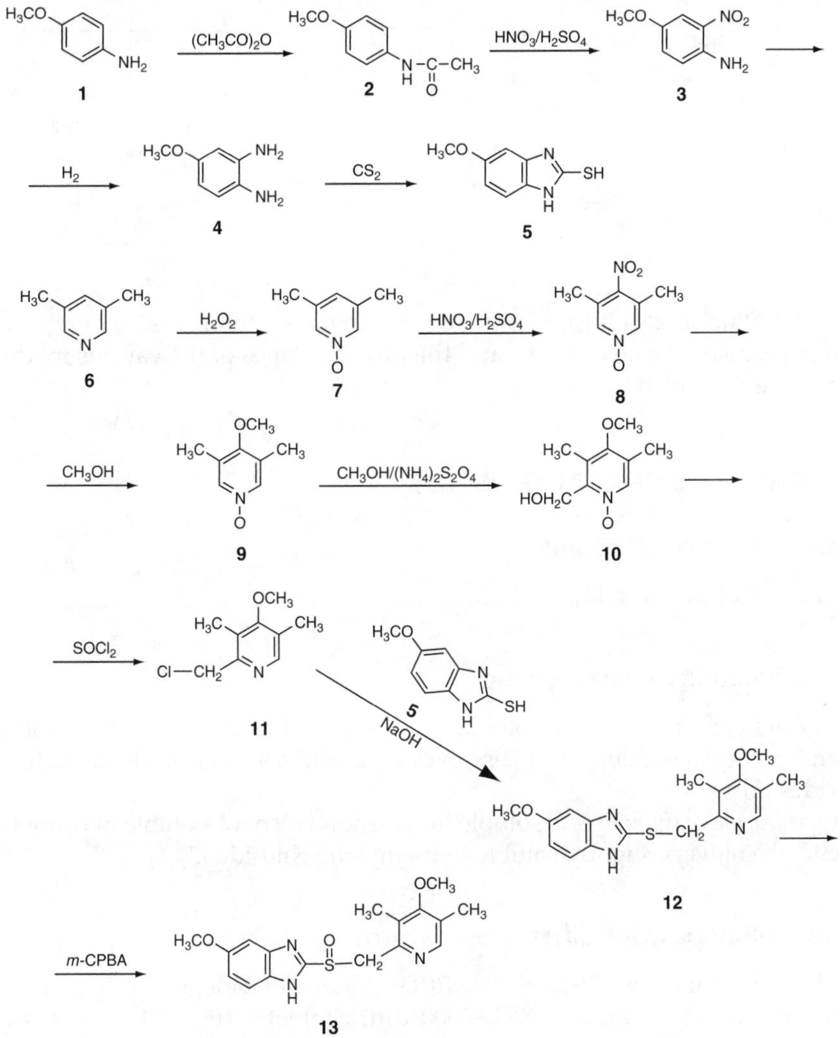

2.8. Omeprazole was prepared [19] by reaction of 4-methoxy-*o*-phenylenediamine **1** with potassium ethyl xanthogenate **2** to give 5-methoxy-2-mercapto-1*H*-benzimidazole **3**. Treatment of compound **3** with 3,5-dimethyl-4-methoxy-2-chloromethyl pyridine **4** in sodium hydroxide gives 5-methoxy-2-[((3,5-dimethyl-4-methoxy-2-pyridinyl)methyl)thio]-1H-benzimidazole **5**. Oxidation of compound **5** with 3-chloroperbenzoic acid produced omeprazole **6**.

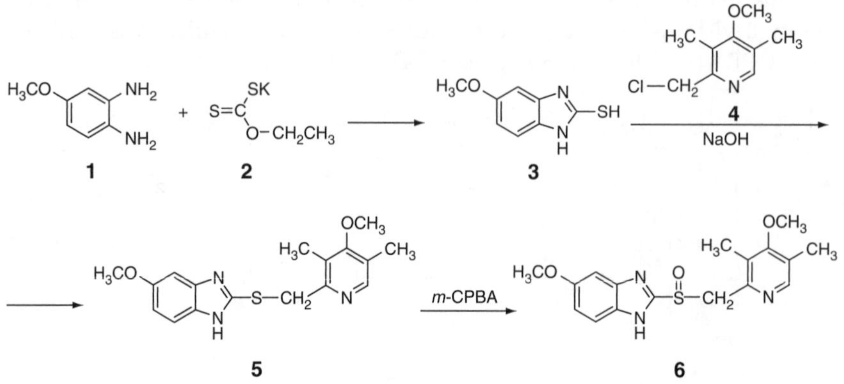

2.9. Rao *et al.* [20] reviewed the synthetic method used to prepare omeprazole. The advantages and the disadvantages of the various methods are described.

3. PHYSICAL CHARACTERISTICS

3.1. Ionization constant

pK_a = 4.61 and 9.08 [4].

3.2. Solubility characteristics

Omeprazole: Very slightly soluble in water, soluble in alcohol, methanol, and methylene chloride. It dissolves in dilute solution of alkali hydroxides [3].
Omeprazole sodium: Freely soluble in water and alcohol, soluble in propylene glycol, very slightly soluble in methylene chloride [3].

3.3. X-Ray powder diffraction pattern

The X-ray powder diffraction (XPRD) pattern of omeprazole was performed using a Simmon XRD-5000 diffractometer (Fig. 4.1). Table 4.1 shows the values for the scattering angles (° 2θ), the interplanar d-spacing (Å), and the relative intensities (%) observed for the major diffraction peaks of a pure sample of omeprazole drug substance.

Omeprazole is known to exist in at least two well-defined polymorphic forms, which have been the subject of patent documentation.

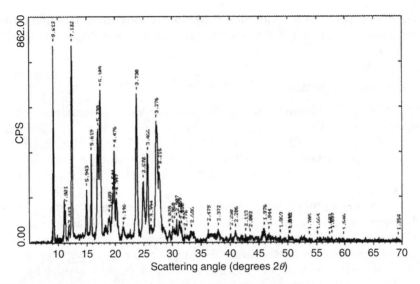

FIGURE 4.1   The X-ray powder diffraction pattern of omeprazole.

TABLE 4.1   The X-ray powder diffraction pattern for omeprazole

Scattering angle (°2 θ)	d-Spacing (Å)	Relative intensity (%)	Scattering angle (°2 θ)	d-Spacing (Å)	Relative intensity (%)
9.192	9.6133	91.05	31.256	2.8594	9.18
11.190	7.9007	14.44	31.587	2.8301	6.71
11.880	7.4433	5.15	32.214	2.7765	3.54
12.400	7.1321	100.00	33.329	2.6861	4.97
14.895	5.9429	23.63	36.204	2.4791	5.67
15.758	5.6192	39.88	37.907	2.3716	5.76
16.910	5.2388	50.92	40.045	2.2497	3.77
17.344	5.1086	68.42	40.874	2.2060	5.28
18.912	4.6887	12.05	42.640	2.1186	3.34
19.533	4.5410	22.00	43.414	2.0826	2.98
19.820	4.4757	41.18	45.896	1.9756	6.04
20.179	4.3969	19.82	46.691	1.9438	4.05
21.413	4.1462	6.28	48.673	1.8692	2.57
23.781	3.7384	66.89	50.200	1.8158	2.05
24.923	3.5697	27.36	50.323	1.8117	2.41
25.678	3.4665	39.81	53.719	1.7049	1.81
26.238	3.3937	7.86	55.145	1.6641	1.77
27.196	3.2763	54.36	57.289	1.6068	1.74
27.725	3.2149	31.37	57.685	1.5967	2.23
29.471	3.0284	5.42	59.747	1.5465	1.65
30.080	2.9684	6.95	69.354	1.3539	1.30
30.727	2.9073	10.24			

The XRPD patterns and defining peak values for these forms are summarized in Figs. 4.2 and 4.3 and in Tables 4.2 and 4.3 (US Patent 6,150,380).

3.4. Crystal structure

Ohishi *et al.* [21] determined the crystal structure of omeprazole. The crystal structure of omeprazole is triclinic, $P1$, $a = 10.686(5)$ Å, $b = 10.608(7)$ Å, $c = 9.666(6)$ Å, $\alpha = 119.75(5)°$, $\beta = 112.02(5)°$, $\gamma = 68.33(4)°$. $V = 859(1)$ Å^3, $Z = 2$, $D_m = 1.332(2)$, $D_x = 1.335$ g/cm^3, Cu Kα, $\lambda = 1.5418$ Å, $\mu = 18.04$ cm^{-1}, $F(0\ 0\ 0) = 364$, $T = 293$ K, $R = 0.057$ for 1962 observed reflections.

The methylsulfinyl group, which adopts a *trans* conformation, links the pyridine and benzimidazole rings in an almost coplanar orientation. Thus the molecule, as a whole, adopts a nearly extended form. Two centrosymmetrically related molecules form a cyclic dimer by intermolecular N–H···O hydrogen bonding, and the dimers are held together by van der Waals contacts between the neighboring aromatic rings in the crystal structure.

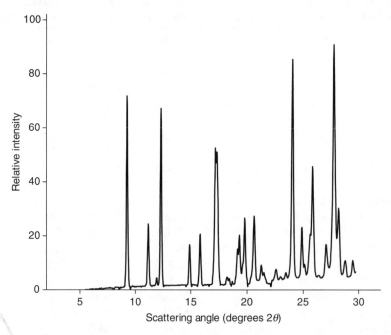

FIGURE 4.2 XRPD pattern of omeprazole, Astra Form-A, scanned and digitized from the pattern disclosed in US patent 6,150,380.

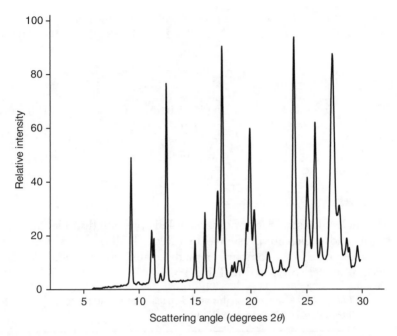

FIGURE 4.3   XRPD pattern of omeprazole, Astra Form-B, scanned and digitized from the pattern disclosed in US patent 6,150,380.

TABLE 4.2   Scattering angles and d-spacing of the 10 most intense peaks (US Patents 6,150,380)

Scattering Angle (degrees 2θ)	d-Spacing ($\mathring{A}$)
9.25	9.557
11.19	7.901
12.32	7.180
14.90	5.941
15.83	5.595
17.18	5.157
20.63	4.302
24.04	3.699
25.84	3.445
27.74	3.214

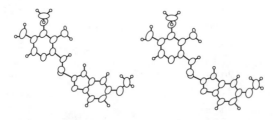

The molecular structure of omeprazole is presented in Fig. 4.4. Bond distances and angles are presented in Table 4.4, all of which are normal within their e.s.d.'s in comparison with related compounds. The final atomic parameters are listed in Table 4.5.

TABLE 4.3 Scattering angles and d-spacing of the 10 most intense peaks (US Patents 6,150,380)

Scattering Angle (degrees 2θ)	d-Spacing (Å)
9.28	9.522
12.46	7.099
15.92	5.561
17.05	5.197
17.45	5.079
19.91	4.455
23.84	3.729
25.02	3.555
25.75	3.457
27.32	3.262

FIGURE 4.4 A stereoscopic view of omeprazole, viewed perpendicular to the pyridine ring [21].

TABLE 4.4 Bond distances (Å) and angles (°) for non-H atoms with e.s.d.'s in parentheses [21]

N(1)–C(2)	1.339(8)	C(10)–N(11)	1.361(7)
N(1)–C(6)	1.319(7)	C(10)–N(14)	1.306(7)
C(2)–C(3)	1.358(9)	N(11)–C(12)	1.396(7)
C(3)–C(4)	1.368(8)	C(12)–C(13)	1.381(7)
C(3)–C(19)	1.53(1)	C(12)–C(15)	1.382(7)
C(4)–C(5)	1.396(7)	C(13)–N(14)	1.376(7)
C(4)–O(20)	1.392(7)	C(13)–C(18)	1.398(8)
C(5)–C(6)	1.387(7)	C(15)–C(16)	1.387(8)
C(5)–C(22)	1.516(8)	C(16)–C(17)	1.377(8)
C(6)–C(7)	1.527(8)	C(16)–O(23)	1.357(7)
C(7)–S(8)	1.815(6)	C(17)–C(18)	1.369(8)
S(8)–O(9)	1.487(4)	O(20)–C(21)	1.421(9)
S(8)–C(10)	1.768(6)	O(23)–C(24)	1.409(8)
C(2)–N(1)–C(6)	117.4(4)	S(8)–C(10)–N(14)	120.7(2)
N(1)–C(2)–C(3)	123.5(4)	N(11)–C(10)–N(14)	115.6(3)
C(2)–C(3)–C(4)	118.0(4)	C(10)–N(11)–C(12)	104.0(3)
C(2)–C(3)–C(19)	121.0(4)	N(11)–C(12)–C(13)	106.1(3)
C(4)–C(3)–C(19)	120.9(4)	N(11)–C(21)–C(15)	130.5(3)
C(3)–C(4)–C(5)	120.9(4)	C(13)–C(12)–C(15)	123.4(3)
C(3)–C(4)–O(20)	121.2(3)	C(12)–C(13)–N(14)	110.8(3)
C(5)–C(4)–O(20)	117.9(3)	C(12)–C(13)–C(18)	120.6(3)
C(4)–C(5)–C(6)	115.6(3)	N(14)–C(13)–C(18)	128.6(3)
C(4)–C(5)–C(22)	120.8(4)	C(10)–N(14)–C(13)	103.6(3)
C(6)–C(5)–C(22)	123.6(4)	C(12)–C(15)–C(16)	114.1(3)
N(1)–C(6)–C(5)	124.5(3)	C(15)–C(16)–C(17)	124.0(4)
N(1)–C(6)–C(7)	115.7(3)	C(15)–C(16)–O(23)	122.4(3)
C(5)–C(6)–C(7)	119.8(3)	C(17)–C(16)–O(23)	113.6(3)
C(6)–C(7)–S(8)	108.7(2)	C(16)–C(17)–C(18)	120.9(4)
C(7)–S(8)–O(9)	105.9(3)	C(13)–C(18)–C(17)	117.0(3)
C(7)–S(8)–C(10)	96.6(3)	C(4)–O(20)–C(21)	114.6(4)
O(9)–S(8)–C(10)	108.0(2)	C(16)–O(23)–C(24)	116.0(4)
S(8)–C(10)–N(11)	123.7(2)		

The molecule takes an extended conformation, in which the pyridine and benzimidazole rings are linked by the methylsulfinyl chain taking a *trans* conformation [C(6)–C(7)–S(8)–C(10) = 179.1(3)°]; the torsion angles N(1)–C(6)–C(7)–S(8), C(5)–C(6)–C(7)–S(8), C(6)–C(7)–S(8)–C(10), and C(7)–S(8)–C(10)–N(14) are −33.6(4)°, 148.3(5)°, 60.9(5)°, and −121.3(5)°, respectively, and the dihedral angle between the aromatic rings is 30.0(2)°. The sulfinyl bond protrudes from the benzimidazole plane.

TABLE 4.5 Fractional atomic coordinates and equivalent isotropic temperature factors ($\mathring{A}^2$) for non-H atoms with e.s.d.'s in parentheses [21]

	x	y	z	$B_{eq}{}^a$
N(1)	0.1549(4)	0.5541(4)	0.9852(5)	6.5(2)
C(2)	0.2194(6)	0.4734(6)	0.0739(7)	7.1(3)
C(3)	0.3401(5)	0.3692(6)	1.0539(7)	6.2(2)
C(4)	0.3972(4)	0.3438(5)	0.9346(6)	6.2(2)
C(5)	0.3343(4)	0.4245(5)	0.8384(6)	5.7(2)
C(6)	0.2131(5)	0.5300(5)	0.8731(6)	5.8(2)
C(7)	0.1411(5)	0.6312(5)	0.7824(7)	6.8(2)
S(8)	−0.0444(1)	0.6751(1)	0.7611(2)	5.94(5)
O(9)	−0.0903(3)	0.5343(3)	0.6413(4)	6.2(1)
C(10)	−0.0916(5)	0.7910(5)	0.6585(6)	5.9(2)
N(11)	−0.0656(4)	0.7435(4)	0.5106(5)	5.8(2)
C(12)	−0.1227(4)	0.8690(5)	0.4758(6)	5.6(2)
C(13)	−0.1783(4)	0.9807(5)	0.6064(6)	5.6(2)
N(14)	−0.1591(4)	0.9295(4)	0.7203(5)	6.2(2)
C(15)	−0.1254(5)	0.8892(5)	0.3440(6)	6.0(2)
C(16)	−0.1882(5)	1.0327(5)	0.3542(6)	6.2(2)
C(17)	−0.2437(5)	1.1469(5)	0.4831(7)	6.8(2)
C(18)	−0.2406(5)	1.1239(5)	0.6119(6)	6.7(2)
C(19)	0.4135(7)	0.2880(9)	1.1666(9)	10.2(4)
O(20)	0.5228(3)	0.2418(4)	0.9104(5)	8.2(2)
C(21)	0.5106(6)	0.0936(6)	0.798(1)	10.3(3)
C(22)	0.3969(6)	0.3954(7)	0.7052(7)	7.9(3)
O(23)	−0.1977(4)	1.0740(4)	0.2377(4)	8.2(2)
C(24)	−0.1520(8)	0.9590(7)	0.0972(7)	8.8(4)

^{a} $B_{eq} = \frac{4}{3}\sum_i \sum_j a_i a_j \beta_{ij}$.

The methoxy group attached to the pyridine ring is almost perpendicular to the ring plane [C(3)–C(4)–O(20)–C(21) = 89.5(6)°, C(5)–C(4)–O(20)–C (21) = −93.4(5)°], while that attached to the benzimidazole ring is almost coplanar with the ring [C(15)–C(16)–O(23)–C(24) = 6.0(5)°, C(17)–C(16)–O(23)–C(24) = −175.4(6)°].

The molecules are arrayed along the (2 2 0) plane corresponding to the strongest intensity among the observed reflections. The two molecules which are related to each other by a center of symmetry form a cyclic dimer with an intermolecular N(11)–H···O(9) hydrogen bond [N (x, y, z)···O(−x, 1 − y, 1 − z) = 2.744(6) Å, H···O = 1.78 (7) Å and angle N–H···O = 169(6)° (see Fig. 4.5A)], and the dimer is stabilized by van der Waals contacts between the pyridine and benzimidazole rings;

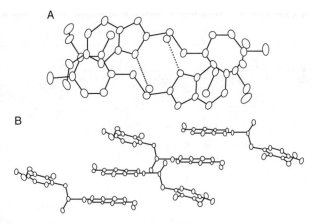

FIGURE 4.5 (A) Cyclic dimer structure formed by N–H· · ·O intermolecular hydrogen bonds represented by dotted lines. (B) Overlapping mode among the neighboring aromatic rings [21].

the average interplanar spacing between the pyridine and benzimidazole rings is 4.13 Å.

On the other hand, the benzimidazole ring in the dimer also forms a stacking interaction with the centrosymmetrically related ring with an average spacing of 3.38 Å (Fig. 4.5B).

3.5. Thermal methods of analysis

3.5.1. Melting range
M.P. 156 °C [3].

3.5.2. Differential scanning calorimetry
The differential scanning calorimetry (DSC) thermogram of omeprazole was obtained using a DuPont 2100 thermal analyzer system. The thermogram shown in Fig. 4.6 was obtained at a heating rate of 10 °C/min, and was run over the range 50–300 °C. Omeprazole was found to melt at 159.65 °C.

3.6. Spectroscopy

3.6.1. Ultraviolet spectroscopy
The ultraviolet (UV) absorption spectrum of omeprazole in methanol (0.0016%, w/v) shown in Fig. 4.7 was recorded using a Shimadzu UV–VIS spectrophotometer 1601 PC. Omeprazole exhibited two maxima at 276 and 302 nm.

Clarke [3] reported the following: Aqueous acid (0.2 M H_2SO_4), 277 and 303 nm; basic, 276 and 305 nm.

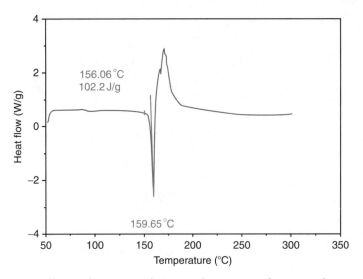

FIGURE 4.6   Differential scanning calorimetry thermogram of omeprazole.

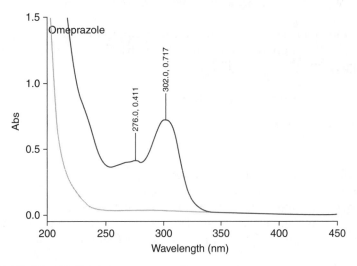

FIGURE 4.7   The UV absorption spectrum of omeprazole.

3.6.2. Vibrational spectroscopy

The infrared (IR) absorption spectrum of omeprazole was obtained in a KBr pellet using a Perkin-Elmer IR spectrophotometer. The IR spectrum is shown in Fig. 4.8, where the principal peaks were observed and the assignments for the major IR absorption bands are listed in Table 4.6.

Clarke [3] reported that principal peaks are at wavenumbers 1625, 1205, 1015 cm^{-1} (KBr disc).

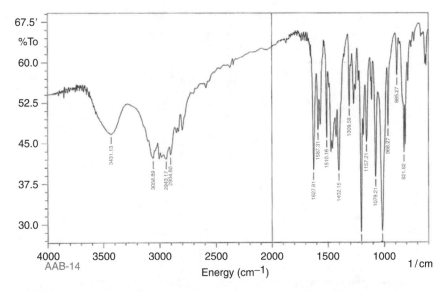

FIGURE 4.8 The infrared absorption spectrum of omeprazole.

TABLE 4.6 Vibrational assignments for omeprazole infrared absorption bands

Frequency (cm^{-1})	Assignment
3431	N–H stretch
3058	Aromatic C–H stretch
2943 and 2904	C–H stretch
1627	C=C stretch
1587	C=N stretch
1510	CH_2 bending
1402 and 1309	CH bending
1157	C=O stretch
1070	C=S stretch
966, 885, and 821	C–H bending

3.6.3. Nuclear magnetic resonance spectrometry

3.6.3.1. ^1H NMR spectrum   The proton nuclear magnetic resonance (^1H NMR) spectrum of omeprazole were obtained using a Bruker Instrument operating at 300, 400, or 500 MHz. Standard Bruker Software was used to execute the recording of DEPT, COSY, and HETCOR spectra. The sample was dissolved DMSO-d_6 and all resonance bands were referenced to tetramethylsilane (TMS) as internal standard. The ^1H NMR spectra of omeprazole are shown in Figs. 4.9–4.12 and the COSY ^1H NMR is

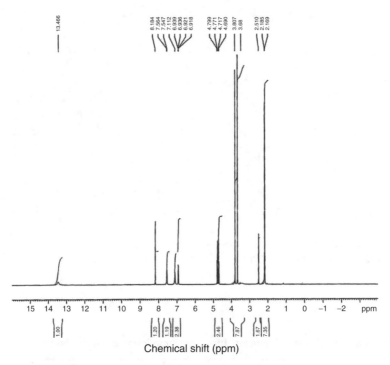

FIGURE 4.9   The ^{1}H NMR spectrum of omeprazole in DMSO-d_6.

shown in Fig. 4.13. The ^{1}H NMR assignments for omeprazole are listed in Table 4.7.

3.6.3.2. ^{13}C NMR spectrum

The ^{13}C NMR spectra of omeprazole were obtained using a Bruker Instrument operating at 75, 100, or 125 MHz. The sample was dissolved in DMSO-d_6 and TMS was added to function as the internal standard. The ^{13}C NMR spectra are shown in Figs. 4.14 and 4.15 and the HSQC and the HMBC NMR are shown in Figs. 4.16 and 4.17, respectively. The DEPT 135 are shown in Figs. 4.18 and 4.19, respectively. The assignments for the observed resonance bands associated with the various carbons are listed in Table 4.8.

Claramunt et al. [22] used a ^{1}H and ^{13}C NMR to study the tautomerisim of omeprazole in solution. The tautomeric equilibrium constant, $K_T = 0.59$ in tetrahydrofuran at 195 K, is in favor of the 6-methoxy tautomer. The assignment of the signals was made by comparison with its two N-methyl derivatives in acetone-d_6 and through theoretical calculations of the absolute shieldings (GIAO/DFT/6-3111++G**).

Claramunt et al. [23] recorded the ^{13}C and ^{15}N CPMAS spectra of solid sample of omeprazole and all signals assigned. The sample consists uniquely of the 6-methoxy tautomer.

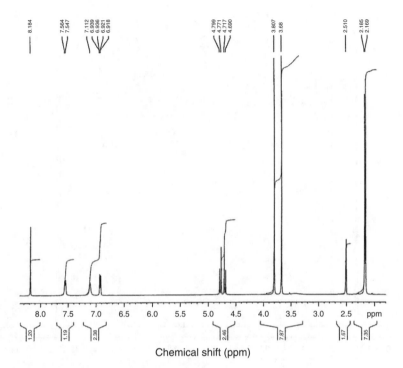

FIGURE 4.10   Expanded ¹H NMR spectrum of omeprazole in DMSO-d_6.

3.7. Mass spectrometry

The mass spectrum of omeprazole was obtained using a Shimadzu PQ-5000 mass spectrometer. The parent ion was collided with helium as the carrier gas. Figure 4.20 shows the mass fragmentation pattern of the drug substance (Table 4.9).

Clarke [3] reported that principal ions are at m/z 151, 136, 121, 120, 180, 297, 77, and 93.

4. METHODS OF ANALYSIS

4.1. Compendial methods

4.1.1. European Pharmacopoeia methods [24]

4.1.1.1. Omeprazole   Omeprazole contains less than 99% and not more than the equivalent of 101% of 5-methoxy-2-[[(*RS*)-(4-methoxy-3,5-dimethyl-pyridine-2-yl)methyl]sulfinyl]-1*H*-benzimidazole, calculated with reference to the dried substance.

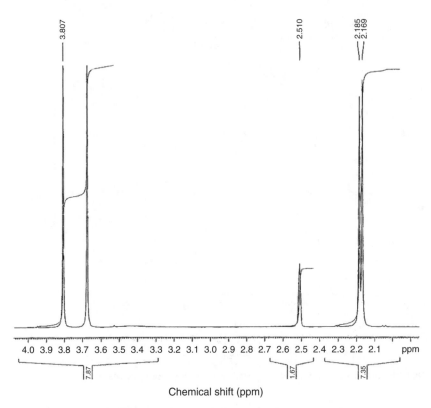

FIGURE 4.11   Expanded ^{1}H NMR spectrum of omeprazole in DMSO-d_6.

- *Identification*

    Test A: Dissolve 2 mg in 0.1 M *sodium hydroxide* and dilute to 100 ml with the same solution. Examined between 230 and 350 nm, according to the general method (2.2.25), the solution shows two absorption maxima, at 276 and 305 nm. The ratio of the absorbance measured at the maximum at 305 nm to that measured at the maximum at 276 nm is 1.6–1.8.

    Test B: Examine by IR absorption spectrophotometry, according to the general method (2.2.24), comparing with the spectrum obtained with *omeprazole CRS*. If the spectra obtained in the solid state show differences, dissolve the substance to be examined and the reference substance separately in *methanol R*, evaporate to dryness and record new spectra using the residues.

    Test C: Examine the chromatograms obtained in the test for omeprazole impurity C. The principal spot in the chromatogram obtained with test solution (b) is similar in position and size to the principal spot in the chromatogram obtained with reference solution (a). Place the plate in a tank saturated with vapor of *acetic acid R*. The spots rapidly turn brown.

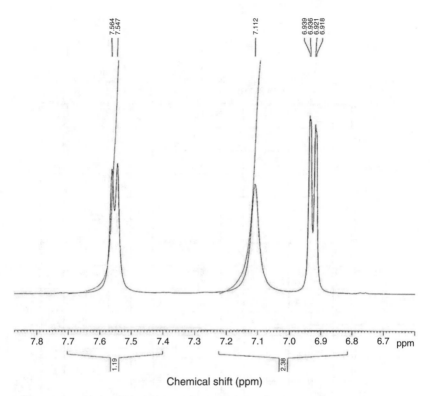

FIGURE 4.12   Expanded ^{1}H NMR spectrum of omeprazole in DMSO-d_6.

- *TESTS*

    Solution S: Dissolve 0.5 g of omeprazole in *methylene chloride R* and dilute to 25 ml with the same solvent.

    Appearance of solution: When this test is carried out according to the general method (2.2.1), solution S is clear.

    Absorbance: When this test is carried out according to the general procedure (2.2.25), the absorbance of solution S measured at 440 nm is not more than 0.1 (this limit corresponds to 0.035% of omeprazole impurity F or omeprazole impurity G).

    Omeprazole impurity C: Examine by thin-layer chromatography (TLC), according to the general procedure (2.2.27), using a TLC *silica gel F$_{254}$ R*.

- *Test solution (a)*. Dissolve 0.1 g of omeprazole in 2 ml of a mixture of equal volumes of *methanol R* and *methylene chloride R*.

- *Test solution (b)*. Dilute 1 ml of test solution (a) to 10 ml with *methanol R*.

- *Reference solution (a)*. Dissolve 10 mg of *omeprazole CRS* in 2 ml of *methanol R*.

- *Reference solution (b)*. Dilute 1 ml of test solution (a) to 10 ml with a mixture of equal volumes of *methanol R* and *methylene chloride R*. Dilute 1 ml of this solution to 100 ml with a mixture of equal volumes of *methanol R* and *methylene chloride R*.

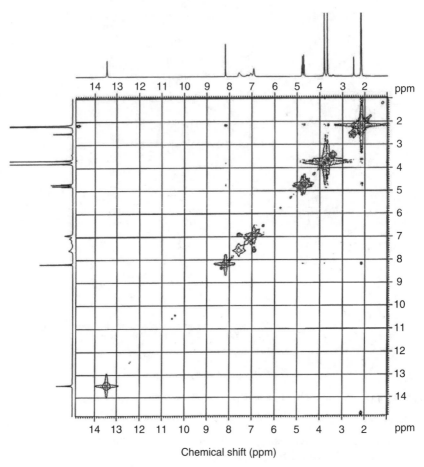

FIGURE 4.13   COSY ^{1}H NMR spectrum of omeprazole in DMSO-d_6.

Apply to the plate 10 μl of each solution. Develop over a path of 15 cm using a mixture of 20 volumes of *2-propanol R*, 40 volumes of *methylene chloride R* previously shaken with *concentrated ammonia R* (shake 100 ml of *methylene chloride R* with 30 ml of *concentrated ammonia R* in a separating funnel; allow the layers to separate and use the lower layer) and 40 volumes of *methylene chloride R*. Allow the plates to dry in air. Examine in UV light at 254 nm. Any spot in the chromatogram obtained with test solution (a) with a higher R_f value than that of the spot due to omeprazole is not more intense than the spot in the chromatogram obtained with reference solution (b) (0.1%).

Related substances. Examine by liquid chromatography, according to the general procedure (2.2.29).

TABLE 4.7   Assignments of the resonance bands in the ¹H NMR spectrum of omeprazole

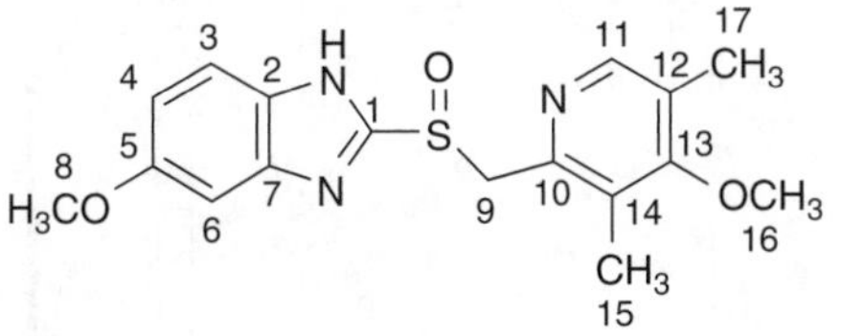

Chemical shift (ppm, relative to TMS)	Number of protons	Multiplicity[a]	Assignment (proton at carbon number)
2.17	3	s	15 or 17
2.19	3	s	15 or 17
2.51	DMSO	s	
3.68	3	s	8 or 16
3.81	3	s	8 or 16
4.68, 4.76	2	2d, $J = 13.5$ Hz	9
6.92–6.94	1	d	4
7.11	1	s	6
7.55–7.56	1	d	3
8.18	1	s	11
13.47	1	s	NH

[a] s, singlet; m, multiplet; d, doublet.

- *Test solution.* Dissolve 3 mg of omeprazole in the mobile phase and dilute to 25 ml with the mobile phase.
- *Reference solution (a).* Dissolve 1 mg of *omeprazole CRS* and 1 mg of *omeprazole impurity D CRS* in the mobile phase and dilute to 10 ml with the mobile phase.
- *Reference solution (b).* Dilute 1 ml of the test solution to 100 ml with the mobile phase. Dilute 1 ml of this solution to 10 ml with the mobile phase.

The chromatographic procedure may be carried out using:

- a stainless-steel column 0.15 m long and 4 mm in internal diameter packed with *octylsilyl silica gel for chromatography R* (5 μm).
- as mobile phase, at a flow-rate of 1 ml/min, a mixture of 27 volumes of *acetonitrile R* and 73 volumes of a 1.4-g/l solution of *disodium hydrogen phosphate R* previously adjusted to pH 7.6 with *phosphoric acid R*.
- as detector a spectrophotometer set at 280 nm.

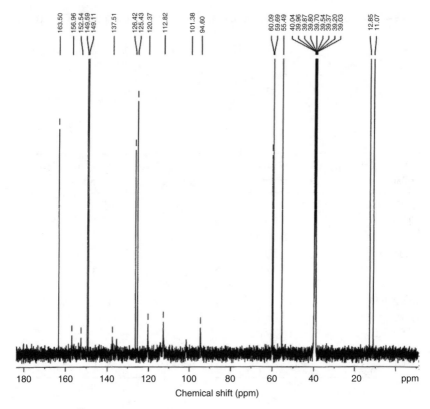

FIGURE 4.14 ^{13}C NMR spectrum of omeprazole in DMSO-d_6.

When the chromatograms are recorded under the prescribed conditions, the retention time of omeprazole is about 9 min and the relative retentions of impurities A, E, D, and B are about 0.4, 0.6, 0.8, and 0.9, respectively. Inject separately 40 μl of each solution and continue the chromatography for three times the retention time of omeprazole. Where applicable, adjust the sensitivity of the system so that the height of the principal peak in the chromatogram obtained with reference solution (b) is at least 15% of the full scale of the recorder. The test is not valid unless the chromatogram obtained with reference solution (a), the resolution between the peaks corresponding to omeprazole impurity D and omeprazole is greater than 3. If necessary, adjust the pH of the mobile phase or the concentration of *acetonitrile R*; an increase in the pH will improve the resolution. The area of any peak due to impurities A, B, D, and E or any other peak, apart from the principal peak, in the chromatogram obtained with the test solution is not greater than the area of the peak in the chromatogram obtained with reference solution (b) (0.1%).

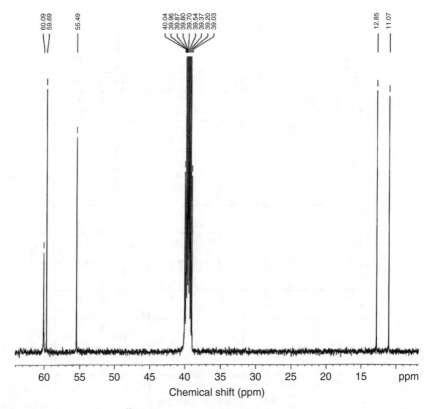

FIGURE 4.15 Expanded ^{13}C NMR spectrum of omeprazole in DMSO-d_6.

Residual solvents. Examine by head-space gas chromatography, according to the general procedure (2.2.28), using the standard addition method. The content of chloroform is not more than 50 ppm and the content of methylene chloride is not more than 100 ppm.

The chromatographic procedure may be carried out using:

— a fused-silica column 30 m long and 0.32 mm in internal diameter coated with a 1.8-μm film of cross-linked *poly-[(cyanopropyl)(phenyl)] [dimethyl] siloxane R*.
— *nitrogen for chromatography R* as the carrier gas.
— a flame-ionization detector.
— a suitable head-space sampler.

Place 0.5 g of omeprazole in a 10-ml vial. Add 4 ml of *dimethylacetamide R* and stopper the vial. Equilibrate the vial at 80 °C for 1 h.

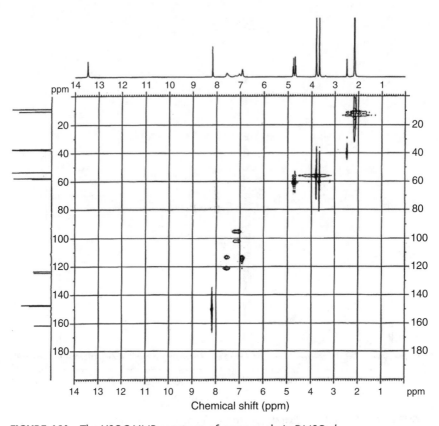

FIGURE 4.16 The HSQC NMR spectrum of omeprazole in DMSO-d_6.

Loss on drying. This test should be carried out according to the general procedure (2.2.32). Maximum 0.2%, determined on 1 g by drying under high vacuum at 60 °C for 4 h.

Sulfated ash. This test should be carried out according to the general procedure (2.4.14). Maximum 0.1%, determined on 1 g.

• *Assay*

Dissolve 1.1 g of omeprazole in a mixture of 10 ml of *water R* and 40 ml of *ethanol R* 96%. Titrate with 0.5 M *sodium hydroxide*, determining the end point potentiometrically as described in the general procedure (2.2.20).

1 ml of 0.5 M *sodium hydroxide* is equivalent to 0.1727 g of $C_{17}H_{19}N_3O_3S$.

• *Storage*

Store in an airtight container, protected from light, at a temperature between 2 and 8 °C.

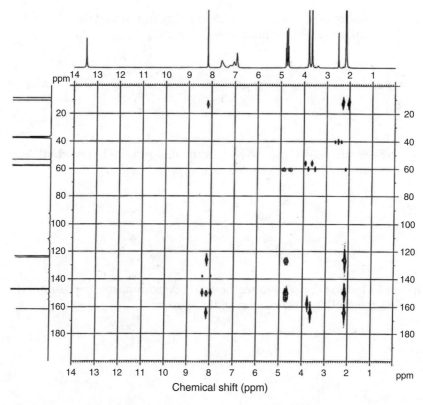

FIGURE 4.17   The HMBC NMR spectrum of omeprazole in DMSO-d_6.

- *Impurities*

  Specified impurities A, B, C, D, E, F, G. Other detectable impurities H, I.

A.   5-Methoxy-1*H*-benzimidazole-2-thiol

B.   2-(*RS*)-[[(3,5-dimethylpyridin-2-yl)methyl]sulphinyl]-5-methoxy-1*H*-benzimidazole

C.   5-Methoxy-2-[[(4-methoxy-3,5-dimethylpyridin-2-yl)methyl]
      sulfanyl]-1*H*-benzimidazole (*ufiprazole*)

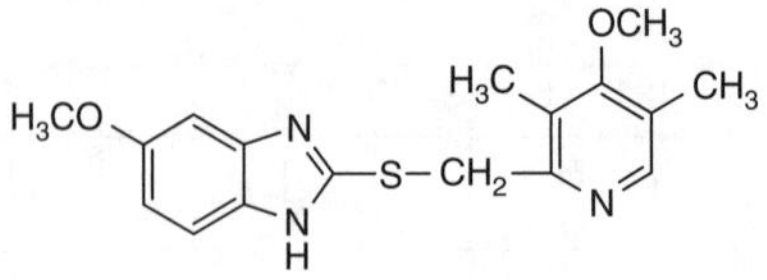

D.   5-Methoxy-2-[[(4-methoxy-3,5-dimethylpyridin-2-yl)methyl]
      sulfonyl]-1*H*-benzimidazole (*omeprazole sulfone*)

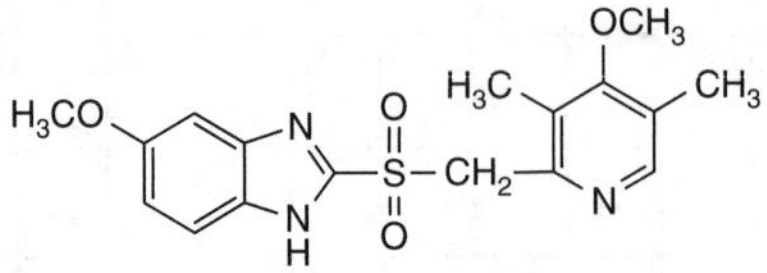

E.   4-Methoxy-2-[[(*RS*)-(5-methoxy-1*H*-benzimidazole-2-yl)sulfinyl]
      methyl]-3,5-dimethyl-pyridine-1-oxide

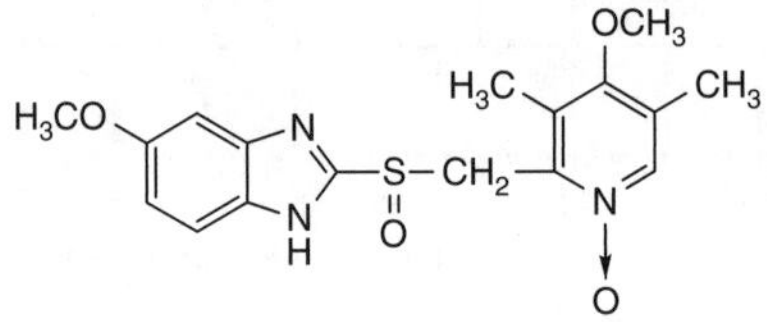

F.   8-Methoxy-1,3-dimethyl-12-thioxopyrido[1′,2′:3,4]imidazo[1,2-*a*]
      benzimidazol-2(12*H*)-one

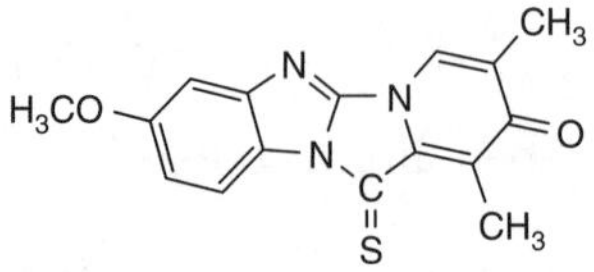

G.   9-Methoxy-1,3-dimethyl-12-thioxopyrido[1′,2′:3,4]imidazo[1,2-*a*]
      benzimidazol-2(12*H*)-one

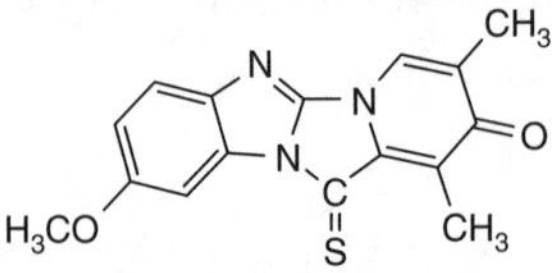

H.   2-[(RS)-[(4-chloro-3,5-dimethylpyridin-2-yl)methyl]sulfinyl]-5-
      methoxy-1H-benzimidazole

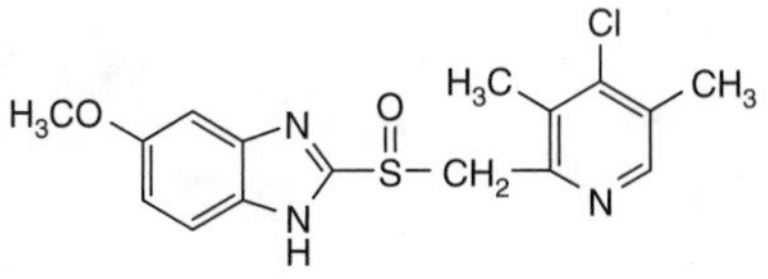

I.   4-Methoxy-2-[[(5-methoxy-1H-benzimidazol-2-yl)sulfonyl]
      methyl]-3,5-dimethyl-pyridine-1-oxide

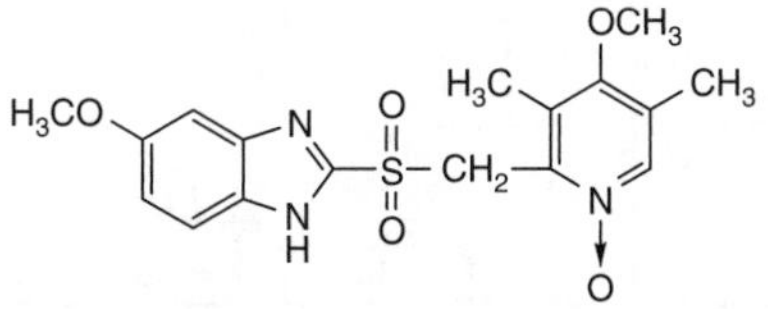

4.1.1.2. Omeprazole sodium

• *Definition*

Omeprazole sodium contains not less than 98% and not more than the equivalent of 101% of sodium 5-methoxy-2[(RS)-[(4-methoxy-3,5-dimethylpyridin-2-yl)methyl]sulfinyl]-1H-benzimidazole,   calculated with reference to the anhydrous substance.

• *Characters*
  A white or almost white powder, hygroscopic, freely soluble in water and in alcohol, soluble in propylene glycol, very slightly soluble in methylene chloride.
• *Identification*
  Test A: Dissolve 2 mg of omeprazole sodium in 0.1 M *sodium hydroxide* and dilute to 100 ml with the same solvent. Examined between 230 and 350 nm, as directed in the general procedure (2.2.25), the solution shows two absorption maxima, at 276 and 305 nm. The ratio of the absorbance measured at the maximum at 305 nm to that measured at the maximum at 276 nm is 1.6–1.8.
  Test B: Examine the chromatograms obtained in the test for omeprazole impurity C. The principal spot in the chromatogram obtained with test solution (b) is similar in position and size to the principal spot in the chromatogram obtained with reference solution (a). Place the plate in a tank saturated with vapor of *acetic acid R*. The spots rapidly turn brown.

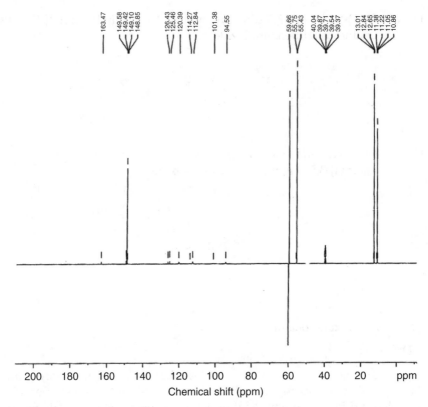

FIGURE 4.18   The DEPT 135 ^{13}C NMR spectrum of omeprazole in DMSO-d_6.

Test C: Ignite 1 g of omeprazole and cool. Add 1 ml of *water R* to the residue and neutralize with *hydrochloric acid R*. Filter and dilute the filtrate to 4 ml with *water R*. 0.1 ml of the solution gives reaction (b) of sodium, this test should be carried according to the general procedure (2.3.1).

• *TESTS*

Solution S: Dissolve 0.5 g of omeprazole sodium in *carbon dioxide-free water R* and dilute to 25 ml with the same solvent.

Appearance of solution: Carry out this test as directed in the general procedure (2.2.1), solution S is clear and not more intensely colored than reference solution B$_6$ (*Method II* in the general procedure (2.2.2)).

pH: This test should be carried out according to the general procedure (2.2.3). The pH of solution S is 10.3–11.3

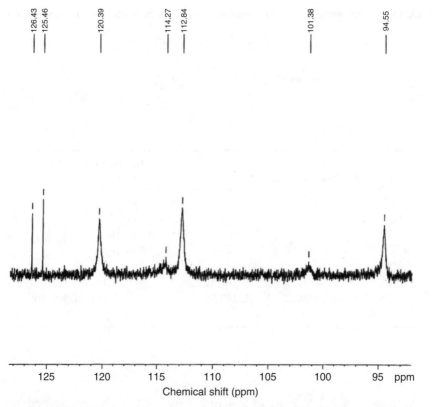

FIGURE 4.19   Expanded DEPT 135 ^{13}C NMR spectrum of omeprazole in DMSO-d_6.

Omeprazole impurity C: Examine by TLC, as directed in the general
procedure (2.2.27), using *silica gel HF$_{254}$ R* as the coating substance.

- *Test solution (a)*. Dissolve 0.1 g of omeprazole sodium in 2 ml of *methanol R*.
- *Test solution (b)*. Dilute 1 ml of test solution (a) to 10 ml with *methanol R*.
- *Reference solution (a)*. Dissolve 9 mg of *omeprazole CRS* in 2 ml of *methanol R*.
- *Reference solution (b)*. Dilute 1 ml of test solution (b) to 100 ml *with methanol R*.

Apply separately to the plate 10 μl of each solution. Develop over a
path of 15 cm using a mixture of 20 volumes of *2-propanol R*, 40 volumes of
methylene chloride R previously shaken with *concentrated ammonia R* (shake
100 ml of *methylene chloride R* with 30 ml of *concentrated ammonia R* in a
separating funnel, allow the layers to separate and use the lower layer)
and 40 volumes of *methylene chloride R*. Allow the plate to dry in air.

TABLE 4.8 Assignments of the resonance bands in the ^{13}C NMR spectra of omeprazole

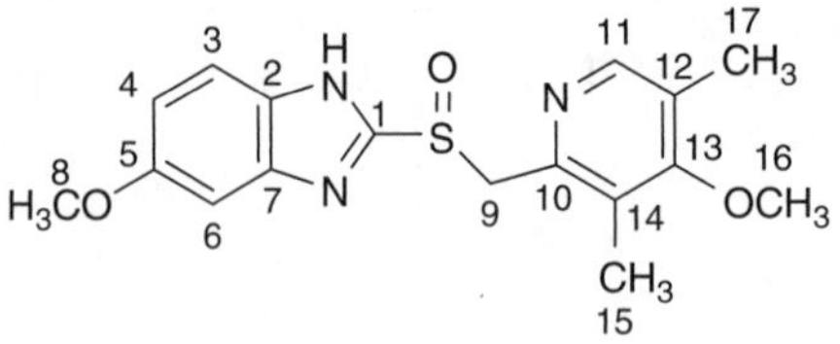

Chemical shift (ppm relative to TMS)	Assignment at carbon number
11.07	15 or 17
12.85	15 or 17
55.49	8 or 16
59.69	8 or 16
60.09	9
94.55, 101.38, 112.84, 114.27, 120.39, 125.46, 126.43, 148.85, 149.10, 149.42, 149.58, 163.47	8 quaternary and 4 protonated carbons

Examine in the UV light at 254 nm. Any spot in the chromatogram obtained with test solution (a) with a higher R_f value than that of the spot corresponding to omeprazole is not more intense than the spot in the chromatogram obtained with reference solution (b) (0.1%).

Related substances. Examine by liquid chromatography, as directed in the general procedure (2.2.29).

- *Test solution.* Dissolve 3 mg of omeprazole sodium in the mobile phase and dilute to 25 ml with the mobile phase.
- *Reference solution (a).* Dissolve 1 mg of *omeprazole CRS* and 1 mg of *omeprazole impurity D CRS* in the mobile phase and dilute to 10 ml with the mobile phase.
- *Reference solution (b).* Dilute 1 ml of the test solution to 100 ml with the mobile phase. Dilute 1 ml of this solution to 10 ml with the mobile phase.

The chromatography may be carried out using:

- a stainless-steel column 0.15 m long and 4 mm in internal diameter packed with *octylsilyl silica gel for chromatography R* (5 μm).

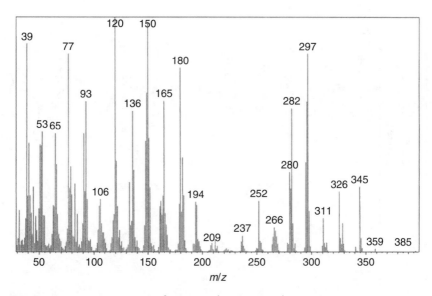

FIGURE 4.20   Mass spectrum of omeprazole in DMSO-d_6.

— as mobile phase at a flow-rate 1 ml/min a mixture of 27 volumes of *acetonitrile R* and 73 volumes of a 1.4-g/l solution of *disodium hydrogen phosphate R*, previously adjusted to pH 7.6 with *phosphoric acid R*.
— as detector a spectrophotometer set at 280 nm.

When the chromatograms are recorded in the prescribed conditions, the retention time of omeprazole is about 9 min and the relative retention time of omeprazole impurity *D* is about 0.8. Inject separately 40 *μl* of each solution and continue the chromatography for three times the retention time of omeprazole. Adjust the sensitivity of the detector so that the height of the principal peak in the chromatogram obtained with reference solution (b) is not less than 15% of the full scale of the recorder. The test is not valid unless the chromatogram obtained with reference solution (a), the resolution between the peaks corresponding to omeprazole impurity *D* and omeprazole is greater than 3. If necessary adjust the pH of the mobile phase or the concentration of *acetonitrile R*, an increase in the pH will improve the resolution. The area of any peak apart from the principal peak in the chromatogram obtained with the test solution is not greater than the area of the peak in the chromatogram obtained with reference solution (b) (0.1%).

Heavy metals: This test should be carried out as directed in the general procedure (2.4.8). One gram of omeprazole sodium complies with limit test C for heavy metals (20 ppm). Prepare the standard using 2 ml of *lead standard solution (10 ppm Pb) R*.

TABLE 4.9 Mass spectral fragmentation pattern of omeprazole

m/z	Relative intensity (%)	Formula	Fragment Structure
345	6	$C_{17}H_{19}N_3O_3S$	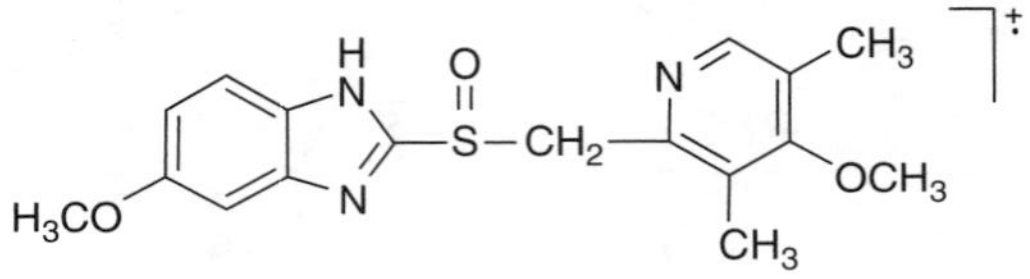
297	6	$C_{17}H_{19}N_3O_2$	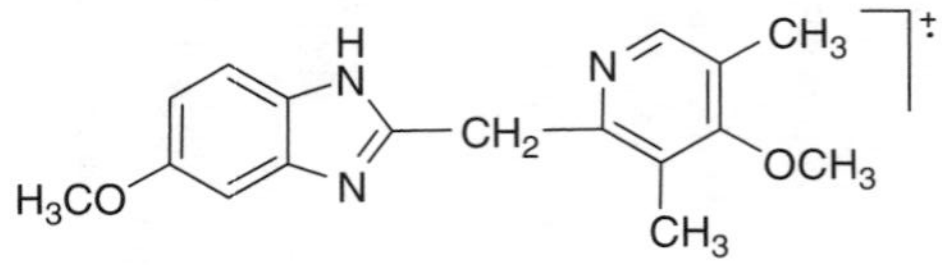
282	61	$C_{16}H_{16}N_3O_2$	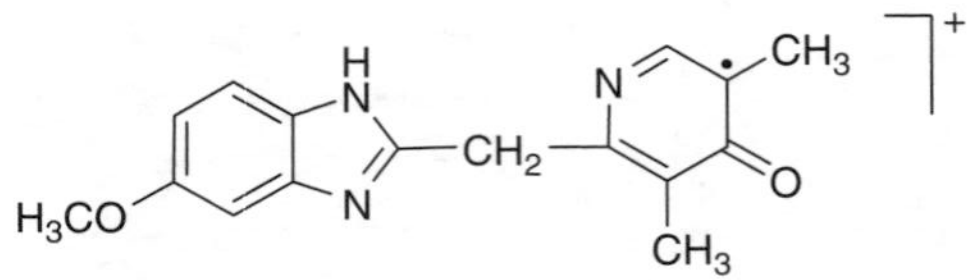
280	34	$C_{16}H_{14}N_3O_2$	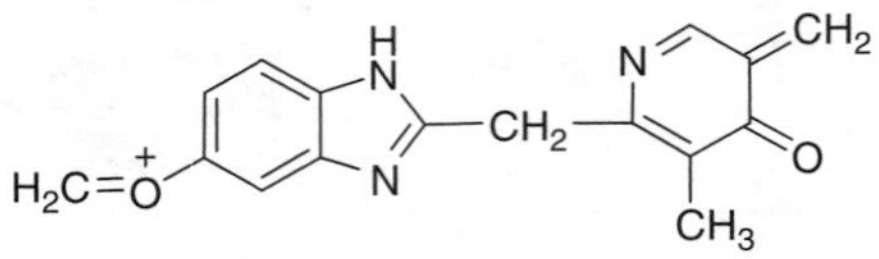

	$C_{16}H_{16}N_3O$	10	266
	$C_{15}H_{15}N_3$	7	237
	$C_9H_9N_2O_2S$	4	209
	$C_8H_7N_2O_2S$	18	195

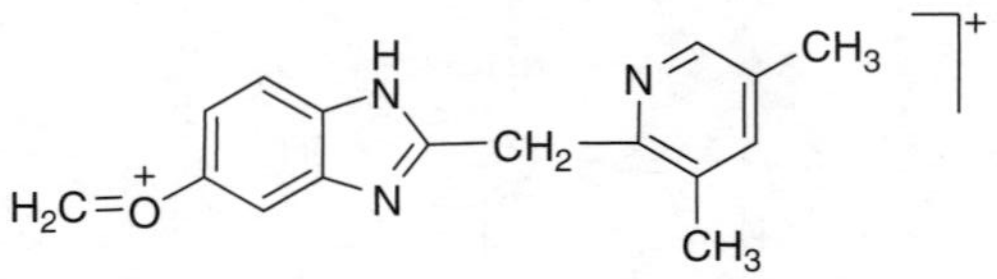

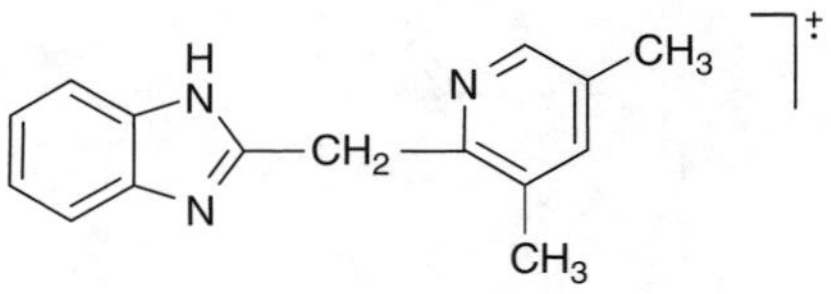

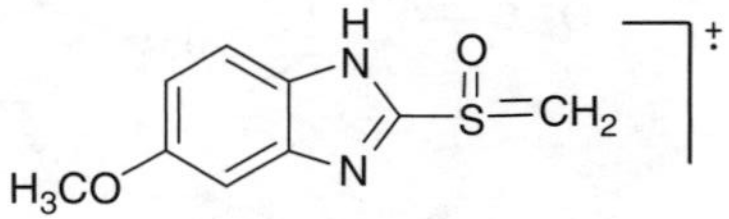

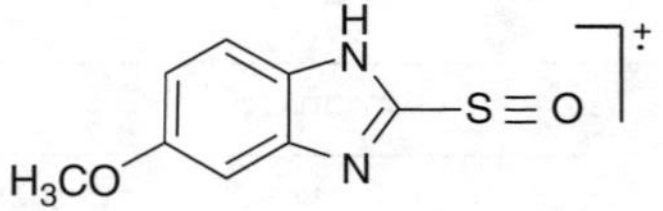

(continued)

TABLE 4.9 (*continued*)

m/z	Relative intensity (%)	Fragment	
		Formula	Structure
194	20	$C_8H_6N_2O_2S$	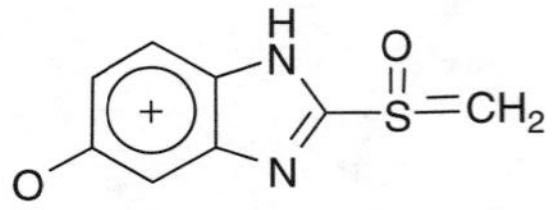 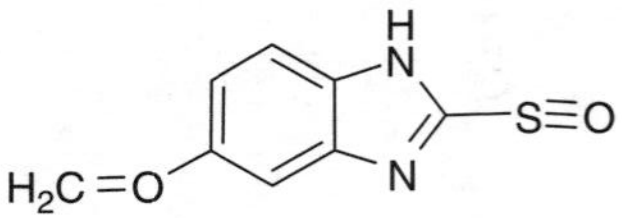
180	79	$C_7H_4N_2O_2S$	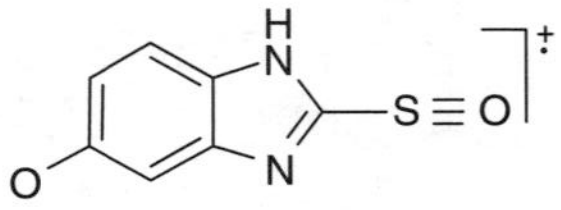
165	64	$C_7H_5N_2OS$	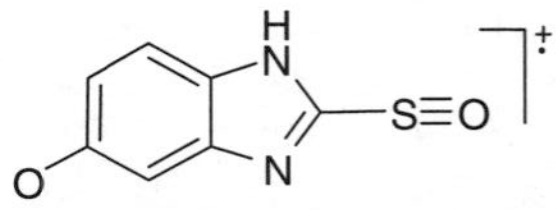

150	98	$C_9H_{12}NO$	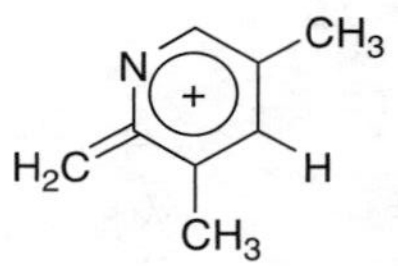
136	60	$C_8H_{10}NO$	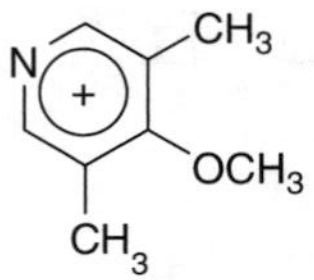
120	100	$C_8H_{10}N$	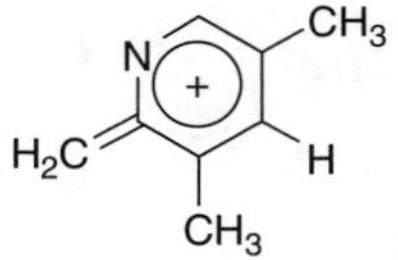
106	22	C_7H_8N	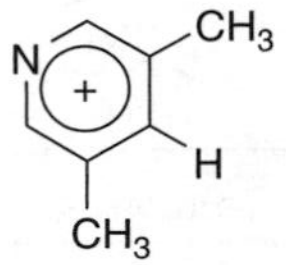

(continued)

TABLE 4.9 (continued)

m/z	Relative intensity (%)	Fragment	
		Formula	Structure
93	64	C_6H_7N	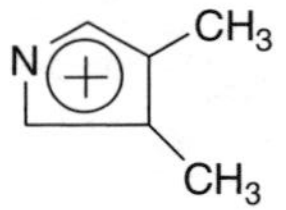
77	85	C_6H_5	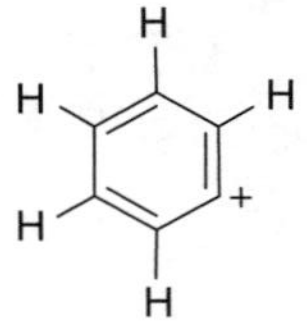
65	50	C_5H_5	
39	89	C_3H_3	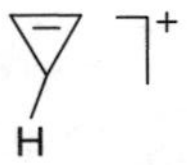

Water: This test should be carried out as directed in the general procedure (2.5.12). 4.5–10%, determined on 0.3 g by the semimicro determination of water.

• *Assay*

Dissolve 0.3 g of omeprazole sodium in 50 ml of *water R*. Titrate with 0.1 M *hydrochloric acid*, determining the end point potentiometrically as directed in the general procedure (2.2.20). One milliliter of 0.1 M *hydrochloric acid* corresponds to 36.74 of $C_{17}H_{18}N_3NaO_3S$.

• *Storage*

Store omeprazole sodium in an airtight container, protected from light.

• *Impurities*

A.   5-Methoxy-1*H*-benzimidazole-2-thiol

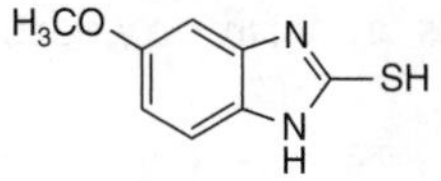

B.   2-[(*RS*)-[(3,5-dimethylpyridin-2-yl)methyl]sulfinyl]-5-methoxy-1*H*-benzimidazole

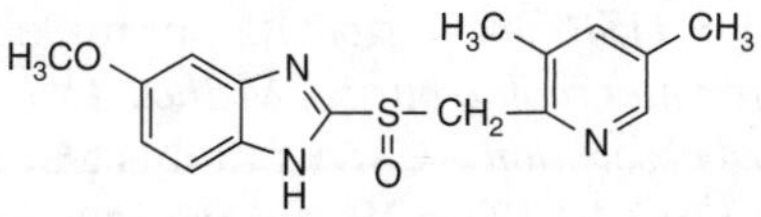

C.   5-Methoxy-2-[[(4-methoxy-3,5-dimethylpyridin-2-yl)methyl]thio]-1*H*-benzimidazole (*ufiprazole*)

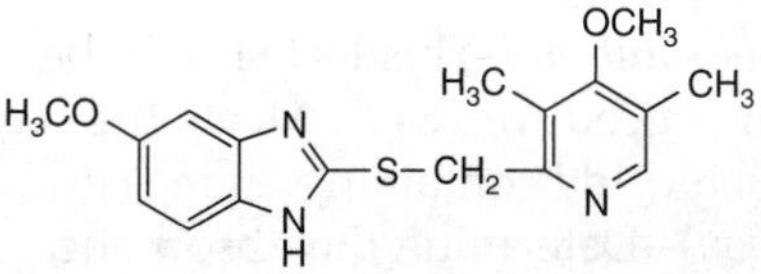

D.   5-Methoxy-2-[[(4-methoxy-3,5-dimethylpyridin-2-yl)methyl]sulfonyl]-1*H*-benzimidazole (*omeprazole sulfone*)

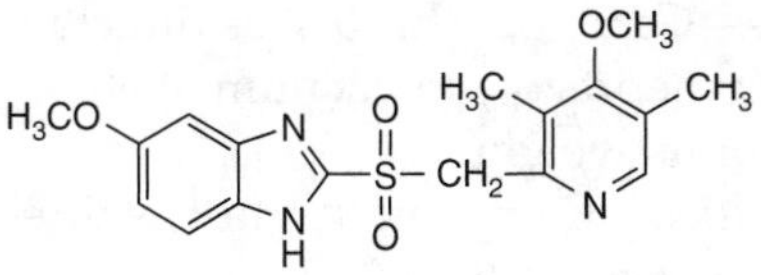

E.   4-Methoxy-2-[[(*RS*)-(5-methoxy-1*H*-benzimidazol-2-yl)sulfinyl]
      methyl]-3,5-dimethyl-pyridine-1-oxide

4.1.2. United States Pharmacopeia (USP) methods [25]

4.1.2.1. Omeprazole Omeprazole contains not less than 98% and not more than 102% of $C_{17}H_{19}N_3O_3S$, calculated on the dried basis.

Packaging and storage: Preserve in tight containers and store in a cold place, protected from moisture.

USP Reference Standards, general procedure ⟨11⟩: *USP Omeprazole RS*.

• *Identification*
   Test A: The R_f value of the principal spot observed in the chromatogram of the *Identification solution* corresponds to that of the principal spot observed in the chromatogram of the *Standard solution* containing 0.15 mg of *USP Omeprazole RS* per ml, obtained as directed in the test for *Chromatographic purity, Method 1*.

   Test B: *Infrared absorption*—Carry out this test as directed in the general procedure ⟨197 K⟩. The IR absorption spectrum of a potassium bromide dispersion of omeprazole previously dried, exhibits maxima only at the same wavelength as that of similar preparation of *USP Omeprazole RS*.

   Completeness of solution—This test should be carried out as directed in the general procedure ⟨641⟩. Meets the requirements, a solution in methylene chloride containing 20 mg/ml being used.

   Color of solution—Determine the absorbance of the solution prepared for the *Completeness of solution* test at 440 nm, in 1-cm cells, using methylene chloride as the blank: the absorbance is not greater than 0.1.

   Loss on drying—Carry out this test as directed in the general procedure ⟨731⟩. Dry omeprazole in vacuum at 60 °C for 4 h: it loses not more than 0.5% of its weight.

   Residue on ignition—Carry out this test as directed in the general procedure ⟨281⟩: not more than 0.1%.

   Heavy metals—Carry out this test as directed in the general procedure ⟨231⟩, *Method II*, 0.002%.

Organic volatile impurities—Carry out this test as directed in the general procedure <467>, *Method IV*: meet the requirements.
Solvents—Use dimethylacetamide.

- *Chromatographic purity*

METHOD 1

Solvents: Prepare a mixture of dichloromethane and methanol (1:1).

Standard solutions: Dissolve an accurately weighed quantity of *USP Omeprazole RS* in *Solvent*, and mix to obtain *Standard solution A* having a known concentration of about 0.5 mg/ml. Dilute this solution quantitatively with *Solvent* to obtain *Standard solution B* and *Standard solution C* having known concentrations of about 0.15 and 0.05 mg/ml, respectively.

Test solution: Prepare a solution of Omeprazole in *Solvent* containing 50 mg/ml.

Identification solution: Dilute a volume of the *Test solution* quantitatively with *Solvent* to obtain a solution containing 0.25 mg/ml.

Procedure: Separately apply 10 μl of the *Test solution*, the *Identification solution*, and each of the *Standard solutions* to a TLC plate (see *Chromatography*, in the general procedure $\langle 621 \rangle$) coated with a 0.25-mm layer of chromatographic silica gel mixture. Allow the spot to dry, and develop the chromatogram in a solvent system consisting of a mixture of ammonia-saturated dichloromethane, dichloromethane, and isopropanol (2:2:1) until the solvent front has moved about three-fourths of the length of the plate.

[**Note:** Prepare ammonia-saturated dichloromethane as follows. Shake 100 ml of dichloromethane with 30 ml of ammonium hydroxide in a separatory funnel, allow the layers to separate, and use the lower layer]. Remove the plate from the developing chamber, mark the solvent front, allow the solvent to evaporate, and examine the plate under short-wavelength UV light: the chromatograms show principal spots at about the same R_f value. Estimate the intensities of any secondary spots observed in the chromatogram of the *Test solution* by comparison with the spots in the chromatograms of the *Standard solutions*: no secondary spot from the chromatogram of the *Test solution* is larger or more intense than the principal spot obtained from *Standard solution B* (0.3%), and the sum of the intensities of all secondary spots obtained from the *Test solution* is not more intense than the principal spot obtained from *Standard solution A* (1%).

METHOD 2

Diluent: Use *Mobile phase*.

Phosphate buffer, *Mobile phase*, *System suitability solution*, and *Chromatographic system*: Proceed as directed in the *Assay*.

Test solution: Dissolve an accurately weighed quantity of omeprazole in *Diluent* to obtain a solution containing about 0.16 mg/ml [**Note:** Prepare this solution fresh].

Procedure: Inject equal volumes (about 40 µl) of the *Test solution* and *Diluent* into the chromatograph, and allow the *Test solution* to elute for not less than two times the retention time of omeprazole. Record the chromatograms, and measure the peak responses. Calculate the percentage of each impurity in the portion of omeprazole taken by the formula:

$$100 \times (r_i/r_s)$$

in which r_i is the peak response for each impurity and r_s is the sum of the responses of all the peaks: not more than 0.3% of any individual impurity is found, and the sum of all impurities is not more than 1%.

* *Assay*

    Phosphate buffer: Dissolve 0.725 g of monobasic sodium phosphate and 4.472 g of anhydrous dibasic sodium phosphate in 300 ml of water, dilute with water to 1000 ml, and mix. Dilute 250 ml of this solution with water to 1000 ml. If necessary, adjust the pH with phosphoric acid to 7.6.

    Mobile phase: Prepare a filtered and degassed mixture of *Phosphate buffer* and acetonitrile (3:1). Make adjustment if necessary (see *system suitability* under *Chromatography*, in the general procedure ⟨621⟩).

    Diluent: Prepare a mixture of 0.01 M sodium borate and acetonitrile (3:1).

    Standard preparation: Dissolve an accurately weighed quantity of *USP Omeprazole RS* in *Diluent*, and dilute quantitatively, and stepwise if necessary, with *Diluent* to obtain a solution having a known concentration of about 0.2 mg/ml.

    Assay preparation: Transfer about 100 mg of Omeprazole, accurately weighed, to a 50-ml volumetric flask, dissolve in and dilute with *Diluent* to volume, and mix. Transfer 5 ml of this solution to a 50-ml volumetric flask, dilute with *Diluent* to volume, and mix.

    System suitability solution: Dilute a volume of *Standard preparation* with *Diluent* to obtain a solution containing about 0.1 mg of *USP Omeprazole RS* per ml.

    Chromatographic system (see *Chromatography*, in the general procedure ⟨621⟩): The liquid chromatography is equipped with a 280-nm detector and a 4.6 mm × 15-cm column that contains 5-µm packing L7. The flow-rate is about 0.8 ml/min. Chromatograph the *System suitability solution*, and record the peak responses as directed for *Procedure*: the capacity factor, k', is not less than 6; the column efficiency is not less than 3000 theoretical plates; the tailing factor is not more than 1.5; and the relative standard deviation (RSD) for replicate injections is not more than 1%.

Procedure: Separately inject equal volumes (about 20 μl) of the *Standard preparation* and the *Assay preparation* into the chromatograph, record the chromatograms, and measure the responses for the major peaks. Calculate the quantity, in mg, of $C_{17}H_{19}N_3O_3S$ in the portion of omeprazole taken by the formula:

$$500 \times C(r_U/r_s)$$

in which C is the concentration (in mg/ml) of *USP Omeprazole RS* in the *Standard preparation*, and r_U and r_s are the peak responses obtained from the *Assay preparation* and *Standard preparation*, respectively.

4.1.2.2. Omeprazole delayed-release capsules
Omeprazole Delayed-Release Capsules contain not less than 90% and not more than 110% of the labeled amount of omeprazole ($C_{17}H_{19}N_3O_3S$).

Packaging and storage: Preserve in tight, light-resistant container. Store between 15 and 30 °C.

Labeling: When more than one *Dissolution test* is given, the labeling states the *dissolution test* used only if *Test 1* is not used.

USP Reference Standards; general procedure ⟨11⟩: *USP Omeprazole RS*.

Identification: The retention time of the major peak in the chromatogram of the *Assay preparation* corresponds to that in the chromatogram of the *Standard preparation*, as obtained in the *Assay*.

Dissolution. Carry out this test as directed in the general procedure ⟨711⟩.

TEST 1
– ACID RESISTANCE STAGE

Medium: 0.1 N hydrochloric acid; 500 ml.
Apparatus 2: 100 rpm.
Time: 2 h.
pH 7.6 Phosphate buffer, Mobile phase, and *Chromatographic system*: Proceed as directed for *Buffer stage*.
Standard solution: Transfer about 50 mg of *USP Omeprazole RS*, accurately weighed, to a 250-ml volumetric flask, dissolve in 50 ml of alcohol, dilute with 0.01 M sodium borate solution to volume, and mix. Transfer 10 ml of this solution into a 100-ml volumetric flask, add 20 ml of alcohol, dilute with 0.01 M sodium borate solution to volume, and mix.
Test solution: After 2 h, filter the *Medium* containing the pellets through a sieve with an aperture of not more than 0.2 mm. Collect the pellets on the sieve, and rinse them with water. Using approximately 60 ml of 0.01 M sodium borate solution, carefully transfer the pellets quantitatively to a 100-ml volumetric flask. Sonicate for about 20 min until the pellets are broken up. Add 20 ml of alcohol to the flask, dilute with 0.01 M sodium borate solution to volume, and mix. Dilute an appropriate amount of this

solution with 0.01 M sodium borate solution to obtain a solution having a concentration of about 0.02 mg/ml. At level L_1, test 6 units. Test 6 additional units at level L_2, and at level L_3, an additional 12 units are tested. Continue testing through the three levels unless the results conform at either L_1 or L_2.

Procedure: Separately inject equal volumes (about 20 μl) of the *Standard solution* and *Test solution* into the chromatograph, record the chromatograms, and measure the responses for the major peaks. Calculate the quantity, in mg, of omeprazole ($C_{17}H_{19}N_3O_3S$) dissolved in the *medium* by the formula:

$$T - CD\,(r_U/r_s)$$

in which T is the labeled quantity (in mg) of omeprazole in the capsule, C is the concentration (in mg/ml) of *USP Omeprazole RS* in the *standard solution*, D is the dilution factor used in preparing the *test solution*, and r_U and r_s are the omeprazole peak responses obtained from *Test solution* and the *Standard solution*, respectively.

Tolerances: Level L_1: no individual value exceeds 15% of omeprazole dissolved. Level L_2: the average of 12 units is not more than 20% of omeprazole dissolved, and no individual unit is greater than 35% of omeprazole dissolved. Level L_3: the average of 24 units is not more than 20% of omeprazole dissolved, not more than 2 units are greater than 35% of omeprazole dissolved, and no individual unit is greater than 45% of omeprazole dissolved.

– BUFFER STAGE

Medium: pH 6.8 phosphate buffer, 900 ml.

Proceed as directed for *Acid resistance stage* with a new set of capsules from the same batch. After 2 h, add 400 ml of 0.235 M dibasic sodium phosphate to the 500 ml of 0.1 N hydrochloric acid medium in the vessel. Adjust, if necessary, with 2 N hydrochloric acid or 2 N sodium hydroxide to a pH of 6.8 ± 0.05.

Apparatus 2: 100 rpm.

At the end of 30 min, determine the amount of $C_{17}H_{19}N_3O_3S$ dissolved in pH 6.8 phosphate buffer by employing the following method:

pH 10.4, 0.235 M Dibasic sodium phosphate: Dissolve 33.36 g of anhydrous dibasic sodium phosphate in 1000 ml of water, and adjust with 2 N sodium hydroxide to a pH of 10.4 ± 0.1.

pH 6.8 Phosphate buffer: Add 400 ml of 0 N hydrochloric acid to 320 ml of pH 10.4, 0.235 M *Dibasic sodium phosphate*, and adjust with 2 N hydrochloric acid or 2 N sodium hydroxide, if necessary, to a pH of 6.8 ± 0.05.

pH 7.6 Phosphate buffer: Dissolve 0.718 g of monobasic sodium phosphate and 4.49 g of dibasic sodium phosphate in 1000 ml of water. Adjust with 2 N hydrochloric acid or 2 N sodium hydroxide, if necessary, to a pH of 7.6 ± 0.1. Dilute 250 ml of this solution with water to 1000 ml.

Mobile phase: Transfer 340 ml of acetonitrile to a 1000-ml volumetric flask, dilute with *pH 7.6 Phosphate buffer* to volume, and pass through a membrane filter having a 0.5-μm or finer porosity. Make adjustments, if necessary (see *System Suitability* under *Chromatography*, in the general procedure ⟨621⟩).

Standard solution 1 (for Capsules labeled 10 mg): Dissolve an accurately weighed quantity of *USP Omeprazole RS* in alcohol to obtain a solution having a known concentration of about 2 mg/ml. Dilute with *pH 6.8 Phosphate buffer* quantitatively, and stepwise if necessary, to obtain a solution having a known concentration of about 0.01 mg/ml. Immediately add 2 ml of 0.25 M sodium hydroxide to 10 ml of this solution, and mix. [**Note:** Do not allow the solution to stand before adding the sodium hydroxide solution.]

Standard solution 2 (for Capsules label 20 mg and 40 mg): Proceed as directed for *Standard solution 1*, except to obtain a solution having a known concentration of about 0.02 mg/ml before mixing with 2 ml of 0.25 M sodium hydroxide.

Test solution 1 (for Capsules containing 10 and 20 mg): Immediately transfer 5 ml of the solution under test to a test tube containing 1 ml of 0.25 M sodium hydroxide. Mix well, and pass through a membrane filter having a 1.2-μm or finer porosity. Protect from light.

Test solution 2 (for Capsules labeled 40 mg): Immediately transfer 5 ml of the solution under test to a test tube containing 2 ml of 0.25 M sodium hydroxide and 5 ml of *pH 6.8 Phosphate buffer*. Mix well, and pass through a membrane filter having a 1.2-μm or finer porosity. Protect from light.

Chromatographic system (see *Chromatography*, in the general procedure ⟨621⟩): The liquid chromatograph is equipped with a 280-nm detector and a 4 mm × 12.5-cm analytical column that contains 5-μm packing L7. The flow-rate is about 1 ml/min. Chromatograph the appropriate *Standard solution* and record the peak responses as directed for *Procedure*: the column efficiency is not less than 2000 theoretical plates, and the RSD for replicate injections is not more than 2%.

Procedure: Separately inject equal volumes (about 20 μl) of the appropriate *Standard solution* and the *Test solution* into the chromatograph, record the chromatograms, and measure the responses for the major peaks. Calculate the quantity, in mg, of omeprazole ($C_{17}H_{19}N_3O_3S$) dissolved by the formula:

$$VCD\,(r_U/r_s)$$

in which V is the volume of *Medium* in each vessel, C is the concentration (in mg/ml) of *USP Omeprazole RS* in the appropriate *standard solution*, D is the dilution factor used in preparing the appropriate *test solution*, and r_U and r_s are the omeprazole peak responses obtained from the appropriate *Test solution* and the *Standard solution*, respectively.

Tolerances: For Capsules labeled 10 and 20 mg, not less than 75% (Q) of the labeled amount of $C_{17}H_{19}N_3O_3S$ is dissolved in 30 min. For capsules labeled 40 mg, not less than 70% (Q) of the labeled amount of $C_{17}H_{19}N_3O_3S$ is dissolved in 30 min. The requirements are met if the quantities dissolved from the product conform to *Acceptance Table*.

TEST 2—If the product complies with this test, the labeling indicates that it meets USP *Dissolution Test 2*.

– ACID RESISTANCE STAGE

Medium: 0.1 N hydrochloric acid; 900 ml.
Apparatus 1: 100 rpm.
Time: 2 h.
Procedure: After 2 h, remove each sample from the basket, and quantitatively transfer into separate volumetric flasks to obtain a solution having a final concentration of about 0.2 mg/ml. Proceed as directed for the *Assay preparation* in the *Assay*, starting with "Add about 50 ml of *Diluent*." Calculate the quantity, in mg, of omeprazole ($C_{17}H_{19}N_3O_3S$) dissolved in the *Medium* by the formula:

$$T - CD(r_U/r_s)$$

in which T is the assayed quantity (in mg) of omeprazole in the capsule, C is the concentration (in mg/ml) of *USP Omeprazole RS* in the *Standard solution*, D is the dilution factor used in preparing the *Test solution*, and r_U and r_s are the omeprazole peak responses obtained from *Test solution* and *Standard solution*, respectively.

Tolerances: It complies with *Acceptance Table* 4.10.

TABLE 4.10   Acceptance Table

Level	Criterion
L_1	The average of the 6 units is not more than 10% of omeprazole dissolved
L_2	The average of the 12 units is not more than 10% of omeprazole dissolved
L_3	The average of the 24 units is not more than 10% of omeprazole dissolved

– BUFFER STAGE

Medium: pH 6.8 0.05 M phosphate buffer; 900 ml (see *Reagents, Indicators, and Solutions*).

Apparatus 1: 100 rpm.

Time: 45 min.

Procedure: Proceed as directed for *Acid resistance stage* with a new set of capsules from the same batch. After 2 h, replace the acid *Medium* with the buffer *Medium* and continue the test for 45 more minutes. Determine the amount of $C_{17}H_{19}N_3O_3S$ dissolved from UV absorbances at the wavelength of maximum absorbance at about 305 nm on portions of the solutions under test passed through 0.2-μm nylon filter, in comparison with a Standard solution having a known concentration of *USP Omeprazole RS* and the same *Medium*.

Tolerances: It complies with *Acceptance Table* 4.10 under *Dissolution*, in the general procedure ⟨711⟩. Not less than 75% (Q) of the labeled amount $C_{17}H_{19}N_3O_3S$ is dissolved in 45 min.

Uniformity of dosage units, general procedure ⟨905⟩: meet the requirements.

- *Chromatographic purity*

  Diluent, Solution A, Solution B, Mobile phase, and *Chromatographic system*: Proceed as directed in the *Assay*.

  Standard solution: Prepare as directed for the *Standard preparation* in the *Assay*.

  Test solution: Use the *Assay preparation*.

  Procedure: Separately inject equal volumes (about 10 μl) of the *Standard solution* and the *Test solution* into the chromatograph, record the chromatograms, and measure all the peak responses. Calculate the percentage of each impurity in the portion of capsules taken by the formula:

$$10 \times (C/A)(1/F)(r_i/r_s)$$

  in which C is the concentration (in μg/ml) of *USP Omeprazole RS* in the *Standard solution*, A is the quantity (in mg) of omeprazole in the portion of capsules taken, as determined in the *Assay*, F is the relative response factor (see Table 4.11 for values), r_i is peak response for each impurity obtained from the *Test solution*, and r_s is the peak response for omeprazole obtained from *Standard solution*. In addition to not exceeding the limits for each impurity in Table 4.11, not more than 2% of total impurities is found.

TABLE 4.11   Limits for impurity

Name	Relative retention time	Relative response factor, F	Limit (%)
Thioxopyrido conversion product[a]	0.33	1.6	0.5
5-Methoxy-1*H*-benzimidazole-2-thiol	0.64	3.1	0.5
Any other individual impurity	–	1.0	0.5

[a] Formed in the solution from two isomers: 1,3-dimethyl-8-methoxy-12-thioxopyrido[1′,2′:3,4]imidazo-[1,2-*a*]benzimidazol-2-(12*H*)-one and 1,3-dimethyl-9-methoxy-12-thioxopyrido[1′,2′:3,4]imidazo[1,2-*a*]benzimidazole-2(12*H*)-one.

- *Assay*

  Diluent: Dissolve 7.6 g of sodium borate decahydrate in about 800 ml of water. Add 1 g of edetate disodium, and adjust with 50% sodium hydroxide solution to a pH of 11 ± 0.1. Transfer the solution to a 2000-ml volumetric flask, add 400 ml of dehydrated alcohol, and dilute with water to volume.

  Solution A: Prepare a filtered and degassed solution of 6 g of glycine in 1500 ml of water. Adjust with 50% sodium hydroxide solution to a pH of 9, and dilute with water to 2000 ml.

  Solution B: Use a filtered and degassed mixture of acetonitrile and methanol (85:15).

  Mobile phase: Use variable mixtures of *Solution A* and *Solution B* as directed for *Chromatograpic system*. Make adjustments if necessary (see *System Suitability* under *Chromatography*, in the general procedure ⟨621⟩).

  Standard preparation: Dissolve, by sonicating, an accurately weighed quantity of *USP Omeprazole RS* in *Diluent*, and dilute quantitatively, and stepwise, if necessary, with *Diluent* to obtain a solution having a known concentration of about 0.2 mg/ml.

  Assay preparation: Weigh and mix the contents of not fewer than 20 Capsules. Transfer an accurately weighed portion of the mixture, equivalent to about 20 mg of omeprazole, to a 100-ml volumetric flask, add about 50 ml of *Diluent*, and sonicate for 15 min. Cool, dilute with *Diluent* to volume, mix, and pass through a membrane filter having 0.45 μm or finer porosity. [**Note:** Bubbles may form just before bringing the solution to volume. Add a few drops of dehydrated alcohol to dissipate the bubbles if they persist for more than a few minutes].

Chromatographic system (see *Chromatography*, in the general procedure ⟨621⟩: The liquid chromatography is equipped with a 305-nm detector and a 4.6 mm × 15-cm column that contains 5-μm base-deactivated packing L7. The flow-rate is about 1.2 ml/min. The chromatograph is programmed as follows:

Time (min)	Solution A (%)	Solution B (%)	Elution
0–20	88 → 40	12 → 60	Linear gradient
20–21	40 → 88	60 → 12	Linear gradient
21–25	88	12	Isocratic

Chromatograph the *Standard preparation*, and record the peak responses as directed for *Procedure*: the column efficiency is not less than 20,000 theoretical plates; the tailing factor is not less than 0.8 and not more than 2; and the RSD for replicate injections is not more than 2%.

Procedure: Separately inject equal volumes (about 10 μl) of the *Standard preparation* and the *Assay preparation* into the chromatograph, record the chromatograms, and measure the peak responses. Calculate the quantity, in mg, of omeprazole ($C_{17}H_{19}N_3O_3S$) in the portion of Capsules taken by the formula:

$$DC(r_U/r_s)$$

in which D is the dilution factor of the *Assay preparation*, C is the concentration (in mg/ml) of *USP Omeprazole RS* in the *Standard preparation*, and r_U and r_s are the peak responses obtained from the *Assay preparation* and the *Standard preparation*, respectively.

4.2. Reported methods of analysis

4.2.1. Spectrophotometry

Dhumal *et al.* [26] described an individual UV spectrophotometric assay method for the analysis of omeprazole from separate pharmaceutical dosage forms. Powdered tablets, equivalent to 50 mg of the drug, were sonicated with 35 ml of 0.1 M sodium hydroxide for 5 min and diluted to 50 ml with 0.1 M sodium hydroxide. The solution was filtered and a 2-ml portion of the filtrate was diluted to 200 ml with 0.1 M sodium hydroxide before the absorbance of the solution was measured at 305 nm versus 0.1 M sodium hydroxide. Beer's law was obeyed for 6–25 μg/ml of omeprazole. Coefficient of variation was 3.1%. Recovery was quantitative.

Sastry *et al.* [27] described four simple and sensitive spectrophotometric methods for the assay of omeprazole in pure and in dosage forms based on the formation of chloroform soluble ion-associated under specified experimental conditions. Four acidic dyes: Suprachen Violet 3B (SV 3B, method A), Tropaeolin 000 (TP 000, method B), Boromocresol Green (BCG, method C), and Azocarmine G (AG, method D) are utilized.

The extracts of the ion-associates exhibit absorption maxima at 590, 420, 500, and 540 nm for methods A, B, C, and D, respectively. Beer's law and the precision and accuracy of the methods are checked by UN reference method.

Ozaltin and Kocer [28] used a derivative spectroscopic method for the determination of omeprazole in pharmaceuticals. Capsule contents were powdered and a sample equivalent to one capsule content was sonicated with 10 ml ethanol, diluted to 100 ml with 0.1 M borate buffer of pH 10 and further diluted as necessary. Spectra were recorded at 50 nm/min with a 3-nm slit width second-order derivative curves were obtained for 200–400 nm using $\Delta\lambda = 31.5$ and $N = 9$. The calibration graph for peak-to-peak measurements between 303 and 310 nm were linear for 0.2–40 μg/ml omeprazole and the RSDs were 1.09–4.55%. Mean recovery was 100.7%. Result agreed with those obtained by polarography.

Tuncel and Dogrukol-Ak [29] developed a flow-through spectrophotometric method for the determination of omeprazole in pharmaceutical preparations containing enteric-coated pellets. Sample was dissolved in 100 ml 0.1 M sodium hydroxide and filtered. Portions were analyzed by flow-through spectrophotometry using a Spectrophoresis 100 system with 75 μm fused-silica capillaries with detection at 305 nm. Samples were pumped through the system for 2 min. Results were compared with those obtained by standard spectrophotometry at 305 nm. Detection limits was 8 μM omeprazole and the calibration graph was linear. The pellet matrix did not interfere. RSD was 1.9% ($n = 6$).

Karlsson and Hermansson [30] used chemometrics for optimization of chiral separation of omeprazole and one of its metabolites on immobilized α1-acid glycoprotein. Plasma was centrifuged at 2500 rpm and a portion (20–50 μl) was injected into a 5-μm Chiral-AGP column (10 cm $\times$ 4 mm) with α1-acid glycoprotein immobilized to silica as a chiral stationary phase and acetonitrile–phosphate buffer of pH 5.7–7.2 as mobile phase (1 ml/min). Detection of omeprazole and its main metabolite, hydroxylated omeprazole, was performed at 302 nm. A statistical model was developed for the optimization of the operational parameters. The experimental data were evaluated with multivariate analyses; column temperature and acetonitrile concentration were the most important variables for the enantioseparations. Complete enantiomeric separation for omeprazole and hydoxylated omeprazole was obtained within 15 min.

El-Kousy and Bebawy [31] described two stability-indicating spectrophotometric methods for the determination of omeprazole in the presence of its photodegradation products. In the first method, omeprazole from capsules or vials were dissolved in acetonitrile/water (1:1) and UV–VIS spectrophotometry used to determine the first-, second-, and third-derivative absorption curves between 200 and 400 nm. The level of omeprazole was assayed from the values of ordinates of the three curves at 290.4,

320.6, and 311.6 nm using calibration curves from standard solutions with concentrations of 3–25 μg/ml. In the second method, omeprazole in tablets was dissolved in chloroform and reacted with 0.2% chloranil solution in chloroform by heating at 70 °C for 20 min. The absorption at 377 nm was measured against a reagent blank. The level of omeprazole was determined against a calibration curve obtained with standard solutions of 8–55 μg/ml.

Wahbi et al. [32] used a spectrophotometric method for the determination of omeprazole in pharmaceutical formulations. The compensation method and other chemometric methods (derivative, orthogonal function, and difference spectrophotometry) have been applied to the direct determination of omeprazole in its pharmaceutical preparations. The method has been validated; the limits of detection was 3.3×10^{-2} μg/ml. The repeatability of the method was found to be 0.3–0.5%. The linearity range is 0.5–3.5 μg/ml. The method has been applied to the determination of omeprazole in its gastro-resistant formulation. The difference spectrophotometric (ΔA) method is unaffected by the presence of acid induced degradation products, and can be used as a stability-indicating assay method.

Karljikovic-Rajic et al. [33] developed a first-derivative UV spectrophotometry, applying the zero-crossing method, for the determination of omeprazole and omeprazole sulfone in methanol/4% ammonia, where sufficient spectral resolution of the drug and corresponding impurity were obtained, using the amplitudes $^1D_{304}$ and $^1D_{307}$, respectively. The method showed good linearity in the ranges (μg/ml) 1.61–17.2 for omeprazole and 2.15–21.50 for omeprazole sulfone and also good accuracy and precision. The experimentally determined values of the limit of detection (μg/ml) were 1.126 and 0.76 for omeprazole and omeprazole sulfone, respectively.

Salama et al. [34] developed and validated a spectrophotometric method for the determination of omeprazole and pantoprazole sodium via their metal chelates. The procedures were based on the formation of 2:1 chelates of both drugs with different metal ions. The colored chelates of omeprazole in ethanol were determined spectrophotometrically at 411, 339, and 523 nm using iron(III), chromium(III), and cobalt(II), respectively. Regression analysis of Beer's plots showed good correlation in the concentration ranges 15–95, 10–60, and 15–150 μg/ml of pure omeprazole using iron(III), chromium(III), and cobalt(II), respectively.

Riedel and Leopold [35] investigated the degradation of omeprazole in organic polymer solutions and aqueous dispersions of enteric-coating polymers by UV spectroscopy. Data were compared with those obtained in a previous high-performance liquid chromatographic (HPLC) study. For comparative purposes the cationic Eudragit RS 100 and the monomeric acid acetic acid were included in this study. The discolorations of the degraded omeprazole solutions were analyzed by visible spectros-copy. UV–VIS spectra were recorded after preparation of the solutions and after 180 min of storage. The change of absorption was calculated as the difference of the absorption values at 305 nm. Degradation of omep-razole depends on the amount of acidic groups in the polymer structure.

Yang et al. [36] studied omeprazole samples from different sources and in different forms spectrophotometrically to obtain pK_a values. In the neutral to alkaline pH region, two consistent pK_a values of 7.1 and 14.7 were obtained from various samples. The assignment of these pK_a values were realized by comparison with the prototropic properties of $N(1)$-methylated omeprazole substituted on the nitrogen at the 1-position of the benzimidazole ring, which was found to have a pK_a value of 7.5. The omeprazole pK_a of 14.7 is assigned to the dissociation of the hydrogen from the 1-position of the benzimidazole ring and the pK_a of of 7.1 is assigned to the dissociation from the protonated pyridine nitrogen of omeprazole. The results presented are at variance with those of earlier work.

4.2.2. Colorimetry

Sastry et al. [37] used a spectrophotometric method for the determination of omeprazole in pharmaceutical formulation. The content of omeprazole capsules were powdered and dissolved in methanol or aqueous sodium hydroxide. Omeprazole was determined by the following:

1. Mixing aqueous sample with 2 ml 10 mM 1,10-phenanthroline solution and 1.5 ml 3 mM ferric chloride, heating on a boiling water bath for 30 min, cooling to room temperature, mixing with 2 ml 20 mM phos-phoric acid, dilution to 10 ml with water and absorbance measurement at 515 nm.
2. Mixing aqueous samples with 1 ml 0.088% N-bromosuccinimide (NBS) and 1 ml 5% acetic acid, dilution to 10 ml with water, equilibration for 20 min, addition of 1 ml 0.3% 4-(methylamino)-phenol sulfate, equili-bration for 2 min, addition of 2 ml 0.2% sulfanilamide, dilution to 25 ml with water, and absorbance measurement at 520 nm within 10–30 min.
3. Mixing methanolic samples with 2 ml methanolic 0.5% p-dimetyyla-mino benzaldehyde and 2 ml sulfuric acid with cooling and agitation, dilution to 10 ml with methanol and absorbance measurement at 420 nm within 20 min. Beer's law was obeyed from 0.5–5, 4–20, and 12.5–125 μg/ml of omeprazole, for the three methods, respectively, and the corresponding molar absorptivities (ε) were 42,800, 6400, and 2380. The relative standard derivations ($n = 6$) were 0.29–0.41% and recov-eries were 98.17–100.1%.

Sastry et al. [38] described four spectrophotometric methods for the determination of omeprazole in bulk form and in pharmaceutical formulations.

Method A: Sample was oxidized with ferric chloride and reacted with 3-methyl-2-benzothiazolinone hydrochloride and the absorbance was measured at 660 nm.

Method B: Sample was oxidized with chloramine T, reacted with *m*-aminophenol and the absorbance was measured at 420 nm.

Method C: Sample was oxidized with excess NBS and the consumed NBS was determined by observing the decrease in color intensity using Celestine blue with measurement of absorbance at 540 nm.

Method D: Sample was reacted with Folin-Ciocalteu reagent and the absorbance was measured at 770 nm.

Beer's law was obeyed for 1–10, 2–32, 0.4–2.4, and 0.8–10 $\mu g/ml$ for methods A, B, C, and D, respectively ($\varepsilon = 21,000$, 11,900, 75,800, and 28,500, respectively). The detection limits were 0.074, 0.104, 0.023, and 0.039 $\mu g/ml$, respectively. The corresponding RSDs were 0.69%, 0.53%, 0.73%, and 0.48%. The method was applied to pharmaceuticals and recoveries were 98.7–100.1%.

4.2.3. Argentometry

Zhang *et al.* [39] carried out studies on the determination of omeprazle by argentometry. Sample (0.3 gm) was dissolved in 20 ml ethanol and 6 ml ammonia reagent (mainly ammonium hydroxide) and 50 ml 0.05 M silver nitrate was added. The mixture was heated at $\leq 50\ ^{\circ}C$ for 15 min, cooled and the solution was diluted to 100 ml with water and filtered. Portions (50 ml) of the filtrate were mixed with 3 ml nitric acid and the mixture was heated until fuming ceased. On cooling, 25 ml water and 2 ml ammonium iron(III) sulfate indicator were added and the excess silver nitrate was titrated against with 50 mM ammonium thiocyanate as titrant until a reddish brown end point was observed, providing an indirect method for determining omeprazole. The average recovery of omeprazole was 99.9% with an RSD was 0.21%. Results were compared with those obtained by HPLC.

4.2.4. Electrochemical analysis

4.2.4.1. Voltammetry Pinzauti *et al.* [40] designed an adsorptive stripping voltammetric method for the determination of omeprazole. The method was optimized using a multivariate procedure and was used to analyze dosage capsules. A 100 μl of sample solution containing 1.4 $\mu g/ml$ omeprazole was added to 10 ml 0.01 M potassium chloride of supporting electrolyte in a voltammetric cell and the accumulation was carried out at 0 V onto a hanging mercury drop electrode (Ag/AgCl reference electrode, Pt wire auxillary electrode) for 68 s at a stirring speed of 400 rev/min. The stirrer was then stopped and the voltammogram was recorded after 10 s by applying a differential-pulse potential scan from -0.7 to -1.5 V at

40 mV/s and 70 mV pulse amplitude. The calibration graph was linear for 2.88–48.9 μg/ml of omeprazole and the detection limit was 2.25 μg/ml. The mean recovery of omeprazole in capsules was 101.9% with an RSD ($n = 5$) of 2%.

Radi [41] used an anodic voltammetric assay method for the analysis of omeprazole and lansoprazole on a carbon paste electrode. The electrochemical oxidations of the drugs have been studied at a carbon paste electrode by cyclic and differential-pulse voltammetry in Britton–Robinson buffer solutions (0.04 M, pH 6–10). The drug produced a single oxidation step. By differential-pulse voltammetry, a linear response was obtained in Britton–Robinson buffer pH 6 in a concentration range from 2×10^{-7} to 5×10^{-5} M for lansoprazole or omeprazole. The detection limits were 1×10^{-8} and 2.5×10^{-8} M for lansoprazole and omeprazole, respectively. The method was applied for the analysis of omeprazole in capsules. The results were comparable to those obtained by spectrophotometry.

Qaisi *et al.* [42] studied the acid decomposition of omeprazole in the absence of thiol using a differential-pulse polarography at the static mercury drop electrode. Reactions were monitored, using differential-pulse polarography at the static mercury drop electrode, in solutions buffered to pH values ranging from 2 to 8. The fast, sensitive, and selective electrochemical technique facilitated to repeat recordings of successive voltammograms [peak current (nA) versus peak potential (volts versus Ag/AgCl saturated with 3 M KCl)]. The differential-pulse polarographic signals of omeprazole and its degradation products, believed to be due sulfur functional group (the principal site for electrode reaction), gave advantages over the previously employed UV detection technique. The latter primarily relied on pyridine and benzimidazole analytical signals, which are common reaction products of proton pump inhibitors in aqueous acidic solutions. After peak identification, the resulting current (nA)–time (s) profiles, demonstrated that omeprazole undergoes degradation to form two main stable compounds, the first is the cyclic sulfenamide (D^+), previously believed to be the active inhibitor of the H^+/K^+-ATPase, the second is omeprazole dimer. This degradation is highly dependent on pH. Unlike previous studies which reported that the lifetime of D^+ is few seconds, the cyclic sulfenamide (D^+) was found to be stable for up to 5–20 min. The results further indicated that omeprazole converts into the cyclic sulfenamide in an irreversible reaction, consequently, D^+ and sulfenic acid (an intermediate which rapidly converts into D^+) were not interconvertible. It is suggested that the sulfenic acid is the active inhibitor *in vivo*. The omeprazole reactions, in the absence of thiol, were not as complicated as were previously reported.

Yan [43] investigated the electrochemical behavior of omeprazole on a glassy carbon electrode by cyclic voltammetry and differential-pulse voltammetry. Omeprazole was found to give a sensitive oxidation peak at +0.74 V in the acetic acid/sodium acetate buffer solution (pH 5.10) under the differential-pulse voltammetric mode. The peak current was linear with the concentration of omeprazole in the range 1–20 mg/l. Based on which, a differential-pulse voltammetric method for the determination of omeprazole with the detection limit of 0.19 mg/l has been developed. The method has been used for determination of omeprazole concentration in omeprazole enteric-coated tablets, the recovery was found to be in the range of 99.3–102%. The mechanism for this electrochemical reaction at the glassy carbon electrode was also discussed in this study.

4.2.4.2. Polarography McClean *et al.* [44] used fluorimetry, UV spectro-photometry, liquid chromatography, and differential-pulse polarography, to study the degradation of omeprazole in 10 mM hydrochloric acid, and the subsequent reactions of the respective degradation products with 2-mercaptoethanol. Omeprazole and its degradation products could also be determined in pharmaceutical formulations or biological fluids, by differential-pulse polarography in Britton–Robinson buffer solution of pH 9. Calibration graphs were linear up to 100 μM and detection limits were 0.07 μM omeprazole and 0.08 μM SK&F 95601. For 0.4 and 100 μM drug the respective RSD ($n = 6$) were 4.28% and 0.55% for omeprazole and 6.11% and 1.15% for SK&F 95601.

Ames and Kovacic [45] studied the electrochemistry of omeprazole, active metabolites and a bound enzyme model, with possible involvement of electron transfer in the antiulcer action of the drug. The active metabolites cyclic sulfenamide and sulfur radical entities, exhibited reduction potentials of −0.3 and −0.2 V, respectively. The value for the bound enzyme model was −0.7 V and that for omeprazole was >−1.4 V. The results lend credence to the hypothesis that electron transfer comprises part of the mode of action in addition to H^+/K^+-ATPase inhibition.

Ozaltin and Temizer [46] used a differential-pulse polarographic method for the determination of omeprazole in pharmaceutical preparations. Various polarographic techniques were investigated and the best results were obtained using differential-pulse polarography in a borate buffer of pH 9. A sensitive well-defined peak was observed at −1.28 V versus Ag/AgCl; no other peaks were observed in the range −0.2 to − 2.0 V. The detection limit was 0.1 μM omeprazole and the calibration graph was linear from 0.2 to 20 μM. The RSD of the calibration plot was 3.92%. The method was applied to two different commercial preparations and was accurate, sensitive, cheap, and easy to apply.

Dogrukol-Ak and Tuncel [47] determined omeprazole in capsules by polarographic techniques. An enteric-coated pellet was mixed with one drop of 1 M sodium hydroxide, made up to 100 ml with deoxygenated water then vigorously shaken. Analysis was carried out with a polarographic system comprising a Polaropulse PRG-5 instrument; dual function EGMA cell, and HG, Pt-wire and Ag/AgCl as working, auxiliary and reference electrodes, respectively. The supporting electrolyte was 0.2 M potassium chloride–ethanol (9:1) in a buffer solution of 0.2 M sodium acetate or 0.2 M sodium dihydrogen phosphate adjusted with 2 M hydrochloric acid or 2 M sodium hydroxide; 10 ml was placed in to the polarographic cell with the sample then purified. Nitrogen was passed through the solution for 10 min. Polarography was carried out by scanning cathodically from 0 to -2000 mV. Results were comparable to with data obtained from a standard spectrophotometric technique. This method was accurate, practical, rapid, and free from interference; and could be applied to routine analysis.

Knoth *et al.* [48] studied the electrochemical behavior of omeprazole with the aid of the direct-current and differential-pulse polarography. Omeprazole was determined in Britton–Robinson buffers pH 7–9 up to a concentration of 10^{-5} M. The mechanism of the reduction process on the dropping mercury electrode is elucidated. With the consumption of two electrons and two protons, omeprazole will be reduced to 5-methoxy-2-[(3,5-dimethyl-4-methoxypyridin-2-yl)methylthio]-1*H*-benzimidazole which will be cleaved with the uptake of two further electrons and two protons into 4-methoxy-2,3,5-trimethyl pyridine and 2-mercapto-5-methoxybenzimidazole.

Oelschlaeger and Knoth [49] described a differential-pulse polaro-graphic procedure for the determination of omeprazole in individual enteric-coated capsules and dry ampoules. The pellet from a hand-opened capsule was disintegrated by ultrasonication for 5 min in 25 ml methanol. For 20–40 mg capsules, 2 and 1 ml of the suspension, respec-tively, was withdrawn and diluted with methanol to 50 ml. After the addition of 9 ml of Britton–Robinson buffer of pH 7, 1 ml of the solution was subjected to differential-pulse polarography from -800 to -1200 mV at 16.67 mV/s. The method can also be applied to methanolic solutions of the contents of single omeprazole ampoules. The method was validated by HPLC and used to check the uniformity of Antra 20 and 40 and Gastroloc capsules and Antra pro infusion ampoules according to the German Pharmacopeia (DAB 1996 V.5.2.).

Belal *et al.* [50] used an anodic polarographic method for the determi-nation of omeprazole and lansoprazole in pure form and in pharmaceuti-cal dosage forms. The study was carried out in Britton–Robinson buffer over the pH range 4.1–11.5. In Britton–Robinson buffer of pH 7, well-defined anodic waves were produced with diffusion–current constant (I_d) of 1.7 ± 0.01 ($n = 6$) and 1.66 ± 0.01 ($n = 8$) for lansoprazole and omep-razole, respectively. The current–concentration plots were rectilinear over the ranges of 4–24, 2–16 μg/ml using direct current (DC_t) mode for lansoprazole and omeprazole, respectively, and over the range 2–18,

0.4–12 μg/ml using the differential-pulse polarographic mode for lansoprazole and omeprazole, respectively. The detection limits ($S/N = 2$) using differential-pulse polarographic modes were 0.2 μg/ml (5.41 $\times$ 10^{-7} M) and 0.05 μg/ml (1.45 $\times$ 10^{-7} M) for lansprazole and omeprazole, respectively. The method was applied to the analysis of the two drugs in their commercial capsules. The average percent recoveries were compared with those obtained by reference methods, with satisfactory standard deviations. The method is simple, accurate, and stability-indicating.

El-Enany et al. [51] studied the alternating current (AC$_t$) polarographic behavior of omeprazole in Britton–Robinson buffers over the pH range 4.1–11.5. In Britton–Robinson buffer of pH 9.6 and 10.5, well-defined AC$_t$ peaks were obtained for omeprazole. The current–concentration plot was rectilinear over the range 0.2–10 μg/ml. The minimum detection limit ($S/N = 2$) was 0.01 μg/ml (2.9 $\times$ 10^{-5} M). The method was applied to the analysis of the drug in its commercial capsules. The average percent recovery was favorably compared to those obtained by reference methods. The pathway for the electrode reaction for the drug involved reduction of the sulfonyl group into the corresponding thiol group at the dropping mercury electrode. The advantages of the method were time saving and more sensitive than other voltammetric method. The method was applied to analysis of lansoprazole.

Cao and Zeng [52] used of an oscillopolarographic method for the determination and the electrochemical behavior of omeprazole. Portions of standard omeprazole solution were treated with 1 ml 1 M ammonia/ammonium chloride at pH 8.9 and the solution was diluted with water to 10 ml. The diluted solution was subjected to single sweep oscillopolarography with measurement of the derivative reduction peak at -1.105 V versus saturated calomel electrode. The calibration graph was linear from 0.5 to 10 μM omeprazole with a detection limit of 0.2 μM. The method was applied to the analysis of omeprazole in capsules with recoveries of 100–118.6% and RSD of 6.78%. The electrochemical behavior of omeprazole at the mercury electrode was also investigated.

4.2.5. Chromatography
4.2.5.1. Thin-layer chromatography Mangalan et al. [53] used of an HPTLC method for the detection and determination of omeprazole in plasma levels. Plasma was extracted three times with dichloromethane at pH 6.5–7 and the combined extracts were evaporated to dryness at 60 °C. The residue was dissolved in dichloromethane and the solution was analyzed by TLC on aluminium-packed plates precoated with Silica gel 60 F$_{254}$ with the upper organic layer of butanol–ammonium hydroxide–water (14:1:15) as mobile phase. The spots were observed by fluorescence quenching under UV light illumination at 280 nm; the total area of each

spot was determined with use a dual wavelength scanner. The calibration graph was rectilinear for 0.1–1 mg of omeprazole. The recovery was 87% and coefficients of variation were 4.1–8.2%.

Ray and Kumar-De [54] described an HPTLC method and a TLC method for the rapid quantification and identification of omeprazole. Ground powder (omeprazole powders, capsules, or tablets) equivalent of 25 mg omeprazole was warmed for 10 min with shaking with 25 ml methanol. After cooling the solution was made up to 50 ml with methanol, mixed and filtered. The filtrate was spotted (2 μl) on to HPTLC plates (20 $\times$ 20 cm) coated with Kieselgel 60 GF_{254} activated at 110 °C for 30 min. The plates were developed to 16 cm with methanol–water (2:1) as mobile phase. After development, the plates were dried in warm air and the spots were visualized at 302 nm. The calibration graph was linear from 2.5 to 10 μg omeprazole and the RSD was $\leq 0.60\%$. The recovery was 99.42%. Results compared well with those obtained by elution of the spots with methanol following TLC and spectrophotometric detection at 302 nm.

Dogrukol-Ak et al. [55] determined omeprazole in pharmaceutical preparations by a TLC densitometric method. Pellets from eneric coated capsules were finely powdered and dissolved in ethanolic 0.05 M potassium hydroxide with sonication. Four microliters of the solution was subjected to TLC on a silica gel FG_{254} plates with chloroform–methanol–25% ammonia (97.5:2.5:1) as mobile phase and densitometric detection of omeprazole ($R_f = 0.46$) at 302 nm. Calibration graphs were linear for 0.42–1.68 μg omeprazole; the detection limit was 25 ng. In the determination of omeprazole in 20 mg Omeprazit, Omeprol, and Losec capsules, the found amounts were 20.2, 20.3, and 19.8 mg omeprazole, respectively, with corresponding RSD 1.9, 1.8, and 1.6% ($n = 8$). The results agree with those of UV spectrophotometry.

Agbaba et al. [56] developed an HPTLC method for the determination of omeprazole, pantoprazole, and their impurities omeprazole sulfone and N-methylpantoprazole in pharmaceutical. The mobile phase chloroform–2-propanol:25% ammonia–acetonitrile (10.8:1.2:0.3:4), enables good resolution of large excesses of the drugs from the possible impurities. Regression coefficients ($r > 0.998$), recovery (90.7–120.0%), and detection limit (0.025–0.05%) were validated and found to be satisfactory. The method is convenient for quantitative analysis and purity control of the compounds.

4.2.5.2. High-performance liquid chromatography   Persson et al. [57] determined omeprazole and three of its metabolites, the sulfone, the sulfide, and the hydroxy metabolite, in plasma and urine by liquid chromatographic methods. The compounds are extracted from the biological sample and the extract is subjected to liquid chromatographic separation, either directly or after evaporation of the organic solvent and dissolution in a polar phase. The effluent from the column is UV-monitored at 302 nm and the quantitative evaluation performed by electronic integrator.

Amantea and Narang [58] used a reversed-phase HPLC method for the quantitation of omeprazole and its metabolites. Plasma was mixed with the internal standard (the 5-methyl analog of omeprazole), dichloromethane, hexane, and 0.1 M carbonate buffer (pH 9.8). After centrifugation, the organic phase was evaporated to dryness and the residue was dissolved in the mobile phase [methanol–acetonitrile–0.025 M phosphate buffer of pH 7.4 (10:2:13)] and subjected to HPLC at 25 °C on a column (15 cm × 4.6 mm) of Beckman Ultrasphere C8 (5 μm) with a guard column (7 cm × 2.2 mm) of Pell C8 (30–40 μm). The mobile phase flow-rate was 1.1 ml/min with detection at 302 nm. The calibration graphs are linear for ≤200 ng/ml, and the limits of detection were 5, 10, and 7.5 ng/ml for omeprazole, its sulfone, and its sulfide, respectively. The corresponding recoveries were 96.42% and 96% and the coefficients of variation ($n = 5$ or 6) were 3.0–13.9%.

Shim et al. [59] developed an HPLC method, with column switching, for the determination of omeprazole in plasma. The plasma samples were injected onto a Bondapak Phenyl/Corasil (37–50 μm) precolumn and polar plasma components were washed with 0.06 M borate buffer. After valve switching, the concentrated drug were eluted in the back-flush mode and separated on a μ-Bondapack C_{18} column with acetonitrile–phosphate buffer as the mobile phase. The method showed excellent precision, accuracy, and speed with detection limit of 0.01 μg/ml. Total analysis time per sample was less than 20 min and the coefficients of variation for intra- and interassay were less than 5.63%. The method has been applied to plasma samples from rats after oral administration of omeprazole.

Balmer et al. [60] separated the two enantiomers of omeprazole on three different stationary phases with immobilized protein, viz, Chiral-AGP with α-1 acid glycoprotein, Ultron ES-OVM with ovomucoid, and BSA–DSC with BSA cross-linked into 3-aminopropyl silica using N-succinimidyl carbonate. The mobile phase (1 ml/min) was phosphate buffer solution with 3–10% 2-propanol as the organic modifier. The enantiomers of omeprazole were separated on Chiralpak AD, an amylose-based chiral stationary phase, with ethanol–hexane (1:4) as mobile phase (1 ml/min).

Kang et al. [61] developed an advanced and sensitive HPLC method for the determination of omeprazole in human plasma. After omeprazole was extracted from plasma with diethylether, the organic phase was transferred to another tube and trapped back with 0.1N sodium hydroxide solution. The alkaline aqueous layer was injected into a reversed-phase C_8 column. Lansoprazole was used as the internal standard. The mobile phase consisted of 30% of acetonitrile and 70% of 0.2 M potassium dihydrogen phosphate, pH 7. Recoveries of the analytes and internal standard were 75.48%. The coefficients of variation of intra- and interday assay were <5.78 and 4.59 for plasma samples. The detection limit in plasma was 2 ng/ml. The method is suitable for the study of the kinetic disposition of omeprazole in the body.

Motevalian et al. [62] developed a rapid, simple, and sensitive HPLC assay method for the simultaneous determination of omeprazole and its major metabolites in human plasma using a solid-phase extraction procedure. Eluent (50 μl) was injected on a μBondapak C_{18} reversed-phase column (4.6 mm × 250 mm, 10 μm). The mobile phase consisted of 0.05 M phosphate buffer (pH 7.5) and acetonitrile (75:25) at a flow-rate of 0.8 ml/min. UV detection was at 302 nm. Mean recovery was greater than 96% and the analytical responses were linear over the omeprazole concentration range of 50–2000 ng/ml. The minimum detection limits were 10, 10, and 15 ng/ml for omeprazole, omeprazole sulfone, and hydroxyomeprazole, respectively. The method was used to determine the plasma concentration of the respective analytes in four healthy volunteers after an oral dose of 40 mg of omeprazole.

Garcia-Encina et al. [63] validated of an automated system using online solid-phase extraction and HPLC with UV detection, to determine omeprazole in human plasma. The extraction was carried out using C_{18} cartridges. After washing, omeprazole was eluted from the cartridge with mobile phase onto an Inertsil ODS-2 column. The developed method was selective and linear for drug concentrations ranging between 5 and 500 ng/ml. The recovery of omeprazole ranged from 88.1% to 101.5% and the limit of quantitation was 5 ng/ml. This method was applied to determine omeprazole in human plasma samples from bioequivalence studies.

Castro et al. [64] reported a comparison between derivative spectrophotometric and liquid chromatographic methods for the determination of omeprazole in aqueous solutions during stability studies. The first derivative procedure was based on the linear relationship between the omeprazole concentration and the first derivative amplitude at 313 nm. The first derivative spectra were developed between 200 and 400 nm ($\Delta\lambda = 8$). This method was validated and compared with the official HPLC method of the USP. It showed good linearity in the range of concentration studied (10–30 μg/ml), precision (repeatability and interday reproducibility), recovery, and specificity in stability studies. It also seemed to be 2.59 times more sensitive than the HPLC method. These results allowed to consider this procedure as useful for rapid analysis of omeprazole in stability studies since there was no interference with its decomposition products.

Persson and Andersson [65] reviewed the unusual effects in liquid chromatographic separations of enantiomers on chiral stationary phases with emphasis on polysaccharide phases. On protein phases and Pirkle phases, reversal of the elution order between enantiomers due to variation and temperature and mobile phase composition has been reported. Most of the nonanticipated observations have dealt with the widely used polysaccharide phases. Reversed retention order and other stereoselective effects have been observed from variation of temperature, organic modifier, and water content in nonpolar organic mobile phases.

Cass et al. [66] used a polysaccharide-based column on multimodal elution for the separation of the enantiomers of omeprazole in human plasma. Amylose tris (3,5-dimethylphenylcarbamate) coated onto APS-Hypersil (5 μm particle size and 120 Å pore size) was used under normal, reversed-phase, and polar-organic conditions for the enantioseparation of six racemates of different classes. The chiral stationary phase was not altered when going from one mobile phase to another. All compounds were enantioresolved within the elution modes with excellent selectivity factor. The separation of the enantiomers of omeprazole in human plasma in the polar-organic mode of elution is described.

Sluggett et al. [67] used an HPLC method with coulometric detection for the determination of omeprazole. A sensitive HPLC method for the analysis of omeprazole and three related benzimidazole with coulometric detection was carried out at +0.8 V using a porous C electrode. The linear range is 0.01–10 μg/ml. The method has a high degree of precision; the RSD of omeprazole at a concentration of 1.06 μg/ml was 0.7% ($n = 4$). The cyclic voltammogram of omeprazole is consistent with the hydrodynamic voltammogram exhibiting a single major irreversible oxidative wave with a peak potential at +1.105 V. The response factors for the four compounds are similar indicating that the oxidative process does not involve the S moiety exclusively. The data are most consistent with oxidation primarily of the benimidazole groups. The method was applied to the determination of omeprazole in a paste formulation.

Dubuc et al. [68] described a rapid HPLC method for the separation and determination of omeprazole extracted from human plasma. Omeprazole and the internal standard (H 168/24) were extracted from plasma samples by solid-phase extraction using a polymeric sorbent-based cartridge. The separation was accomplished under reversed phase conditions using an Eclipse XDB-C8 Rapid Resolution (4.6 × 50 mm) column. The mobile phase consisted of 23% acetonitrile and 77% of 30.4 mM disodium hydrogen phosphate and 1.76 mM potassium dihydrogen phosphate solution, pH 8.4, in which a gradient elution was used to linearly change solvent composition to 33% acetonitrile and 67% phosphate buffer during the first minute. Absorbance was monitored at 302 nm for omeprazole and at 294 nm for the internal standard and the total analysis time was 4 min.

Gonzalez et al. [69] presented a new simple and reliable HPLC method for measuring omeprazole and its two main metabolites in plasma. Omeprazole, hydroxyomeprazole, and omeprazole sulfone were extracted from plasma samples with phosphate buffer and dichloromethane–ether (95:5). HPLC separation was achieved using an Ultrasphere ODS C_{18} column. The mobile phase was acetonitrile–phosphate buffer (24:76, pH 8) containing nonylamine at 0.015%. Retention times were 9.5 min for omeprazole, 3.25 min for hydroxyomeprazole, 7.4 min for omeprazole sulfone, and 6.27 min for internal standard (phenacetin). Detection (UV at 302 nm) of analytes was linear in the range from 96 to 864 ng/ml.

Cheng et al. [70] used a microdialysis technique coupled to a validated microbore HPLC system to monitor the levels of protein-unbound omeprazole in rat blood, brain, and bile, constructing the relationship of the time course of the presence of omeprazole. Microdialysis probes were simultaneously inserted into the jugular vein toward right atrium, the brain striatum, and the bile duct of the male Sprague–Dawley rats for biological fluid sampling after the administration of omeprazole (10 mg/kg) through the femoral vein. The concentration–response relationship from the present method indicated linearity ($r^2 > 0.995$) over a concentration range of 0.01–50 μg/ml for omeprazole. Intra- and interassay precision and accuracy of omeprazole fell well within the predefined limit of acceptability. Following omeprazole administration, the blood-to-brain coefficient of distribution was 0.15, which was calculated as the area under the concentration versus time curve in the brain divided by the area under the curve in blood. The blood-to-bile coefficient of distribution was 0.58.

Cass et al. [71] described a direct injection HPLC method, with column-switching, for the determination of omeprazole enantiomers in human plasma. A restricted access media of bovine serum albumin octyl column has been used in the first dimension for separation of the analyte from the biological matrix. The omeprazole enantiomers were eluted from the restricted access media column onto an amylose tris (3,5-dimethylphenylcarbamate) chiral column by the use of a column-switching valve and the enantioseparation was performed using acetonitrile–water (60:40) as eluent. The analytes were detected by their UV absorbance at 302 nm. The validated method was applied to the analysis of the plasma samples obtained from 10 Brazilian volunteers who received a 40-mg oral dose of racemic omeprazole and was able to quantify the enantiomers of omeprazole in the clinical samples analyzed.

Schubert et al. [72] developed and validated a liquid chromatographic method for the determination of omeprazole in powder for injection and in pellets. The analyses were performed at room temperature on a reversed-phase C_{18} column of 250 mm $\times$ 4.6 mm (5 μm). The mobile phase, composed of methanol–water (90:10) was pumped at a constant flow-rate of 1.5 ml/min. Detection was performed on a UV detector at 301 nm. The method was validated in terms of linearity, precision, accuracy, and ruggedness.

Orlando and Bonato [73] presented a practical and selective HPLC method for the separation and quantification of omeprazole enantiomers in human plasma. C_{18} solid-phase extraction cartridges were used to extract the enantiomers from plasma samples and the chiral separation was carried out on a Chiralpak AD column protected with a CN guard column, using ethanol–hexane (70:30) as the mobile phase, at a flow-rate of 0.5 ml/min. The detection was carried out at 302 nm. The method is linear in the range of 10–1000 ng/ml for each enantiomer, with a quantification limit of 5 ng/ml. Precision and accuracy, demonstrated by within-day and between-day assays, were lower than 10%.

Rezk *et al.* [74] developed and validated a reversed-phase HPLC assay method for the simultaneous quantitative determination of omeprazole and its three metabolites in human plasma. The method provides excellent chromatographic resolution and peak shape for the four components and the internal standard within a 17-min run time. The simple extraction method results in a clean baseline and relatively high extraction efficiency. The method was validated over the range of 2–2000 ng/ml. The resolution and analysis for the four analytes; omeprazole, hydroxyomeprazole, omeprazole sulfone, and omeprazole sulfide and the internal standard utilized a Zorbax C18 (15 cm × 3 mm, 5 μm) with a Zorbax C18 (12.5 cm × 4.6 mm) guard column. The mobile phase consisted of two components. Mobile phase A was 22 mM phosphate monobasic, adjusted to a pH of 6 with diluted sodium hydroxide. This solution was filtered through a 0.45-μm membrane filter, then mixed as 900 ml buffer to 100 ml methanol. Mobile phase B was composed of 100 ml of the phosphate buffer as mobile phase A, mixed with 800 ml of acetonitrile, 100 ml of methanol, and 100 μl of trifluoroacetic acid with an initial flow-rate of 0.55 ml/min and detection at 302 nm.

Shimizu *et al.* [75] described a column-switching HPLC method for the simultaneous determination of omeprazole and its two main metabolites, 5-hydroxyomeprazole and omeprazole sulfone, in human plasma. Omeprazole and its two metabolites and lansoprazole as an internal standard were extracted from 1 ml of alkalinized plasma sample using diethyl ether–dichloromethane (45:55). The extract was injected into a column I (TSK-PW precolumn, 10 μm, 35 mm × 4.6 mm) for cleanup and column II (Inertsil ODS-80A column, 5 μm, 150 mm × 4.6 mm) for separation. The mobile phase consisted of phosphate buffer–acetonitrile (92:8, pH 7) for cleanup and phosphate buffer–acetonitrile–methanol (65:30:5, pH 6.5) for separation, respectively. The peak was detected with a UV detector set at a wavelength of 302 nm, and total time for chromatographic separation was approximately 25 min. The validated concentration ranges of this method were 3–2000 ng/ml for omeprazole, 3–50 ng/ml for 5-hydroxyomeprazole, and 3–1000 ng/ml for omeprazole sulfone. Mean recoveries were 84.3% for omeprazole, 64.3% for 5-hydroxyomeprazole, and 86.1%

for omeprazole sulfone. Intra- and interday coefficient variations were less than 5.1 and 6.6 for omeprazole, 4.6% and 5.0% for hydroxyomeprazole and 4.6% and 4.9% for omeprazole sulfone at the different concentrations. The limits of quantification were 3 ng/ml for omeprazole and its metabolites. This method was suitable for use in pharmacokinetic studies in human volunteers.

Zarghi et al. [76] developed an HPLC method, using a monolithic column, for quantification of omeprazole in plasma. The method is specific and sensitive with a quantification limit of 10 ng/ml. Sample preparation involves simple, one-step extraction procedure, and analytical recovery was complete. The separation was carried out in reversed-phase conditions using a Chromolith Performance (RP-18e, 100 × 4.6 mm) column with an isocratic mobile phase consisting of 0.01 mol/l disodium hydrogen phosphate buffer–acetonitrile (73:27) adjusted to pH 7.1. The wavelength was set at 302 nm. The calibration curve was linear over the concentration range 20–1500 ng/ml. The coefficients of variation for intra- and interday assay were found to be less than 7%.

Jia et al. [77] validated an HPLC method without solvent extraction and using UV detection at 302 nm for the determination of omeprazole in rat plasma. Plasma sample after pretreatment with acetonitrile to effect deproteinization were dried under nitrogen at 40 °C and reconstituted with mobile phase. The apparatus used was an Agilent 1100 quaternary pump, with a variable wavelength detector, thermostated autosampler, and column thermostat. A Hypersil ODS_2 C_{18} column (250 mm × 4.6 mm, 5 μm) was fitted with a Phenomenex guard column packed with octadecyl C_{18}. The mobile phase comprised 50 mM potassium dihydrogen phosphate buffer (pH 7.1, contained 0.7% triethylamine) and acetonitrile (75:25), the detection wavelength was 302 nm. Analyses were run at a flow-rate of 1 ml/min at 25 °C and the samples were quantified using peak areas. The standard calibration curve for omeprazole was linear ($r^2 = 0.999$) over the concentration range of 0.02–3 μg/ml. The intra- and interday assay variability range was 4.8–9.2% and 5.2–10.3% individually. This method has been applied to a pharmacokinetic study of omeprazole in rats.

Pearce and Lushnikova [78] used semipreparative HPLC method for the isolation of the three omeprazole metabolites produced by the fungi. Incubation of Cunninghamella elegans ATCC 9245 and omeprazole allowed putative fungal metabolite to be isolated in sufficient quantities for structural elucidation. The metabolites structures were identified by a combination of LC/MS and NMR spectrometric experiments. These isolates are used as reference standards in the confirmatory analysis of mammalian metabolites of omeprazole. In the LC/MS and LC/MS/MS analysis, components were separated by reversed-phase HPLC on a Hypersyl HyPurity (15 cm × 4.6 mm, 5 μm) column. The mobile phase consisted of 10 mM ammonium hydroxide in water (solvent A) and 1 mM ammonium hydroxide in 90% methanol (solvent B). A linear, one-step solvent gradient was applied, changing an initial composition of 85% solvent A to a final composition of 10% solvent A in 14 min. The injection volume was 10 μl, the solvent flow-rate was 0.25 ml/min and the total run time for each sample was 15 min. Metabolite isolation was carried out by reversed-phase HPLC on a semipreparative scale. A Hypersil HyPurity Elite C_{18} column (15 cm $\times$ 10 mm, 5 μm) column was used. The mobile phase consisted of 10 mM ammonium hydroxide in water (solvent A) and 1 mM ammonium hydroxide in 90% methanol (solvent B). A linear, three-step solvent gradient was applied, maintaining an initial composition of 75% solvent A for 1 min, changing to 40% solvent A over the next 15 min, then to 30% solvent A in the next 2 min and finally to 10% solvent A in the next 2 min. Injection volume was 2 ml, the solvent flow-rate was 1 ml/min., the UV detection wavelength was at 302 nm and the total run time for each sample loading was 21 min.

El-Sherif et al. [79] developed and validated a reversed-phase HPLC method for the quantitative determination of omeprazole and two other proton pump inhibitors in the presence of their acid-induced degradation products. The drugs were monitored at 280 nm using Nova-Pak C_{18} column and mobile phase consisting of 0.05 M potassium dihydrogen phosphate–methanol–acetonitrile (5:3:2). Linearity range for omeprazole was 2–36 μg/ml. The recovery of omeprazole was 100.50 $\pm$ 0.8%, and the minimum detection was 0.54 μg/ml. The method was applied to the determination of pure, laboratory prepared mixtures, and pharmaceutical dosage forms. The results were compared with the official USP method for omeprazole.

Linden et al. [80] developed a simple HPLC-diode array detector method using a reverse phase column and isocratic elution for the simultaneous determination of omeprazole, 5-hydroxyomeprazole, and omeprazole sulfone. The method was used to study CYP_2C_{19} and CYP_3A_4 genetic polymorphisms using omeprazole as the probe drug in a group of Brazilian volunteers. Omeprazole, 5-hydroxyomeprazole, and omeprazole sulfone were extracted from plasma samples with Tris buffer pH 9.5 (0.2 mol/l) and ethyl acetate. HPLC separation was achieved using a Shim-Pack RP-18e (15 cm $\times$ 4.6 mm, 5 μm) column with acetonitrile–phosphate buffer, pH 7.6 (24:76) as mobile phase and total run time for 15 min. Retention times were 2.7 min for internal standard, sulpiride, 4.1 min for 5-hydroxyomeprazole, 11.6 min for omeprazole, and 12.6 min for omeprazole sulfone. Detection (UV at 302 nm) of analytes was linear in the range from 25 to 1000 ng/ml. Extraction recoveries were in the range of 64.3–73.2% for all analytes. A group of 38 Brazilian healthy volunteers was phenotyped with this method, after a single oral dose of 20 mg of omeprazole. The method presented adequate accuracy and precision, with a limit of quantification of 25 ng/ml for omeprazole and metabolites, which allowed the identification of ultra-rapid metabolizers for both CYP_2C_{19} and CYP_3A_4 and took advantage of the selective identification offered by diode-array detectors.

Sivasubramanian and Anilkumar [81] described a simple reversed-phase HPLC method for the determination of omeprazole and domperidone from tablet formulations. The analysis was carried out on a Hypersil ODS C_{18} (15 cm $\times$ 4.6 mm, 5 μm) column using a mobile phase of methanol– 0.1 M ammonium acetate, pH 4.9 (60:40). The flow-rate and run time were 1 ml/min and 10 min, respectively. The eluent was monitored at 280 nm. The method was reproducible, with good resolution between omeprazole and domperidone. The detector response was linear in the concentration range of 10–60 μg/ml for omeprazole.

Murakami et al. [82] developed and validated a sensitive HPLC technique to quantify omeprazole in delayed release tablets. The analysis was carried out using a RP-C_{18} column with UV–VIS detection at 280 nm. The mobile phase was diluted with phosphate buffer (pH 7.4) and acetonitrile (70:30) at a flow-rate of 1.5 ml/min. The parameters used in the validation process were linearity, range, quantification limit, accuracy, specificity, and precision. The retention time of omeprazole was about 5 min with symmetrical peaks. The linearity in the range of 10–30 μg/ml presented a correlation coefficient of 0.9995. The excipients in the formulation did not interfere with the analysis and the recovery was quantitative. Results were satisfactory and the method proved to be adequate for quality control of omeprazole delayed-release tablets.

Silva et al. [83] separated omeprazole and other chiral drugs on a tartardiamide-based stationary phase commercially named Kromasil CHI-TBB. The effect of temperature on the chromatographic separation of the chiral drugs using the Kromasil CHI-TBB stationary phase was determined quantitatively so as to contribute toward the design for the racemic mixtures of the named compound using chiral column. A decrease in the retention and selectivity factors was observed, when the column temperature increased. Van't Hoff plots provided the thermodynamic data. The variation of the thermodynamic parameters enthalpy and entropy are clearly negative meaning that the separation is enthalpy controlled. The chiral column (25 cm $\times$ 1 cm) used was Kromasil CHI-TBB. The column was packed with 16 μm particle diameter and 100 Å of internal pore diameter of Kromasil silica which is covalent bonded with O,O'-di(4-tert-butyl-benzoyl)-N,N'-diallyl-L-tartardiamide. A mobile phase of n-hexane–isopropanol–triethylamine–acetic acid (98:2:0.15:0.05) was used. Omeprazole (0.15 g/l) solutions were prepared using this mobile phase. The solutions and the mobile phases were filtered in a Millipore filter system (0.45 μm) and degasified in a Cole Parmer 8892 ultrasonic bath. The experiments were carried out using a single chromatographic column in an HPLC system equipped with a Waters 1525 dual pump, a Waters 2487 dual absorbance UV–VIS detector, temperature controller, manual injector, and digital data acquisition system. The chromatograms of omeprazole and the other chiral drugs were obtained by small pulse experiments (20 μl) after a time interval necessary to the stabilization of the system at four different flow-rates (1–4 ml/min). The chromatographic experiments were performed at 25, 35, and 45 °C. Detection was carried out at 302 nm for omeprazole.

Belaz et al. [84] separated the enantiomers omepraozle and other proton pump inhibitors by HPLC at multimilligram scale on a polysaccharide-based chiral stationary phase using normal and polar organic conditions as mobile phase. The values of the recovery and production rate were significant for each enantiomer; better results were achieved using a solid-phase injection system. The chiroptical characterization of the compounds was performed using a polarimeter and a circular dichroism detector. The preparative HPLC system consisted of a Shimadzu liquid chromatographic-6AD pump, a Rheodyne 7725 injector fitted with a 200-μl loop or a cylindrical stainless-steel precolumn coated with Teflon for the injections for samples, and a 10-AVvp variable wavelength UV–VIS detector with a CMB SCL-10 AVvp interface. The columns were prepared at the UFSCar Laboratory [85–87]. The tris-3,5-dimethylphenylcarbamate and tris(S)-1-phenylethylcarbamate of amylose were coated on to APS-Neocleosil (500 Å, 7 μm, 20%, w/w) and packed into a stainless-steel 20 cm × 0.7 cm column for semipreparative chromatography (CSP 1 and 2) and into a 15 cm × 0.46-cm column for analytical separation (CPS 3 and 4, respectively). Amylose tris-(S)-1-phenylcarbamate coated on to APS-Hypersil (120 Å, 5 μm, 25%, w/w) packed into a stainless-steel 20 cm × 0.4 cm semipreparative column (CPS 5) was used. A Shandon HPLC packing pump was employed for column packing. The mobile phase for omeprazole was methanol at a flow-rate of 3 ml/min and detection at 302 nm on column 1 and the mobile phase was methanol at a flow-rate of 1 ml/min on column 3 were used for the semipreparative and analytical chromatographic separation of omeprazole, respectively.

Sultana et al. [88] developed a reversed-phase HPLC method for the simultaneous determination of omeprazole in Risek® capsules. Omeprazole and the internal standard, diazepam, were separated by Shim-pack CLC-ODS (0.4 × 25 cm, 5 μm) column. The mobile phase was methanol–water (80:20), pumped isocratically at ambient temperature. Analysis was run at a flow-rate of 1 ml/min at a detection wavelength of 302 nm. The method was specific and sensitive with a detection limit of 3.5 ng/ml at a signal-to-noise ratio of 4:1. The limit of quantification was set at 6.25 ng/ml. The calibration curve was linear over a concentration range of 6.25–1280 ng/ml. Precision and accuracy, demonstrated by within-day, between-day assay, and interoperator assays were lower than 10%.

The *in vitro* availability of omeprazole in presence of manganese, cobalt, nickel, copper, and zinc was studied by this method. Recovery of omeprazole in presence of various metals was from 41% and 74%.

Rambla-Alegre *et al.* [89] reported a chromatographic procedure that uses micellar mobile phases of sodium dodecyl sulfate and propanol buffered at pH 7 and a C_{18} column for the determination of omeprazole and its metabolites, omeprazole sulfone, and hydroxyomeprazole, in urine and serum samples. Direct injection and UV detection set at 305 nm was used. Omeprazole and its metabolites were eluted in less than 11 min with no interference by the protein band or endogenous compounds. Adequate resolution was obtained with a chemometric approach, in which the retention factor and shape of the chromatographic peaks were taken into account. The chromatographic system was equipped with a quaternary pump, thermostated autosampler tray, and column compartments, and a diode-array detector (190−700 nm). Separation was performed in a reversed phase Kromasil C_{18} column thermostated at 25 °C. The mobile Phase composition was 0.08 M sodium dodecyl sulfate, 10% propanol, 0.01 M sodium dihydrogen phosphate at pH 7. The flow-rate, injection volume, and UV wavelength were 1 ml/min, 20 μl, and 305 nm, respectively. Under these conditions, the total analysis time for omeprazole and its main metabolites was less than 11 min. The analytical parameters including linearity ($r = >0.9998$) intra- and interday precision (RSD) 06−7.9% and 0.14−4.7%, respectively) and robustness were studied in the validation of the method for the three compounds. The limit of detection and quantification were less than 6 and 25 ng/ml, respectively. Recoveries in micellar medium, plasma, and urine matrices were in the 98−102 range. The method was applied to the determination of omeprazole and its metabolites in physiological samples. Omeprazole was also analyzed in pharmaceutical formulations.

Other HPLC methods [90–115] are listed in Table 4.12.

4.2.5.3. Liquid chromatography–mass spectrometry
Woolf and Matuszewski [116] described a liquid chromatography–tandem mass spectrometric method for the simultaneous determination of omeprazole and 5-hydroxyomeprazole in human plasma. Omeprazole and its 5-hydroxy-metabolite plus 2-[(4-methoxy-3-methyl-2-pyridinylmethyl)sulfinyl]-1*H*-benzimidazole as internal standard were separated from plasma by solid-phase extraction on to a Waters Oasis cartridge (60 mg bed) and elution with methanol, followed after solvent evaporation by transfer into mobile phase. For HPLC, a column (5 cm × 4.6 mm) of Zorbax XDB C_{18} silica (3 μm) was used with 35 μl sample injection, a mobile phase (1 ml/min) of 10 mM ammonium hydroxide in aqueous 21% acetonitrile adjusted to pH 8.5 with formic acid. Transfer to triple quadrupole PE Sciex API III$^+$ tandem mass spectrometer was via nebulizer at 500 °C for positive

TABLE 4.12 HPLC conditions of the methods used for the determination of omeprazole

Column	Mobile phase and [flow-rate]	Detection (nm)	Remarks	References
Waters Rad-Pak A C$_{18}$	1% triethylamine solution in aqueous 60% methanol adjusted to pH 7 with phosphoric acid	302	Analysis of omeprazole and its sulfone and sulfide metabolites in human plasma and urine	[90]
LiChrosorb Si 60 or Polygosil C$_{18}$	Methanol–aqueous ammonia–dichloromethane for omeprazole and its sulfone in plasma or acetonitrile–phosphate buffer for the hydroxy metabolite in plasma and all compounds in urine	302	Analysis of omeprazole and metabolites in plasma and urine. Drug and metabolites are extracted in CH$_2$Cl$_2$	[91]
3 cm × 4.6 mm of Spheri-5 RP-8 guard column and a column 15 cm × 4 mm of LiChrosorb RP-8 (5 or 7 μm)	Oxygenated and N-deoxygenated phosphate buffer solution at pH 7.6 containing 32.5% acetonitrile [1 ml/min]	d.c. mercury-drop detection or silver/silver chloride electrode	Oxygen effects in amperometric liquid chromatography detection of the drug at a mercury electrode	[92]

(continued)

TABLE 4.12 *(continued)*

Column	Mobile phase and [flow-rate]	Detection (nm)	Remarks	References
15 cm × 4.5 mm of Polygosil C_{18} (5 μm) and a guard column 3 cm × 4.6 mm of Spheri-5 RP-18	Acetonitrile–phosphate buffer solution (pH 7.7) from 25% of acetonitrile (maintained for 3 min) to 40% (in 1 min; maintained for 6 min) and decreased to 25% (in 5 min). [1.5 ml/min]	302	Fully automated gradient-elution LC assay of the drug and two of its metabolites	[93]
20 cm × 4.6 mm of triphenylcarbomethylcellulose immobilized on 3-aminoproyl silica	Phosphate buffer (pH 6.6)–propanol [1 ml/min]	229 and 280	Resolution of enantiomers of omeprazole and its analogs by LC on triphenylcarbamoyl cellulose-based stationary phase	[94]
15 cm × 4.6 mm of Nucleosil C_{18} (5 μm) or 10 cm × 5 mm of Novapack C_{18} (4 μm)	Acetonitrile–phosphate buffer solution (3:7 or 2:3) [1 ml/min]	280	Peak distortion in the column LC analysis of the drug dissolved in borax buffer	[95]

25 cm × 5 mm of Shimadzu Nucleosil C_8 operated at 35 °C	Acetonitrile–phosphate buffer pH 7.6 (17:33). [1 ml/min]	280	Studies on the quantitative determination of omeprazole	[96]
25 cm × 4.6 mm of Capcell Pak C_{18} SG 120 (5 μm)	Acetonitrile–0.05 M phosphate buffer pH 8.5. [0.8 ml/min]	302	Simultaneous analysis of the drug and its metabolites in plasma and urine by RPHPLC with an alkaline-resistant polymer-coated C_{18} column	[97]
NH_2, diol, CN, C_{18}, Si-60 columns thermostated at 40 °C	Carbon dioxide (3.5 grade) [1.2 ml/min] plus 60 μl/min of 1% (v/v) methanolic triethylamine mobile phase modifier	300	Packed-column supercritical fluid chromatography of the drug and related compounds	[98]
12.5 cm × 4 mm of Supersphere SI-60 (4 μm) with a guard column (1.5 cm × 3 mm) of Brownlee Aquapore Silica (7 μm)	Dichloromethane–5% ammonium hydroxide in methanol–isopropanol (191:8:1) [1.5 ml/min]	302	HPLC assay for human liver microsomal omeprazole metabolism	[99]

(continued)

TABLE 4.12 (continued)

Column	Mobile phase and [flow-rate]	Detection (nm)	Remarks	References
25 cm × 6.4 mm of YWG C_{18} (10 μm) operated at 25 °C	Methanol–water–triethylamine–phosphoric acid (3850:1650:25:4) [1 ml/min]	302	Analysis of omeprazole and its analogs by HPLC	[100]
25 cm × 4.6 mm of Capcell Pak C_{18} SG120 operated at 30 °C	Acetonitrile–0.05 M sodium phosphate buffer of pH 8.4 (13:37) [0.8 ml/min]	302	Development and preliminary application of an HPLC assay for omeprazole metabolism in human liver microsomes	[101]
15 cm × 4 mm of Resolvosil BSA-7 (5 μm)	50 mM phosphate buffer pH 7 containing 0.05–1% of propanol [1.5 ml/min]	302	Enantioselective HPLC analysis of omeprazole in human plasma	[102]
15 cm × 4.6 mm of Chiracel OJ-R (5 μm) with 1 cm × 6 mm precolumn (25–40 μm) LiChroprep PR-2 and 4 mm × 4 mm of LiChrospher 100 RP-18 (5 μm) guard column	25% aqueous acetonitrile or 50 mM sodium chlorate–acetonitrile (3:1) [0.5 ml/min] Water [0.5 ml/min]	286	Direct HPLC separation of enantiomers of omeprazole and other benzimidazole sulfoxides using CBCS phases in reversed-phase mode	[103]

Column	Mobile phase	m/z	Application	Reference
15 cm × 4.6 mm of Shim Pack CLC-C$_{18}$ (5 μm)	Methanol–0.05 M phosphate buffer of pH 5.5 (1:1) [1 ml/min]	302	Studies on chromatographic optimization and its application in pharmacokinetics research	[104]
15 cm × 6 mm of 5 M-Shim-pack CLC-ODS operated at room temperature	Aqueous 63% methanol containing 1% triethylamine, pH adjusted to 7 with 85% phosphoric acid [1 ml/min]	302	Analysis of the drug as capsules in human plasma by RPHPLC	[105]
25 cm × 4.6 mm of Nucleosil 120-5 C$_{18}$ at 37 °C with a precolumn 10 cm × 4.6 mm of similar material	0.1 M dipotassium hydrogen phosphate of pH 7.8–methanol (53:47) containing 40 mg/l of azide [1.2 ml/min]	302	Analysis of the drug in human plasma by HPLC	[106]
25 cm × 4.6 mm of Zorbax C-8 (5 μm)	Acetonitrile–8 mM disodium hydrogen orthophosphate buffer pH 7.5 (7:13) [1 ml/min]	302	RPHPL chromatographic assay of omeprazole in plasma	[107]

(continued)

TABLE 4.12 (continued)

Column	Mobile phase and [flow-rate]	Detection (nm)	Remarks	References
Omnipac Pax-500 fitted with a C$_{18}$ reversed-phase guard column	0.1 M sodium phosphate–methanol–acetonitrile (3:1:1) adjusted to pH 2.3 with 85% phosphoric acid [0.7 ml/min]	254	Simple HPLC analysis of omeprazole in human plasma and gastric fluid	[108]
C$_{18}$ column	50 mM phosphate buffer in acetonitrile (22–50% in 43 min followed by 15 min equilibration	302	Analysis of omeprazole and its metabolite in human plasma by HPLC	[109]
20 cm × 4.6 mm of Hypersil ODS 2 (5 μm)	Methanol–water–glacial acetic acid–triethylamine (120:80:1:1) [1 ml/min]	302	Analysis of omeprazole and its pharmacokinetics in human plasma by an improved HPLC	[110]
15 cm × 4.6 mm of Zorbax Eclipse XDB-C$_8$ (5 μm)	Phosphate buffer adjusted to pH 7 with phosphoric acid–acetonitrile (7:3) [2 ml/min]	280	Effect of various salts on stability of omeprazole as determined by HPLC	[111]

Column	Mobile phase	Detection (nm)	Description	Reference
15 cm × 4 mm of ET Resolvosil-BSA-7	0.5 M phosphate buffer of pH 7.9/2% propanol [1 ml/min]	250	Direct optical resolution of racemic sulfoxide by HPLAC	[112]
15 cm × 4.6 mm of Crestpak C$_{18}$ (5 μm) preceded by a refillable guard column packed with Perisorb RP-18 (30–40 μm)	0.05 M disodium hydrogen phosphate buffer–acetonitrile (65:35) adjusted to pH 6.5 [1 ml/min]	302	Improved HPLC analysis of omeprazole in human plasma	[113]
10 cm × 4 mm of Chiral-AGP	10 mM sodium phosphate buffer of pH 6.5 containing 10% acetonitrile	210	Omeprazole chiral separation chiral chromatography	[114]
3.5 cm × 4.6 mm of MF Ph-1 precolumn and 25 cm × 1.5 mm of Capcell Pak C$_{18}$ UG 120 (5 μm) and 3.5 cm × 2 mm of Capcell Pak C$_{18}$ UG 120 (5 μm)	Buffer–acetonitrile (90:10) for precolumn and buffer–acetonitrile (60:40) for the analytical column	302	Assay of omeprazole and its sulfone by semimicrocolumn LC with mixed-function precolumn in human plasma samples	[115]

chemical ionization mass spectrometery with Argon collision-induced dissociation and monitoring at m/z 214 for 5-hydroxy omeprazole, 198 for omeprazole and 147 for the internal standard. The method was validated for 10–500 ng/ml of the drug and its metabolite in plasma with ($n = 5$) an accuracy of 98.2–102.1% for the drug and 95.9–103.1% for the metabolite and precision RSD ranging from 1.8% to 4.7% for the drug and 2.8% to 6.7% for the metabolite across the calibration range.

Stenhoff et al. [117] determined enantiomers of omeprazole in blood plasma by normal-phase liquid chromatography and detection by atmospheric-pressure ionization tandem mass spectrometry. The enantioselective assay of omeprazole is using normal-phase liquid chromatography on a Chiralpak AD column and detection by mass spectrometry. Omeprazole is extracted by a mixture of dichloromethane and hexane and, after evaporation, redissolution and injection, separated into its enantiomers on the chiral stationary phase. Detection is made by a triple quadrupole mass spectrometer, using deuterated analogs and internal standards. The method enables determination in plasma down to 10 nmol/l and shows excellent consistency suited for pharmacokinetic studies in man.

Kanazawa et al. [118] performed a chiral separation of omeprazole on a chiral column with circular dichroism detection and LC/MS. A good resolution of enantiomers was obtained. The column used for the chiral separation was Chiralpak AD-RH column (4.6 mm $\times$ 150 mm) using phosphate buffer and (or ammonium acetate) acetonitrile as an eluent. After a single oral dose of omeprazole (20 mg), the plasma concentrations of the separate enantiomers of omeprazole were determined for 3.5 h after drug intake. This study is useful because of the part polymorphism plays in the therapeutic effectiveness of omeprazole and other proton pump inhibitors during the treatment of acid-related diseases. This study demonstrates the stereospecific analysis of omeprazole in human plasma as a probe drug of CYP_2C_{19} phenotyping.

Martens-Lobenhoffer et al. [119] used chiral HPLC-atmospheric pressure photoionization tandem mass-spectrometric method for the enantioselective quantification of omeprazole and its main metabolites in human serum. The method features solid-phase separation, normal phase chiral HPLC separation, and atmospheric pressure photoionization tandem mass spectrometry. The internal standards serve stable isotope labeled omeprazole and 5-hydroxy omeprazole. The HPLC part consists of Agilent 1100 system comprising a binary pump, an autosampler, a thermostated column component, and a diode array UV–VIS detector. The enantioselective chromatographic separation took place on a ReproSil Chiral-CA 5 μm 25 cm $\times$ 2 mm column, protected by a security guard system, equipped with a 4 mm $\times$ 2-mm silica filter insert. The analytes were detected by a Thermo Scientific TSQ Discovery Max triple quadrupole mass spectrometer, equipped with an APPI ion source with a krypton UV-lamp. System control and data handling were carried out by the Thermo Scientific Xcalibur software. Solid-phase extraction of the samples was performed on OASIS HLB 1 ml extraction column containing 30 mg sorbet. After injection of 10 μl of the prepared samples, enantioselective chromatographic separation was achieved by HPLC normal phase gradient elution. The mobile phase A consisted of 2-propanol–acetic acid–diethylamine (100:4:1) and mobile phase B was pure hexane. At a flow-rate of 0.35 ml/min, the gradient started with a composition of 10:90 A:B, fraction of A was increased to 15% in the next 10 min and was held constantly for 1 min. Subsequently, a washing step with 25% A for 1 min was performed. After the washing step, the mobile phase composition was turned back to starting conditions. The column temperature was held constant at 20 °C. A divert valve directed the HPLC effluent without splitting to the mass spectrometer in the run-time window of 5–15.9 min, otherwise to the waste container. In the mass spectrometer detector, ions were formed by photoionization using krypton light source radiating two emission lines with energies of 10 and 10.6 eV. Vaporizer and capillary temperatures were set to 300 and 220 °C, respectively. Nitrogen served as sheath and AUX gas, with flow settings of 41 and 8 arbitrary units, respectively. Under these conditions, the analytes were ionized exclusively to $[M + H]^+$ parent ions. Prior to detection, collision induced fragmentation of the parent ions was achieved with argon serving as collision gas at a pressure of 1 mTorr. The calibration functions are linear in the range 5–750 ng/ml for the omeprazole enantiomers, and omeprazole sulfone, and 2.5–375 ng/ml for 5-hydroxyomeprazole enantiomers, respectively. Intra- and interday RSDs are 7% for omeprazole and 5-hydroxy omeprazole enantiomers and 9% for omeprazole sulfone.

Macek *et al.* [120] developed a method to quantitate omeprazole in human plasma using liquid chromatography–tandem mass spectrometry. The method is based on the protein precipitation with acetonitrile and a reversed-phase liquid chromatography performed on an octadecylsilica column (55 × 2 mm, 3 μm). The mobile phase consisted of methanol–10 mM ammonium acetate (60:40). Omeprazole and the internal standard, flunitrazepam, elute at 0.80 ± 0.1 min with a total run time 1.35 min. Quantification was through positive-ion made and selected reaction monitoring mode at m/z 346.1 → 197.9 for omeprazole and m/z 314 → 268 for flunitrazepam, respectively. The lower limit of quantification was 1.2 ng/ml using 0.25 ml of plasma and linearity was observed from 1.2 to 1200 ng/ml. The method was applied to the analysis of samples from a pharmacokinetic study.

4.2.5.4. High-performance liquid chromatography–mass spectrometry
Weidolf and Covey [121] described the application of the ionspray interface for liquid chromatography and atmospheric-pressure ionization mass spectrometry to samples obtained in a study on the metabolism of

omeprazole. In this study, [^{34}S]omeprazole was utilized for the stable isotope cluster technique. Over 40 metabolites in a sample of partially purified rat urine were resolved by gradient elution liquid chromatography with ionspray atmospheric pressure ionization mass spectrometric detection and each of them produced molecular ion 1:1 clusters (MH$^+$ and [MH + 2]$^+$). The chromatographic fidelity of the total-ion current was excellent. The endogenous matrix of the sample was quite low, allowing a background-substracted averaged mass spectrum of the entire total-ion current trace to produce a metabolite mass profile depicting all the molecular ion 1:1 clusters in the sample. From this mass profile, it was possible to obtain direct information concerning oxygenation and conjugation reactions of the parent compound.

Weidolf and Castagnoli [122] reported a detailed analysis of the product ion spectrum generated from the protonated molecule under electrospray ionization (ESI)-MS/MS conditions using a triple quadrupole mass spectrometer for omeprazole. Unambiguous molecular composition data of the fragment ions were obtained with the aid of regioselectively ^{14}C-, ^{34}S-, and ^{18}O-labeled analogs. Attempts have been made to provide rationale pathways for the formation of the fragment ions from four protonated omeprazole species. These results will facilitate the characterization of the complex metabolic fate of omeprazole in humans, which involve the excretion of at least 50 metabolites.

Naidong et al. [123] demonstrated a novel approach in 96-well solid-phase extraction by using normal phase LC/MS/MS methods with low aqueous/high organic mobile phases, which consisted of 70–95% organic solvent, 5–30% water, and small amount of volatile acid or buffer. While the commonly used solid-phase extraction elution solvents (acetonitrile and methanol) have stronger elution strength than a mobile phase on reversed-phase chromatography. Analytical methods for omeprazole and other polar compounds in biological fluids were developed and optimized.

Kanazawa et al. [124] determined omeprazole and its metabolites in human plasma by liquid chromatography–three-dimensional quadrupole mass spectrometry with a sonic spray ionization interface. The analytical column was YMC-Pack Pro C$_{18}$ (5 cm × 2 mm) using acetonitrile–50 mM ammonium acetate (pH 7.25) (1:4) at a flow-rate of 0.2 ml/min. The drift voltage was 30 V. The sampling aperture was heated at 110 °C and shield temperature was 230 °C. In the mass spectrum, the molecular ions of omeprazole, hydroxyomeprazole, and omeprazole sulfone were clearly observed as base peaks. The method is sufficiently sensitive and accurate for pharmacokinetic studies of omeprazole.

Jensen et al. [125] investigated an HPLC/ICP-MS (inductively coupled plasma mass spectrometry) with sulfur-specific detection, as a method for obtaining metabolite profiles for omeprazole administered as a 1:1

mixture of ^{32}S- and ^{34}S-labeled material. Analysis based on the monitoring of the chromatographic eluent at either m/z 32 or 34 was not successful due to insufficient sensitivity caused by interferences from polyatomic ions. Reaction of sulfur with oxygen in the hexapole collision cell, combined with monitoring at m/z 48 (for ^{32}S) or m/z 50 (for ^{34}S), provided a facile method for metabolite profiling. Detection of m/z 48 was superior in sensitivity to detection of m/z 50.

Tolonen et al. [126] described a simple and efficient method for the determination of labile protons in drug metabolites using postcolumn infusion of deuterium oxide in LC/MS experiments with ESI and time-of-flight (TOF)-MS. The number of exchangeable protons in the analytes; the hydroxyl, amine, thiol, and carboxylic acid protons can easily be determined by comparing the increase in m/z values after H/D-exchange occurring online between an HPLC column and electrospray ion source. The hydroxyl metabolites and sulfur/nitrogen oxides with the same accurate mass can be distinguished. A good degree of exchange was obtained in repeatable experiments. Only a low consumption of deuterium oxide is needed in a very easy and rapidly set-up procedure. This method is applied to the study of metabolites of omeprazole in human and mouse in vitro samples, together with exact mass data obtained from TOF-MS experiments.

Wang et al. [127] developed an analytical method for the determination of omeprazole in human plasma, based on LC–MS. The analyte and the internal standard sildenafil are extracted from plasma by liquid–liquid extraction using diethyl ether–dichloromethane (60:40) and separated by reversed-phase HPLC using acetontile–methanol–10 mM ammonium acetate (37.5:37.5:25) as mobile phase. Detection is carried out by multiple reaction monitoring on a Q TRAP LC/MS/MS system. The method has a chromatographic run time of 3.5 min and is linear within the range 0.50–800 ng/ml. Intra- and interday precision expressed as RSD ranged from 0.4% to 8.5% and from 1.2% to 6.8%, respectively. Assay expressed as relative error was < 5.7%. The method has been applied in a bioequivalence study of two capsule formulations of omeprazole.

Frerichs et al. [128] developed and validated a method for the quantitation of omeprazole and hydroxyomeprazole from one 250 μl sample of human plasma using HPLC coupled to tandem mass spectrometry. The method was validated for a daily working range of 0.4–100 ng/ml, with limits of detection between 2 and 15 pg/ml. The interassay variation was less than 15% for all analytes at four control concentrations and the samples were stable for three freeze–thaw cycles under the analysis conditions and 24 h in the postpreparative analysis matrix. The method was used to analyze samples in support of clinical studies probing the activity of the cytochrome P-450 enzyme system.

Song and Naidong [129] analyzed omeprazole and 5-hydroxyomepra-
zole in human plasma using hydrophilic interaction chromatography
with tandem mass spectrometry. Omeprazole and its metabolite 5-
hydroxy omeprazole and the internal standard desoxyomeprazole were
extracted from 0.05 ml of human plasma using 0.5 ml of ethyl acetate in a
96-well plate. A portion (0.1 ml) of the ethyl acetate extract was diluted
with 0.4 ml of acetonitrile and 10 μl was injected onto a Betasil silica
column (5 cm $\times$ 3 mm, 5 μm) and detected by atmospheric pressure
ionization 3000 and 4000 with positive electrospray ionization. Mobile
phase with linear gradient elution consists of acetonitrile, water, and
formic acid (from 95:5:0.1 to 73.5:26.5:0.1 in 2 min). The flow-rate was
1.5 ml/min with total run time of 2.75 min. The method was validated for
a low limit of quantitation at 2.5 ng/ml for both analytes. The method was
also validated for specificity, reproducibility, stability, and recovery.

Hultman *et al.* [130] developed a LC/MS/MS method for the quantitative determination of esomeprazole and its two main metabolites 5-hydro-
xyesomeprazole and omeprazole sulfone in 25 μl human, rat, or dog
plasma. The analytes and their internal standards were extracted from
plasma into methyl *tert*-butyl ether–dichloromethane (3:2). After evaporation and reconstitution of the organic extract, the analytes were separated
on a reversed-phase liquid chromatography column and measured by
atmospheric-pressure positive ionization mass spectrometry.

Hofmann *et al.* [131] described a sensitive method for the simultaneous
determination of omeprazole and its major metabolites 5-hydroxyome-
prazole and omeprazole sulfone in human plasma by HPLC-electrospray
mass spectrometry. Following liquid–liquid extraction HPLC separation
was achieved on a Prontosil AQ, C_{18} column using a gradient with 10 mM
ammonium acetate in water (pH 7.25) and acetonitrile. The mass spectrometer was operated in the selected ion monitoring mode using the
respective $MH^{(+)}$ ions, m/z 346 for omeprazole, m/z 362 for 5-hydroxyo-
meprazole, and omeprazole sulfone and m/z 300 for the internal standard
(2-[[(3,5-dimethylpyridin-2-yl)methyl]thio]-1H-benzimidazole-5-yl)
methanol. The limit of quantification was 5 ng/ml for 5-hydroxyomepra-
zole and 10 ng/ml for omeprazole and omeprazole sulfone using 0.25 ml
of plasma. Intra- and interassay variability was below 11% over the whole
concentration range from 5 to 250 ng/ml for 5-hydroxyomeprazole and
from 10 to 750 ng/ml for omeprazole and omeprazole sulfone. The
method was used for the determination of pharmacokinetic parameters
of esomeprazole and the two major metabolites after a single dose and
under steady state conditions.

4.2.5.5. Supercritical fluid chromatography   Toribio *et al.* [132] used supercritical fluid chromatography for the enantiomeric separation of omeprazole. The drug was separated at semipreparative scale on a

polysaccharide-based chiral stationary phase. A modular supercritical fluid chromatograph was adapted to operate at semipreparative scale on a Chiralpak AD (25 cm × 10 mm) column was used. The effect of two organic modifiers (ethanol and isopropanol) was studied, and different injection volumes and concentrations of omeprazole racemic mixture were evaluated to obtain high enantiomeric purities and production rates. Better results were achieved using concentration overloading instead of volume overloading. The recoveries decreased when the requirements of enantiomeric purity or the load increased, but it was possible to recover 100% of both enantiomers at an enantiomeric purity higher than 99.9% under some loading conditions, like injecting 1 and 2 ml of a solution of 3 g/l. As far as production rates are concerned, the best result for S-(−)-omeprazole at that purity (27.2 mg/h) was achieved with sample concentrations of 10 g/l and the injection of 2 ml, while a volume of 4 ml was better in the case of R-(+)-omeprazole (20.5 mg/h).

4.2.5.6. Electrophoresis McGrath *et al.* [133] used capillary zone electrophoresis to study the migration behavior of selected 1,4-benzodiazepines and metabolites over the pH range 2–12 exhibiting the ability to determine, pK_a values using this technique. The method was applied to the assay of a variety of pharmaceutical formulations which contain omeprazole, metronidazole, and 1,4-benzodiazepines and was compared with alternative analytical techniques such as reversed-phase HPLC, capillary gas chromatography, and automated differential-pulse polarography. Limit of detection of capillary zone electrophoresis and alternative techniques are compared for these molecules. The selectivity of capillary electrophoresis was demonstrated for the separation of four benzodiazepines using capillary zone electrophoresis with 20 mM citric acid + 15% methanol, and miceller electrokinetic capillary chromatography with 75 mM sodium dodecyl sulfate in 6 mM sodium tetraborate—12 mM disodium hydrogen phosphate + 5% methanol, and compared with other topical analytical techniques in terms of retention times, capacity factors, and efficiencies.

Eberle *et al.* [134] separated the enantiomers of omeprazole and structurally related drugs by capillary zone electrophoresis with bovine serum albumin as chiral selector. The separations were carried out on a fused silica column (60 cm × 50 μm, 50 cm to detector) with a buffer consisting of 100-μM-bovine serum albumin and 7% 1-propanol in 10 mM potassium phosphate pH 7.4. Electrokinetic injection was at 5–8 kV for 7 s. An applied voltage of 300 V/cm was used. Detection was at 290 nm. Detection limits were 0.04 mg/ml for the analytes studied.

Altria *et al.* [135] used a validated capillary electrophoresis method for the analysis of omeprazole among other acidic drugs and excipients. The results of validation experiments for the capillary electrophoretic separation of water-soluble and -insoluble pharmaceutical compounds at 30 °C with 15 mM borax as run buffer and detection at 200 nm are presented. Hewlett-Packard bubble cell and Beckman instruments were used with fused-silica column (34 and 27 cm, respectively ×75 mm) operated at 7 and 6.5 kV, respectively. When peak area ratios and internal standards sodium-β-naphtoxy acetate and aminobenzoic acid were used to determine 100 ppm omeprazole, the R.S.D. was 0.34–1.31% ($n = 10$). Typical detection and determination limits were 0.4 and 1.2–1.7 mg/l, respectively. Calibration graphs were linear from 50 to 150 mg/l of omeprazole.

Bonato and Paias [136] developed two sensitive and simple assay procedures based on HPLC and capillary electrophoresis for the enantioselective analysis of omeprazole in pharmaceutical formulations. Racemic omeprazole and (S)-omeprazole were extracted from commercially available tablets using methanol–sodium hydroxide 2.5 mol/l (90:10). Chiral HPLC separation of omeprazole was obtained on a ChiralPak AD column using hexane–ethanol (40:60) as the mobile phase and detection at 302 nm. The resolution of omeprazole enantiomers by capillary electrophoresis was carried out using 3% sulfated β-cyclodextrin in 20 mmol/l phosphate buffer, pH 4 and detection at 202 nm.

Lin and Wu [137] established a simple capillary zone electrophoresis method for the simultaneous analysis of omeprazole and lansoprazole. Untreated fused-silica capillary was operated using a phosphate buffer (50 mM, pH 9) under 20 kV and detection at 200 nm. Baseline separation was attained within 6 min. In the method validation, calibration curves were linear over a concentration range of 5–100 μM, with correlation coefficients 0.9990. RSD and relative error were all less than 5% for the intra- and interday analysis, and all recoveries were greater than 95%. The limits of detection for omeprazole and lansoprazole were 2 μM ($S/N = 3$, hydroxynamic injection 5 s). The method was applied to determine the quality of commercial capsules. Assay result fell within 94–106%.

Berzas Nevado et al. [138] developed a new capillary zone electrophoresis method for the separation of omeprazole enantiomers. Methyl-β-cyclodextrin was chosen as the chiral selector, and several parameters, such as cyclodextrin structure and concentration, buffer concentration, pH, and capillary temperature were investigated to optimize separation and run times. Analysis time, shorter than 8 min was found using a background electrolyte solution consisting of 40 mM phosphate buffer adjusted to pH 2.2, 30 mM β-cyclodextrin and 5 mM sodium disulfide, hydrodynamic injection, and 15 kV separation voltage. Detection limits were evaluated on the basis of baseline noise and were established 0.31 mg/l for the omeprazole enantiomers. The method was applied to pharmaceutical preparations with recoveries between 84% and 104% of the labeled contents.

Olsson *et al.* [139] developed and validated a nonaqueous capillary electrophoresis method for the enantiomeric determination of omeprazole and 5-hydroxyomeprazole. Heptakis-(2,3-di-O-methyl-6-O-sulfo)-β-cyclodextrin was chosen as the chiral selector in an ammonium acetate buffer acidified with formic acid in methanol. Parameters such as cyclodextrin concentration, concentration of buffer electrolyte, voltage, and temperature were studied to optimize both the enantioresolution and migration times. An experimental design was utilized for method optimization, using software Modde 5. Validation of the method showed good linearity, which was tested over a concentration range of 2.5–500 μM. The regression coefficients for S-omeprazole, S-5-hydroxyomeprazole, R-omeprazole, and R-5-hydroxyomeprazole were between 0.996 and 0.997. The limits of detection for the four enantiomers were in the range from 45 to 51 μM and the limits of quantification were between 149 and 170 M with UV detection at 301 nm.

Perez-Ruiz *et al.* [140] developed a sensitive method for the determination of omeprazole and its metabolites, hydroxylomeprazole, and omeprazole sulfone using automated solid-phase extraction and micellar electrokinetic capillary chromatography. The method involves an automated solid-phase extraction procedure and capillary electrophoresis with UV detection. Omeprazole, hydroxyomeprazole, and omeprazole sulfone could be separated by micellar electrokinetic capillary chromatography using a background electrolyte composed of 20 mM borate buffer and 30 mM sodium dodecyl sulfate, pH 9.5. The isolation of omeprazole and its metabolites from plasma was automatically accomplished with an original solid-phase extraction procedure using surface-modified styrene–divinyl benzene polymer cartridges. Good recovery data and satisfactory precision values were obtained. Responses were linear for the three analytes, from 0.08 to 2 μg/ml of plasma. Intra- and interday precision values of about 1.6% RSD ($n = 10$) and 2.5% RSD ($n = 36$), respectively, were obtained. The method is highly robust and no breakdown of the current or capillary blockages was observed during several weeks of operation. The method was applied to the determination of omeprazole in pharmaceutical preparations and for the analysis of plasma samples obtained from three volunteers who received oral doses of omeprazole.

Olsson and Blomberg [141] enantioseparated omeprazole and its metabolite 5-hydroxyomeprazole using open tubular capillary electrochromatography with immobilized avidin as chiral selector. The separation was performed with open tubular capillary electrochromatography. The protein avidin was used as the chiral selector. Avidin was immobilized by a Schiffs base type of reaction where the protein was via glutraldehyde covalently bonded to the amino-modified wall of a fused-silica capillary, 50 μm i.d. Both racemates were baseline resolved. Resolution was 1.9 and 2.3, respectively, using ammonium acetate buffer, pH 8.5, 5% methanol, with UV detection. These values of resolution using open tubular capillary electrochromatography are higher than earlier published results regarding chiral separation of omeprazole and 5-hydroxyomeprazole on packed capillary electrochromatography. The number of theoretical plates also indicated good separation efficiency.

5. PHARMACOKINETICS AND METABOLISM

5.1. Pharmacokinetics

Regardh *et al.* [142] studied the pharmacokinetics of omeprazole in mouse, rat, dog, and man. The drug is rapidly absorbed in all species. The systemic availability is relatively high in dog and in man provided the drug is protected from acidic degradation in stomach. In man the fraction of the oral dose reaching the systemic circulation was found to increase from an average of 40.3–58.2 when the dose was raised from 10 to 40 mg, suggesting some dose-dependency in this parameter. Omeprazole distributes rapidly to extravascular sites. Omeprazole is bound to about 95% to proteins in human plasma. The drug is eliminated almost completely by metabolism and no unchanged drug has been recovered in the urine in the species studied. Two metabolites, the sulfone and sulfide of omeprazole, have been identified and quantified in human plasma.

Regardh *et al.* [143] found that about 54% of an oral dose of omeprazole, administered by young healthy subjects, is available to the systemic circulation. The distribution of the drug after an intravenous dose was consistent with localization of a major fraction of the drug in the extracellular water, with about 25% restricted to the blood. Omeprazole was rapidly cleared and possessed the characteristics of a high clearance drug; insignificant amounts of ^{14}C-omeprazole were excreted by the kidneys, though metabolites were excreted very rapidly. Six different metabolites were reported and the major one being hydroxyomeprazole.

Naesdal *et al.* [144] studied the pharmacokinetics of ^{14}C-omeprazole and its metabolites after single intravenous and oral doses of 20–40 mg, respectively, to 12 patients with chronic renal insufficiency. Blood samples for determination of total radioactivity, omeprazole, hydroxyomeprazole, sulfone, and sulfide were taken for 24 h. Urine was collected over 96 h for determination of total radioactivity and during the first 24 h for additional assay of omeprazole and metabolites. The mean systemic availability was 70% and the mean plasma $t_{1/2}$ of omeprazole was 0.6 h.

Cederberg *et al.* [145] reported that omeprazole has to be protected from exposure to the acidic gastric juice when given orally. Following a single oral dose of buffered suspension, omeprazole is rapidly absorbed with peak plasma concentrations within 0.5 h. The volume of distribution is 0.3 l/kg corresponding to the volume of extracellular water. In contrast to the long duration antisecretory action, omeprazole is rapidly eliminated from plasma. The half-life is less than 1 h, and omeprazole is almost entirely cleared from plasma with 3–4 h. Omeprazole is completely metabolized in the liver. The two major plasma metabolites are the sulfone and hydroxyomeprazole, neither of which contributes to the antisecretory activity. About 80% of a given dose is excreted in the urine, and the remainder via the bile.

Regardh *et al.* [146] studied the pharmacokinetics of omeprazole, hydroxyomeprazole, omeprazole sulfone, and other metabolites, in eight young healthy subjects following an acute intravenous and oral dose of 10 and 20 mg of ^{14}C-labeled drug, respectively. The oral dose was given as a buffered solution. Two subjects exhibited essentially higher and more sustained plasma levels of omeprazole than the others. This was due to a higher bioavailability, lower clearance, and longer $t_{1/2}$ of omeprazole in these two subjects. Maximum concentration, 0.7–4.6 μmol/l, was reached between 10 and 25 min after oral dosing. Omeprazole was rapidly distributed to extravascular sites. Low systemic clearance of omeprazole was associated with a decreased formation rate of hydroxyomeprazole and other metabolites. Omeprazole sulfone formation seemed to be less affected. The excretion of hydroxymeprazole during the first 12 h varied between 4.6% and 15.5% of a given dose.

Andersson *et al.* [147] studied the influence of dose on the kinetics of omeprazole and two of its metabolites, hydroxyomeprazole and the sulfone. Ten healthy subjects were given omeprazole 10 and 40 mg intravenously and 10, 40, and 90 mg orally. No significant dose-related difference in parameter calculated from the intravenous experiments was detected. Following the oral solutions, there was a dose-dependent increase in the systemic availability, probably due to saturable first-pass elimination.

Sohn *et al.* [148] examined the kinetic variables of omeprazole and its two primary metabolites in plasma, 5-hydroxyomeprazole and omeprazole sulfone, and the excretion profile of its principal metabolite in urine, 5-hydroxyomeprazole, in eight extensive metabolizers and eight poor metabolizers. Each subject received a postoral dose of 20 mg of omeprazole as an enteric-coated formulation, and blood and urine samples were collected up to 24 h postdose. Omeprazole and its metabolites were measured by HPLC with UV detection. The mean omeprazole area under the concentration–time curve, elimination half-life, and apparent postoral clearance were significantly greater, longer, and lower, respectively, in the poor metabolizers than in the extensive metabolizers. The mean cumulative urinary excretion of 5-hydroxyomeprazole up to 24 h postdose was significantly less in the poor metabolizers than in the extensive metabolizers.

Landahl *et al.* [149] studied the pharmacokinetics of omeprazole and its metabolites in eight healthy elderly volunteers using ^{14}C-omeprazole. In another six healthy elderly volunteers, the pharmacokinetics of omeprazole were studied using unlabeled drug. Each volunteer received single doses of omeprazole intravenously, 20 mg, and orally, 40 mg, as solutions in a randomized crossover design. The plasma concentrations and urinary excretion of omeprazole and metabolites were followed for 24 and 96 h, respectively. The results indicate that the average metabolic capacity of omeprazole is decreased in the elderly compared with that found in earlier studies of healthy young individuals. This was reflected in an increase in bioavailability from 56% to 76%, a reduction in mean systemic clearance by approximately 50% (0.25 l/min) and a prolongation of the mean elimination half-life from 0.7 to 1 h, compared with the young.

Andersson *et al.* [150] studied the pharmacokinetics of omeprazole and its metabolites following single doses, in eight patients with liver cirrhosis. Each patient participated in two experiments in which ^{14}C-omeprazole was administered either intravenously, 20 mg, or in an oral solution, 40 mg, in a randomized crossover design. Plasma concentrations of omeprazole and two of its identified metabolites, as well as total radioactivity were followed for 24 h. Urinary excretion was followed for 96 h. The mean elimination half-life of omeprazole in the patients with cirrhosis was 2.8 h and the mean total plasma clearance was 67 ml/min (4.02 l/h); corresponding values from separate studies in young healthy volunteers were 0.7 h and 594 ml/min (35.64 l/h). Almost 80% of a given dose was excreted as urinary metabolites in both patients and young volunteers.

Okada *et al.* [151] determined omeprazole and its metabolites in human plasma as a probe for CYP_2C_{19} phenotype. The drug is metabolized in the liver to varying degree by several cytochrome P450 (CYP) isoenzymes which are further categorized into subfamilies of related polymorphic gene products. The metabolism of omeprazole is dependent on CYP_3A_4 and CYP_2C_{19}. Omeprazole is metabolized to two major metabolites, 5-hydroxyomeprazole (CYP_2C_{19}) and omeprazole sulfone (CYP_3A_4). Minor mutations in CYP_2C_{19} affect its activity in the liver and the metabolic and the pharmacokinetic profiles of omeprazole. The frequency of CYP_2C_{19} poor metabolizers in population of Asian descent has been reported to range from 10% to 20%. This study demonstrates determination of omeprazole in human plasma as a probe drug of CYP_2C_{19} phenotyping. The method allows the quantitation of omeprazole and its metabolite in human plasma after the administration of therapeutic dose of the drug. The analytical column used for LC/MS was YMC-Pack Pro C_{18} (5 cm $\times$ 2 mm) and operated at 25 °C. The mobile phase was acetonitrile–ammonium acetate at a flow-rate of 0.2 ml/min. The drift voltage was 30 V. The sampling aperture was heated at 110 °C and shield temperature was 230 °C. The column used for chiral separation by HPLC was Chiralpak AD-RH column (15 cm × 4.6 mm) using phosphate buffer/acetonitrile as the eluent and operated at 40 °C. The flow-rate was 0.5 ml/min and detection was 302 nm.

Abelo *et al.* [152] studied the stereoselective metabolism of omeprazole by human cytochrome P450 enzymes. This study demonstrates the stereo-selective metabolism of the optical isomers of omeprazole in himan liver microsomes. The intrinsic clearance of the formation of the hydroxyl metabolite from *S*-omeprazole was 10-fold lower than that from *R*-omep-razole. However, the intrinsic clearance value for the sulfone and 5-*O*-desmethyl metabolites from *S*-omeprazole was higher than that from *R*-omeprazole. The sum of the intrinsic clearance of the formation of all three metabolites was 14.6 and 42.5 μl/min/mg protein for *S*- and *R*-omeprazole, respectively. This indicates that *S*-omeprazole is cleared more slowly than *R*-omeprazole *in vivo*. The stereoselective metabolism of the optical isomers is mediated primarily by cytochrome P450 (CYP)2C19, as indicated by studies using cDNA-expressed enzymes. This is the result of a considerable higher intrinsic clearance of the 5-hydroxy metabolite formation for *R*- and *S*-omeprazole. For *S*-omeprazole, CYP_2C_{19} is more important for 5-*O*-desmethyl formation than for 5-hydroxylation. Predictions of the intrinsic clearance using data from cDNA-expressed enzymes suggest that CYP_2C_{19} is responsible for 40% and 87% of the total intrinsic clearance of *S*- and *R*-omeprazole, respectively, in human liver micro-some. According to experiments using cDNA-expressed enzymes, the sulfoxidation of both optical isomers is metabolized by a single isoform, CYP_3A_4. The intrinsic clearance of the sulfone formation by CYP_3A_4 is 10-fold higher for *S*-omeprazole than for *R*-omeprazole, which may contribute to their stereoselective disposition. The results of this study show that both CYP_2C_{19} and CYP_3A_4 exhibit a stereoselective metabolism of omeprazole. CYP_2C_{19} favors 5-hydroxylation of the pyridine group of *R*-omeprazole, whereas the same enzyme mainly 5-*O*-demethylates *S*-omepyrazole in the benzimidazole group. Sulfoxidation mediated by CYP_3A_4 highly favors the *S*-form.

Pique *et al.* [153] examined the pharmacokinetics of omeprazole dur-ing intravenous infusion in patients with varying degrees of liver dys-function. Thirteen patients, five males and eight females with a mean age of 59 years with proved hepatic cirrhosis, classified according to Child-Pugh criteria as A ($n = 5$), B ($n = 4$), or C ($n = 4$). Each patient received an 80 mg bolus of omeprazole over 30 min followed by a continuous infusion of 8 mg/h for 47.5 h. Blood sample was taken frequently throughout the infusion and during the subsequent 24-h washout period for determination of omeprazole and its metabolites. Data were evaluable for 12 patients. For omeprazole, the mean total area under the plasma concentration–time curve was 286.5 μmol h/l. Peak plasma concentration was 14.9 μmol/l and terminal elimination half-life was 4.1 h; these values were higher than those observed historically in control patient populations. Concentrations of the metabolite omeprazole sulfone were also increased, but there was a decrease in concentrations of hydroxyomeprazole. Exposure to omeprazole following intravenous administration was higher in patients with liver dysfunction than in normal population. However, in patients with severely impaired liver function, the omeprazole plasma concentration did not change by more than 100% and the drug was well tolerated.

Kita et al. [154] have undertaken a study to help predict the optimal dosage of omeprazole for extensive metabolizers in the anti-H. pylori therapy. Seven healthy Japanese subjects, classified based on the CYP_2C_{19} genotype into extensive metabolizers ($n = 4$) and poor metabolizers ($n = 3$), participated in this study. Each subject received a single oral dose of omeprazole 20, 40, and 80 mg, with at least a 1-week washout period between each dose. Plasma concentrations of omeprazole and its two metabolites were monitored for 12 h after each dose of medication. After each dose was administered, the pharmacokinetic profiles of omeprazole and its two metabolites were significantly different between extensive metabolizers and poor metabolizers. The area under the plasma concentration–time curve of omeprazole in extensive metabolziers was disproportionally increased 3.2- or 19.2-fold with dose escalation from 20 to 40 to 80 mg omeprazole, respectively. In contrast, the area under the plasma concentration–time curve of omeprazole was proportionally increased with the higher dose in poor metabolizers. The area under the plasma concentration–time curve of omeprazole after 20 mg administration to poor metabolizers was almost equal to the area under the plasma concentration–time curve in extensive metabolizers after 80 mg administration. The recommended dose of omeprazole for extensive metabolizers is a maximum of 80 mg $\times$ 2/day based on pharmacokinetic considerations.

Omeprazole is a racemate, from which the R- and S-isomers are isolated as reported by Kendall [155]. Both of these isomers convert to the same inhibitor of the H^+/K^+-ATPase and produce the same reduction in the gastric acid secretion. The S-isomer, esomeprazole, is metabolized more slowly and reproducibly than the R-isomer of omeprazole and therefore produces higher plasma concentrations for longer and, as a result, inhibits gastric acid production more effectively and for longer. Esomeprazole has the pharmacological properties of a more effective form of treatment for disorders related to gastric acid secretion.

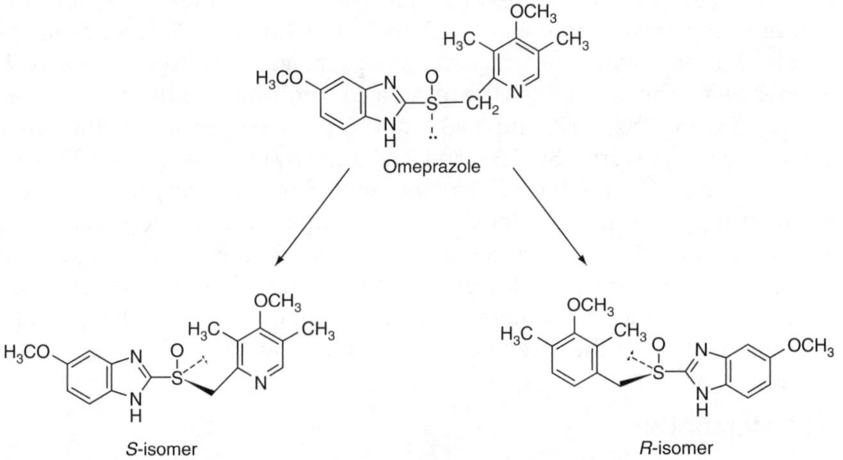

Omeprazole

S-isomer   R-isomer

Kumar *et al.* [156] carried out a study aiming to determine the pharmacokinetics of omeprazole in different degrees of liver cirrhosis and in patients with extrahepatic portal vein obstruction (EHPVO), compared with healthy volunteers. Ten healthy volunteers, 30 patients with cirrhosis of the liver, divided into three groups of 10 depending on severity (according to Child-Pugh classification A, B, and C) and 10 patients with EHPVO participated in this study. The subjects received an omeprazole 20 mg capsule after an overnight fast. Blood samples were collected at 0, 0.5, 1, 1.5, 2, 2.5, 3, 6, 9, and 24 h after drug administration. Omeprazole level in plasma was estimated by reversed-phase HPLC. The elimination half-life was significantly increased to 2.38 ± 0.16, 3.26 ± 0.12, 3.58 ± 0.31, and 2.59 ± 0.22 h in patients with different grades of cirrhosis A, B, and C and also in patients with EHPVO, respectively, compared with 1.054 ± 0.1 h in healthy volunteers. It was concluded that the metabolism of omeprazole was significantly impaired in both liver cirrhosis and EHPVO in comparison with healthy volunteers.

Hassan-Alin *et al.* [157] investigated the pharmacokinetics of S-omeprazole, R-omeprazole, and racemic omeprazole following single and repeated oral doses of 20 and 40 mg of each compound in healthy male and female subjects. Twelve subjects received 20 mg and another 12 subjects received 40 mg of S-omeprazole, R-omeprazole, and racemic omeprazole as oral solutions once daily for 5 days, separated by washout periods of at least 10 days. Blood samples were taken for analysis predose and at selected time points during a 12-h period following drug administration on study day 1 and day 5. Pharmacokinetic parameters of S-omeprazole, R-omeprazole, and racemic omeprazole and the two main metabolites, 5-hydroxyomeprazole and omeprazole sulfone, were calculated using noncompartmental analysis. Following the 20-mg dose of each compound, values of the total area under the plasma concentration–time curve were 1.52, 0.62, and 1.04 μmol h/l for S-omeprazole, R-omeprazole, and racemic omeprazole, respectively, on day 1. Respectively, the area under the plasma concentration–time curve values on day 5 were 2.84, 0.68, and 1.63 μmol h/l. Corresponding values after the 40-mg doses were 3.88, 1.39, and 2.44 μmol h/l on day 1 and 9.32, 1.80, and 5.79 μmol h/l and day 5. Treatment with S-omeprazole, 20 and 40 mg, resulted in higher area under the concentration–time curve values than either R-omeprazole or racemic omeprazole after both single and repeated doses due to a lower metabolic rate of S-omeprazole than R-omeprazole and consequently, racemic omeprazole. S-Omeprazole, R-omeprazole, and the racemate were well tolerated.

5.2. Metabolism

Helander *et al.* [158] reported that radioactive omeprazole was given intravenously or orally to mice, and the distribution of the drug was investigated at various intervals by scintillation counting and by autoradiography. The half-life for radioactivity in the stomach was 14 h versus 30–36 h in the liver, kidneys, and blood. At 16 h after the drug was given, the radioactivity in the stomach was 10 times higher than that in the liver and kidneys, and 100 times that in the blood. Whole-body autoradiography showed sustained high levels of radioactivity only in the gastric mucosa. Light microscopic autoradiographic investigations of gastric mucosa from mice killed 1 or 16 h after the drug was given, revealed radioactivity in the parietal cells. By electron microscopy of gastric mucosa from the mouse killed 16 h after omeprazole injection, the isotope label was found mainly over the secretory surface and the tubulo-vesicles. At these locations H^+/K^+-ATPase has previously been demonstrated, and it is suggested that omeprazole or its metabolites binds to this enzyme.

Hoffmann *et al.* [159] studied the metabolic disposition of ^{14}C omeprazole in dogs, rats, and mice after the administration of pharmacologically active, single oral doses of the drug in buffer solutions (pH 9). Averages of 38% (dogs), 43% (rats), and 55% (mice) of the radiolabeled doses were excreted in the urine in 72 h. Most of the remaining dose was recovered in the feces. Omeprazole was extensively metabolized in all species studied and the metabolites were eliminated rapidly. No unchanged drug could be detected in the urine samples (less than 0.1% of dose). In each species at least 10 metabolites were detected in urine (pH 9) by gradient elution reversed-phase HPLC. Based on liquid chromatographic retention data, the metabolic patterns were very complex and exhibited some quantitative differences between species. Bile was collected from rats and from chronic bile-fistulated dogs. Biliary excretion was a major route of elimination of omeprazole metabolites.

Hoffmann [160] identified the main urinary metabolites of omeprazole after an oral dose to rats and dogs. The structures of seven urinary metabolites of omeprazole following high oral doses to rats and dogs were determined by combining different analytical and spectroscopic techniques including derivatization and stable isotopes. Omeprazole was metabolized by aromatic hydroxylation at position 6 in the benzimidazole ring followed by glucuronidation. There was also oxidative O-dealkylation of both methoxy groups, and aliphatic hydroxylation of a pyridine methyl group followed by oxidation to the corresponding carboxylic acid. Due to the experimental design, implying no pH control of collected samples, all metabolites were isolated as sulfides.

Renberg *et al.* [161] identified two main urinary metabolites of ^{14}C omeprazole in humans. The excretion and metabolism of ^{14}C omeprazole given orally as a suspension was studied in 10 healthy male subjects. An average of 79% of the dose was recovered in the urine in 96 h, with most of the radioactivity (76% of dose) being eliminated in the first 24 h. Pooled urine (0–2 h) from five subjects, containing about 47% of the dose, was analyzed by reversed-phase gradient elution liquid chromatography with radioisotope detection. Omeprazole was completely metabolized to at least six metabolites. The two major metabolites were extensively purified by liquid chromatography and their structures were determined by mass spectrometry with derivatization and use of stable isotopes, ^{1}H NMR, and comparison with synthetic references. They were formed by hydroxylation of a methyl group in the pyridine ring, followed by further oxidation of the alcohol to the corresponding carboxylic acid. Both metabolites retained the sulfoxide group of omeprazole, rendering them as unstable as the parent compound at pH less than 7. They accounted for approximately 28% hydroxyomeprazole, and 23% (omeprazole acid) of the amount excreted in the 0–2-h collection interval. Based on *in vitro* studies, with the synthetic metabolites in isolated gastric glands, it is unlikely that metabolite 1 and metabolite 2 will contribute to the pharmacological effect of omeprazole in human.

Weidolf *et al.* [162] established the metabolic route of omeprazole involving glutathione through identification of end products excreted in the urine of rats after oral administration of 400 μmol/kg of a mixture of ^{3}H- and ^{14}C-omeprazole. The labeled positions enabled facile tracing of metabolites that were formed through fission of omeprazole, producing ^{3}H-pyridine and ^{14}C-benzimidazole metabolites. The structures of the metabolites were established by HPLC thromospray mass spectrometry and MS/MS. Two of the metabolites were isolated and characterized by ^{1}H NMR studies. The fact that the N-acetylcysteine derivative of the benzimidazole was one of the end products indicated that the initial reaction involved glutathione. Three metabolites reflecting the fate of the pyridine moiety were identified. Their proposed formation route is *via* initial reduction to the pyridylmethylthiol compound followed by *S*-methylation and *S*-oxidation to the corresponding sulfoxide or sulfone.

Chiba *et al.* [163] studied the oxidative metabolism of omeprazole in 14 human liver microsomes in relation to the 4'-hydroxylation capacity of *S*-mephenytoin. The formation of 5-hydroxyomeprazole and omeprazole sulfone from omeprazole exhibited a biphasic kinetic behavior, indicating that at least two distinct enzymes are involved in either of the metabolic pathways of omeprazole. These findings suggest that *S*-mephenytoin 4'-hydroxylase is an enzyme primarily responsible for the 5-hydroxylation of omeprazole and further metabolism of omeprazole sulfone, but not for the sulfoxidation of omeprazole in human liver microsomes.

Zhao and Lou [164] studied the metabolism of omeprazole to its two major metabolites, hydroxyomeprazole and omeprazole sulfone, in rat liver microsomes by a reversed-phase HPLC assay. The formation of metabolites of omeprazole depended on incubation time, substrate concentration, microsomal protein concentration, and was found to be optimal at pH 7.4. The $V_{\max}$ and K_m of omeprazole hydroxylation in the rat liver microsomal preparation were 2033 nmol/(min mg protein), and 46.8 μmol/l, respectively. The effects of seven drugs on omeprazole metabolism were tested. Mephenytoin, five benzodiazepines and pavaverine caused inhibition of omeprazole metabolism.

Meyer [165] studied the metabolic interactions of omeprazole with other drugs by cyctochrome P450 enzymes. Omeprazole was extensively metabolized by several human cytochromes P450, most prominently by mephenytoin hydroxylase (CYP_2C_{19}) and nifedipine hydroxylase (CYP_3A_4). The substrates and inhibitors of CYP_2C_{19} and CYP_3A_4 and the known genetic polymorphism of CYP_2C_{19} explain the interactions of omeprazole with carbamazepine, diazepam, phenytoin, and theophylline or caffeine.

Pearce and Lushnikova [78] incubated *C. elegans* ATCC 9245 and omeprazole and isolated putative fungal metabolites in sufficient quantities for structural elucidation. The metabolites were isolated by using semipreparative HPLC and the structures were identified by a combination of LC/MS and NMR experiments. Metabolites are shown in Scheme 4.1.

Tyrbing *et al.* [166] studied the stereoselective disposition of omeprazole and its formed 5-hydroxy metabolite in five poor metabolizers, and five extensive metabolizers of 5-mephenytoin. After a single oral dose of omeprazole (20 mg), the plasma concentrations of the separated enantiomers of the parent drug and the 5-hydroxy metabolite were determined for 10 h after drug intake. In poor metabolizers, the area under the plasma concentration versus time curve [AUC(0–8)] of (+) omeprazole was larger

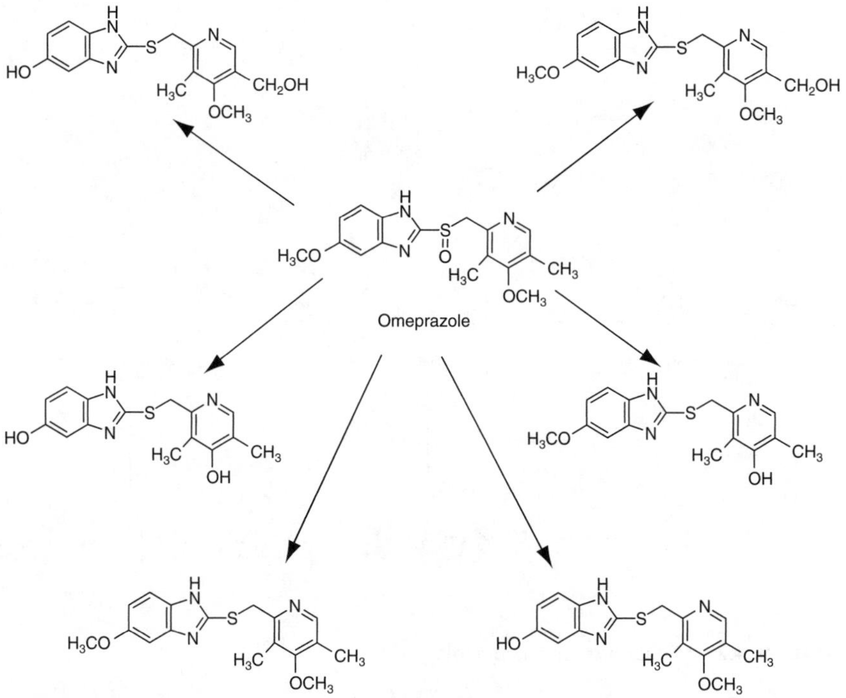

SCHEME 4.1 Biotransformation pathway of omeprazole induced by *C. elegans* ATCC 9245 [78].

and that of the 5-hydroxy metabolite of this enantiomer was smaller than the AUC (0–8) values in extensive metabolizers.

Jensen *et al.* [125] investigated an HPLC/ICP-MS with sulfur-specific detection, as a method for obtaining metabolite profiles for omeprazole administered as 1:1 mixture of ^{32}S- and ^{34}S-labeled material. Analysis based on the monitoring of the chromatographic eluent at either m/z 32 or m/z 34 was not successful due to insufficient sensitivity caused by interferences from polyatomic ions. However, reaction of sulfur with oxygen in the hexapole collision cell, combined with monitoring at m/z 48 (for ^{32}S) or m/z 50 (for ^{34}S), provided a facile method for metabolite profiling. Detection at m/z 48 was superior in sensitivity to detection of m/z 50.

Tolonen *et al.* [167] described a simple and efficient method for determination of labile protons in drug metabolites using postcolumn infusion of deuterium oxide in LC/MS experiments with ESI and TOF-MS. The number of exchangeable protons in analytes; hydroxyl, amine, thiol, and carboxylic acid protons can easily be determined by comparing the increase in m/z values after H/D-exchange occurring on line between

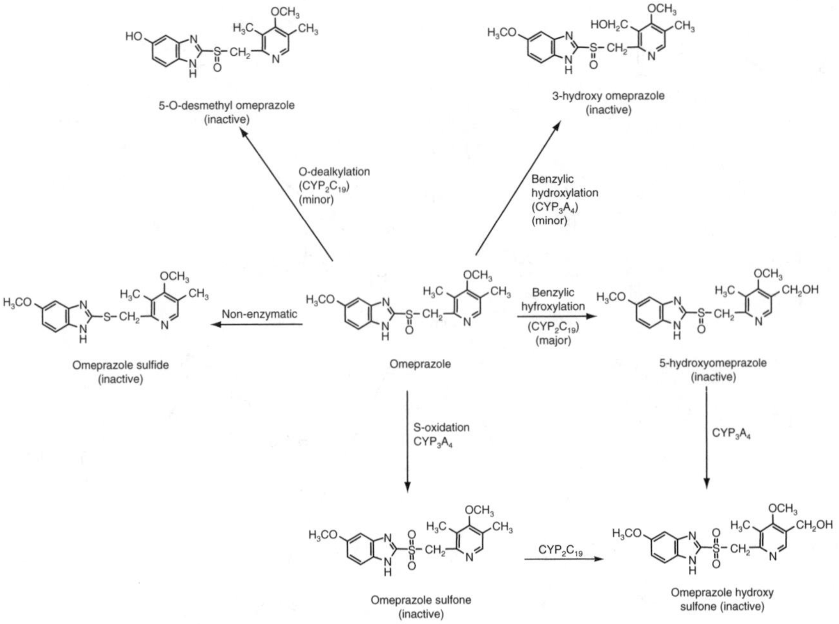

SCHEME 4.2   Metabolism of omeprazole [168].

an HPLC column and electrospray ion source. The hydroxyl metabolites and sulfur/nitrogen oxide with the same accurate mass can be distinguished. A good degree of exchange was obtained in repeatable experiments. Only the low consumption of deuterium oxide is needed in a very easy and rapidly set-up procedure. The method is applied in the study of metabolites of omeprazole in human and mouse *in vitro* samples, together with the exact mass data obtained from TOF-MS experiments.

Roche [168] pointed out that omeprazole and other proton pump inhibitors bind strongly to serum proteins and are extensively metabolized by the CYP450 family of enzymes. The CYP_2C_{19} isoform is important in converting parent structures to inactive metabolites, although CYP_3A_4 also plays a role in the proton pump inhibitors biotransformation. The metabolic degradation pathways for omeprazole are shown in Scheme 4.2.

6. MECHANISM OF ACTION

Lindberg *et al.* [169] proposed a mechanism of action for omeprazole, the inhibitor of the gastric H^+/K^+-ATPase, which is responsible for the gastric acid production and located in the secretory membranes of the parietal cell. Omeprazole itself is not an active inhibitor of this enzyme, but it is transformed within the acid compartments of the parietal cell into the active inhibitor, close to the enzyme. Omeprazole **1** is transformed to the sulfenamide isomer **4**. The reaction is reversible and goes via the spiro intermediate **2** and the sulfenic acid **3**. The spiro intermediate **2** is a dihydrobenzimidazole, with a tendency to undergo aromatization to form the sulfenic acid **3** by a C–S bond cleavage. The reaction of **4** with β-mercaptoethanol forms the disulfide **5**. The adduct **5** reacts with molecule of β-mercaptoethanol in a base-catalyzed reaction to form the sulfide **8**, probably via the unstable mercaptan **7** resulting from the S–S bond cleavage during the simultaneous formation of the disulfide of β-mercaptoethanol. The mechanism of action is illustrated in Scheme 4.3.

Puscas *et al.* [170] suggested that omeprazole has a dual mechanism of action: H+/K+-ATPase inhibition and gastric mucosa carbonic anhydrase enzyme inhibition and that these enzymes may be functionally coupled.

Roche [168] presented Scheme 4.4 to illustrate the activation and the reaction pathway of omeprazole.

7. STABILITY

Quercia *et al.* [171] studied the stability of omeprazole 2 mg/ml in an extemporaneously prepared oral liquid. The contents of five 20-mg omeprazole capsules were mixed with 50 ml of 8.4% sodium bicarbonate solution in a Luer-Lok syringe. Three vials of this liquid were prepared for storage at 24, 5, and − 20 °C. A 3-ml sample of each was taken initially and on days 1, 2, 3, 4, 6, 8, 10, 12, 14, 18, 22, 26, and 30 and assayed by HPLC. The liquids stored at 5 and − 20 °C did not change color during the study period, but the color of the liquid stored at 24 °C changed from white to brown.

DiGiacinto *et al.* [172] determined the stability of omeprazole suspensions at ambient and refrigerated temperatures using HPLC. The contents of omeprazole capsules were suspended in separate flasks containing sodium bicarbonate 8.4% to concentrations of 3 and 2 mg/ml, respectively. The contents of each flask were drawn into six amber-colored oral syringes, with one-half of the syringes stored at 22 °C and the other half at 4 °C. Omeprazole concentrations were determined by a stability-indicating HPLC assay at baseline and at 4, 8, 12, and 24 h, and on days 4, 7, 14, 21, 30, 45, and 60 after mixing. Omeprazole was considered stable if it retained 90% of the baseline drug concentration. Omeprazole was stable for up to 14 days at 22 °C and 45 days at 4 °C.

Palummo *et al.* [173] comparatively evaluated the stability of capsules containing 20 mg of omeprazole, in enteric-coated pellets, from seven pharmaceutical laboratories on Argentine market. The stability test was

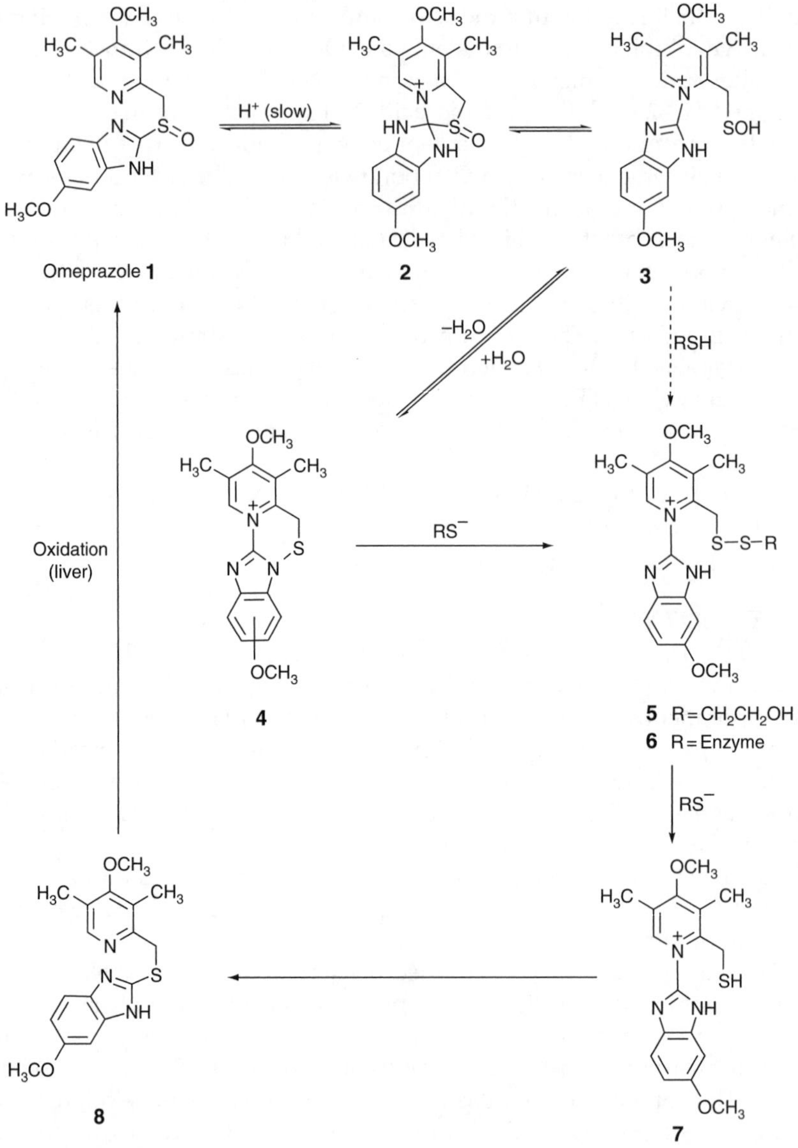

SCHEME 4.3 Mechanism of action of omeprazole [169].

performed under the conditions indicated by the international conference of harmonization: 40 °C, 75% HR, with and without light, during a 6-month period. The remaining content of omeprazole, total percentage of impurities and percentage of released active principle *in vitro*, were determined by HPLC. The organoleptic characteristics of the pellets were

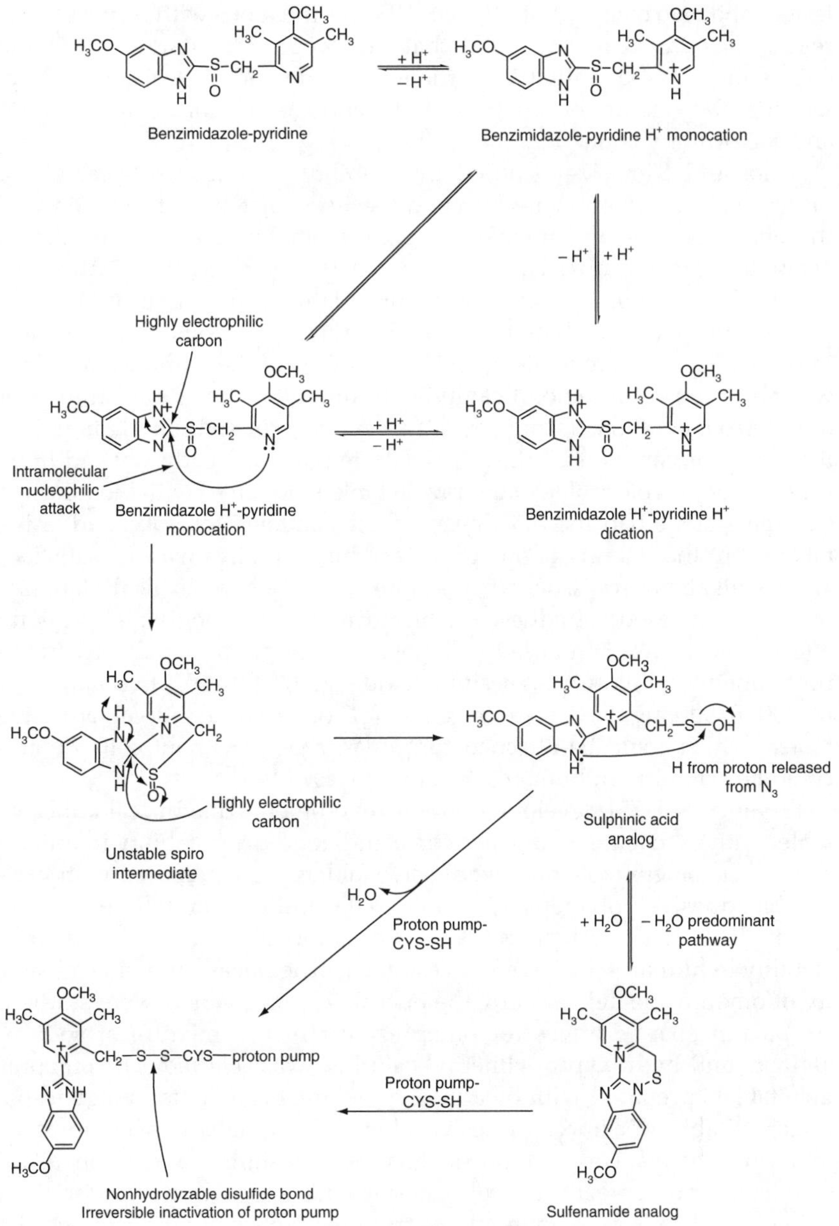

SCHEME 4.4 Omeprazole activation and reaction pathway [168].

visually examined. The results obtained at 6 months indicate that, from the seven products studied, four were found to have a content of omeprazole higher than 90% of the labeled amount, in both lighting conditions tested, and also comply with the USP 23 specifications with respect to the release *in vitro*. It is concluded that the progressive darkening of the pellets indicates, quantitatively, the level of degradation of the product and that the stability of omeprazole depends on the correct formulation and the primary container.

Choi and Kim [174] studied the stability of omeprazole tablets in human saliva. Omeprazole buccal adhesive tablet was developed and the absorption of omeprazole solutions from human oral cavity was evaluated and the physicochemical properties such as the bioadhesive forces of various omeprazole tablet formulations composed of bioadhesive polymers and alkali materials, and the stability of omeprazole tablets in human saliva were investigated. About 23% of the administered dose was absorbed from the oral cavity at 15 min after the administration of omeprazole solutions (1 mg/15 ml). A mixture of sodium alginate and hydroxypropylmethylcellulose was selected as the bioadhesive additive for the omeprazole tablet. Omeprazole tablets prepared with bioadhesive polymers alone had the bioadhesive forces suitable for buccal adhesive tablets, but the stability of omeprazole in human saliva was not satisfied. Among alkali materials, only magnesium oxide could be an alkali stabilizer for omeprazole buccal adhesive tablets due to its strong water-proofing effect. Two tablets composed of omeprazole/sodium alginate/hydroxypropylmethylcellulose/magnesium oxide (20/24/6/50, mg/tab) and (20/30/0/50, mg/tab) were suitable for omeprazole buccal adhesive tablets which could be attached to human cheeks without collapse and could be stabilized in human saliva for at least 4 h.

Yong *et al.* [175] developed an effective omeprazole buccal adhesive tablet with excellent bioadhesive force and good drug stability in human saliva. The omeprazole buccal adhesive tablets were prepared with various bioadhesive polymers, alkali materials, and croscarmellose sodium. Their physicochemical properties, such as bioadhesive force and drug stability in human saliva, were investigated. The release and bioavailability of omeprazole delivered by the buccal adhesive tablets were studied. As bioadhesive additives for omeprazole tablet, a mixture of sodium alginate and hydroxypropylmethyl cellulose was selected. The omeprazole tablets prepared with bioadhesive polymers alone had bioadhesive forces suitable for a buccal adhesive tablet, but the stability of omeprazole in human saliva was not satisfactory. Magnesium oxide is an alkali stabilizer for omeprazole buccal adhesive tablets. Croscarmellose sodium enhanced the release of omeprazole from the tablets but it decreased the bioadhesive forces and stability of omeprazole tablets in human saliva.

Leitner and Zollner [176] compared pH stabilities of the intravenous formulations of omeprazole. The solutions prepared according to the official instructions were exposed to four different light conditions. Both manufacturers state reference solutions for acceptable discoloration.

Those were prepared according to the European Pharmacopeia. Discoloration of solutions was evaluated as criterion for the stability of the proton pump inhibitors prodrugs. Optical alterations under different light conditions were compared and documented photographically with standardized illumination. Intensity and spectral composition of light as well as temperature had only minor influence on discoloration as measure of degradation of the prodrugs. Both formulations for injection fulfill the specifications for pH stability within the stated time frames mentioned in the summaries of product characteristics. After 1 h omeprazole injectable formulation showed a substantial optical discoloration. After 6 h these changes were more intense than the reference solution β-1,3-glucanase 5 (BG5) allows.

Moschwitzer et al. [177] developed an intravenously injectable chemically stable aqueous omeprazole formulation using nanosuspension technology. The feasibility of omeprazole stabilization using the DissoCubes technology and the optimal production parameters for a stable, highly concentrated omeprazole nanosuspension were studied. The HPLC analysis has proved the predominance of the nanosuspension produced by high-pressure homogenization in comparison to an aqueous solution. After 1 month of production, no discoloration or drug loss was recognizable when the nanosuspension was produced at 0 °C. It can be stated that the production of nanosuspensions by high-pressure homogenization is suitable for preventing degradation of labile drugs.

Riedel and Leopold [178] developed two reversed-phase HPLC methods to investigate the degradation of omeprazole in organic polymer solutions and aqueous dispersions of enteric-coating polymers (Eudragit L-100, S-100, CAP, HP-55, HPMCAS-HF, -LF, and shellac). The overall goal of the study was to determine the influence of the polymer structure on the degradation of omeprazole, whether the acid structure of the enteric-coating polymers caused an instability of the drug. It was investigated whether a difference in omeprazole degradation could be detected between organic polymer solutions and aqueous dispersions. pK_a values of the polymers and pH values of the aqueous dispersions were determined to see whether there was a correlation with the extent of degradation of omeprazole induced by enteric polymers. As the polymers containing phthalate moieties are very susceptible to hydrolysis, the influence of free phthalic acid on omeprazole stability was investigated. The degradation kinetics of omeprazole in organic polymer solutions were determined. Omeprazole degradation is more pronounced in aqueous polymer dispersions than in organic polymer solutions. The influence of organic polymer solutions on the stability of omeprazole depends on the amount of acidic groups in the polymeric structure.

Burnett and Balkin [179] investigated the stability and viscosity of preparations of a commercially available, flavored, immediate-release powder for oral suspension (omeprazole–sodium bicarbonate) during refrigerator and room temperature storage. Omeprazole–sodium bicarbonate 20-mg packets were suspended to initial omeprazole concentrations of 0.6 and 2 mg/ml, and omeprazole–sodium bicarbonate 40-mg packets were suspended to initial omeprazole concentrations of 1.2, 2, 3, and 4 mg/ml. Suspensions were stored at 4 °C in darkness (refrigerated) or 22–25 °C (room temperature) in light for 1 week. A third set of suspensions was stored refrigerated for 1 month. Omeprazole's stability was quantified after 0, 6, 12, 24, 48, and 168 h in 1-week samples and after 0, 7, 14, 21, and 28 days in 1-month samples using high-pressure liquid chromatography. Viscosities of refrigerated suspensions were measured after 0, 1, and 7 days. Refrigerated suspension retained 98% and 96% of their initial omeprazole concentrations after 1 week and 1 month, respectively. Stability at room temperature suspensions was concentration dependent. After 1 week, the 0.6- and 1.2-mg/ml suspensions retained 87.2% and 93.1% of their respective initial omeprazole concentrations, whereas the 2-, 3-, and 4-mg/ml suspensions retained 97% of their initial omeprazole concentrations.

8. REVIEW

Bosch *et al.* [180] reviewed the analytical methodologies that have been used for the determination of omeprazole. The drug has been determined in formulation and in biological fluids by a variety of methods such as spectrophotometry, HPLC with UV detection, and liquid chromatography coupled with tandem mass spectrometry. The overview includes most relevant analytical methodologies used for the determination of the drug since the origin.

ACKNOWLEDGMENT

The author thanks Mr. Tanvir A. Butt, Pharmaceutical Chemistry Department, College of Pharmacy, King Saud University, for his secretarial assistance in preparing this manuscript.

REFERENCES

[1] S. Sweetman (Ed.), Martindale, The Complete Drug Reference, Pharmaceutical Press, Electronic version 2007.
[2] M.J. O'Neil (Ed.), The Merck Index, 14th ed., Merck & Co., Inc., NJ, USA, 2006, p. 1179.
[3] A.C. Moffat, M.D. Osselton, B. Widdop (Eds.), Clarke's Analysis of Drugs and Poisons, third ed., Vol. 2, Pharmaceutical Press, London, 2004, pp. 1366–1368.
[4] J.H. Block, J.M. Beale Jr. (Eds.), Wilson and Gisvold's Textbook of Organic Medicinal and Pharmaceutical Chemistry, 11th ed., Lippincott Williams & Wilkins, 2004, p. 954.

[5] P. Richardson, C.J. Hawkey, W.A. Stack, Drugs 56 (1998) 307–335.
[6] H.D. Langtry, M.I. Wilde, Drugs 56 (1998) 447–486.
[7] R.R. Berardi, L.S. Welage, Am. J. Health-Syst. Pharm. 55 (1998) 2289–2298.
[8] G.J.E. Brown, N.D. Yeomans, Drug Safety 21 (1999) 503–512.
[9] B.L. Erstad, Ann. Pharmacother. 35 (2001) 730–740.
[10] M. Robinson, J. Horn, Drugs 63 (2003) 2739–2754.
[11] R. Dekel, C. Morse, R. Fass, Drugs 64 (2004) 277–295.
[12] A.E. Brandstrom, B.R. Lamm, U.S. Patent 4,620,008. (Aktiebolaget Hassle), October 28, 1986.
[13] C. Slemon, B. Macel 1995, US Patent 5,470,983. (Torcan Chemical Ltd., Canada) November 28, 1995.
[14] H. Cotton, T. Elebring, M. Larsson, L. Li, H. Sorensen, S. von Unge, Tetrahedron: Asymmetry 11 (2000) 3819–3825.
[15] Aktiebolag Haessle. Jpn, Kokai Tokkyo Koho JP 62-72666 (Refs. 24 and 25. In: Rao *et al.*; Ref. 20).
[16] J.E. Baldwin, R.M. Adlington, N.P. Crouch, US Patent 6043371, 2000..
[17] S.P. Singh, S.M.J. Mukarram, D.G. Kulkami, M. Purohit, U.S. Patent 6,245,913 B1 (Wockhardt Europe Limited), June 12, 2001.
[18] X.L. Liu, Shanxi Med. Univ. J. 33 (2002) 330–332.
[19] http://www.stn-international.de/archive/presentations/online_information03/psspa.pdf.
[20] G.W. Rao, W.X. Hu, Z.Y. Yang, Chin. J. Synth. Chem. 10 (2002) 297–301.
[21] H. Ohishi, Y. In, T. Ishida, M. Inoue, F. Sato, M. Okitsu, Acta Cryst. C45 (1989) 1921–1923.
[22] R.M. Claramunt, C. Lopes, I. Alkorta, J. Elguero, R. Yang, S. Schulman, Magn. Reson. Chem. 42 (2004) 712–714.
[23] R.M. Claramunt, C. Lopes, J. Elguero, Arkivoc V (2006) 5–11.
[24] British Pharmacopoeia 2005, Vol. 2, The Stationary Office, London, 2005, pp. 1457–1460.
[25] The United States Pharmacopeia, USP 31, Vol. 3, The United States Pharmacopeial Convention, Rockville, MD, 2008, pp. 2850–2853.
[26] S.N. Dhumal, P.M. Dikshit, I.I. Ubharay, B.M. Mascarenhas, C.D. Gaitonde, Indian Drugs 28 (1991) 565–567.
[27] C.S.P. Sastry, P.Y. Naidu, S.S.N. Murthy, Indian J. Pharm. Sci. 59 (1997) 124–127.
[28] N. Ozaltin, A. Kocer, J. Pharm. Biomed. Anal. 16 (1997) 337–342.
[29] M. Tuncel, D. Dogrukol-Ak, Pharmazie 52 (1997) 73–74.
[30] A. Karlsson, S. Hermansson, Chromatographia 44 (1997) 10–18.
[31] N.M. El-Kousy, L.I. Bebawy, J. AOAC Int. 82 (1999) 599–606.
[32] A.A. Wahbi, O. Abdel Razak, A.A. Gazy, H. Mahgoub, M.S. Moneeb, J. Pharm. Biomed. Anal. 30 (2002) 1133–1142.
[33] K. Karljikovic-Rajic, D. Novovic, V. Marinkovic, D. Agbaba, J. Pharm. Biomed. Anal. 32 (2003) 1019–1027.
[34] F. Salama, N. El-Abasawy, S.A.A. Razeq, M.M.F. Ismail, M.M. Fouad, J. Pharm. Biomed. Anal. 33 (2003) 411–421.
[35] A. Riedel, C.S. Leopold, Pharmazie 60 (2005) 126–130.
[36] R. Yang, S.G. Schulman, P.J. Zavala, Anal. Chim. Acta 481 (2003) 155–164.
[37] C.S.P. Sastry, P.Y. Naidu, S.S.N. Murty, Indian Drugs 33 (1996) 607–610.
[38] C.S.P. Sastry, P.Y. Naidu, S.S.N. Murty, Talanta 44 (1997) 1211–1217.
[39] A.H. Zhang, F. Wang, X.J. Chen, L.K. Wu, Yaowu Fenxi-Zazhi 16 (1996) 194–195.
[40] S. Pinzauti, P. Gratteri, S. Furlanetto, P. Mura, E. Dreassi, R. Phan-Tan-Luu, J. Pharm. Biomed. Anal. 14 (1996) 881–889.
[41] A. Radi, J. Pharm. Biomed. Anal. 31 (2003) 1007–1012.
[42] A.M. Qaisi, M.F. Tutunji, L.F. Tutunji, J. Pharm. Sci. 95 (2005) 384391.

[43] Jin-Long Yan, J. Appl. Sci. 6 (2006) 1625–1627.
[44] S. McClean, E. O'Kane, V.N. Ramachandran, W.F. Smyth, Anal. Chim. Acta 292 (1994) 81–89.
[45] J.R. Ames, P. Kovacic, J. Electroanal. Chem. 343 (1992) 443–450.
[46] N. Ozaltin, A. Temizer, Electroanalysis 6 (1994) 799–803.
[47] D. Dogrukol-Ak, M. Tuncel, Pharmazie 50 (1995) 701–702.
[48] H. Knoth, H. Oelschlager, J. Volke, J. Ludvik, Pharmazie 52 (1997) 686–691.
[49] H. Oelschlaeger, H. Knoth, Pharmazie 53 (1998) 242–244.
[50] F. Belal, N. El-Enany, M. Rizk, J. Food Drug Anal. 12 (2004) 102–109.
[51] N. El-Enany, F. Belal, M. Risk, J. Biochem. Biophys. Methods 70 (2008) 889–896.
[52] E.X. Cao, Y.H. Zeng, Fenxi Kexue Xuebao 17 (2001) 110–113.
[53] S. Mangalan, R.B. Patel, B.K. Chakravarthy, J. Planar. Chromatogr. Modern TLC 4 (1991) 492–493.
[54] S. Ray, P. Kumar-De, Indian Drugs 31 (1994) 543–547.
[55] D. Dogrukol-Ak, Z. Tunalier, M. Tuncel, Pharmazie 53 (1998) 272–273.
[56] D. Agbaba, D. Novovic, K. Karljikovic-Rajic, V. Marinkovic, J. Planar. Chromatogr. Modern TLC 17 (2004) 169–172.
[57] B.A. Persson, P.O. Lagerstrom, I. Grundevik, Scand. J. Gastroenterol. Suppl. 108 (1985) 71–77.
[58] M.A. Amantea, P.K. Narang, J. Chromatogr. 426 (1988) 216–222.
[59] S.H. Shim, S.J. Bok, K.I. Kwon, Arch. Pharm. Res. 17 (1994) 458–461.
[60] K. Balmer, B.A. Persson, P.O. Lagerstroem, J. Chromatogr. 660 (1994) 269–273.
[61] W.K. Kang, D.S. Kim, K.I. Kwon, Arch. Pharm. Res. 22 (1999) 86–88.
[62] M. Motevalian, G. Saeedi, F. Keyhanfar, L. Tayebi, M. Mahmoudian, Pharm. Pharmacol. Commun. 5 (1999) 265–268.
[63] G. Garcia-Encina, R. Farran, S. Puig, L. Martinez, J. Pharm. Biomed. Anal. 21 (1999) 371–382.
[64] D. Castro, A.M. Moreno, S. Torrado, J.L. Lastres, J. Pharm. Biomed. Anal. 21 (1999) 291–298.
[65] B.A. Persson, S. Andersson, J. Chromatogr. 906 (2001) 195–203.
[66] Q.B. Cass, A.L.G. Degani, N. Cassiano, J. Liq. Chromatogr. Relat. Technol. 23 (2000) 1029–1038.
[67] G.W. Sluggett, J.D. Stong, J.H. Adams, Z. Zhao, J. Pharm. Biomed. Anal. 25 (2001) 357–361.
[68] M.C. Dubuc, C. Hamel, M.S. Caubet, J.L. Brazier, J. Liq. Chromatogr. Relat. Technol. 24 (2001) 1161–1169.
[69] H.M. Gonzalez, E.M. Romero, T.de J. Chavez, A.A. Peregrina, V. Quezada, C. Hoyo-Vadillo, J. Chromatogr. 780 (2002) 459–465.
[70] F.C. Cheng, Y.F. Ho, L.C. Hung, C.F. Chen, T.H. Tsai, J. Chromatogr. 949 (2002) 35–42.
[71] Q.B. Cass, V.V. Lima, R.V. Oliveira, N.M. Cassiano, A.L. Degani, J. Pedrazzoli, J. Chromatogr. B 798 (2003) 275–281.
[72] A. Schubert, A.L. Werle, C.A. Schmidt, C. Codevilla, L. Bajerski, R. Chiappa, et al., J. AOAC Int. 86 (2003) 501–504.
[73] R.M. Orlando, P.S. Bonato, J. Chromatogr. B 795 (2003) 227–235.
[74] N.L. Rezk, K.C. Brown, A.D. Kashuba, J. Chromatogr. B 844 (2006) 314–321.
[75] M. Shimizu, T. Uno, T. Niioka, N. Yaui-Furukori, T. Takahata, K. Sugawara, J. Chromatogr. B 832 (2006) 241–248.
[76] A. Zarghi, S.M. Foroutan, A. Shafaati, A. Khoddam, Arzneimittelforschung 56 (2006) 382–386.
[77] H. Jia, W. Li, K. Zhao, J. Chromatogr. B. 837 (2006) 112–115.
[78] G.M. Pearce, M.V. Lushnikova, J. Mol. Catal. B: Enzym. 41 (2006) 87–91.

[79] Z.A. El-Sherif, A.O. Mohamed, M.G. El-Bardicy, M.F. El-Tarras, Chem. Pharm. Bull. 54 (2006) 814–818.

[80] R. Linden, A.L. Ziulkoski, M. Wingert, P. Tonello, A.A. Souto, J. Braz. Chem. Soc. 18 (2007) 733–740.

[81] L. Sivasubramanian, V. Anilkumar, Indian J. Pharm. Sci. 69 (2007) 674–676.

[82] F.S. Murakami, A.P. Cruz, R.N. Pereira, B.R. Valente, M.A.S. Silva, J. Liq. Chromatogr. Relat. Technol. 30 (2007) 113–121.

[83] I.J. de Silva Jr., J.P. Sartor, P.C.P. Rosa, V. de Veredas, A.G. Barreto Jr., C.C. Santana, J. Chromatogr. 1162 (2007) 97–102.

[84] K.R.A. Belaz, M. Coimbra, J.C. Barreiro, C.A. Montanari, Q.B. Cass, J. Pharm. Biomed. Anal. 47 (2008) 81–87.

[85] S.A. Matlin, M.E. Tiritan, Q.B. Cass, D.R. Boyd, Chirality 8 (1996) 147–152.

[86] M.E. Tiritan, Q.B. Cass, A. Del Alamo, S.A. Matlin, S.J. Grieb, Chirality 10 (1998) 573–577.

[87] S.A. Matlin, M.E. Tiritan, A.J. Crawford, Q.B. Cass, D.R. Boyd, Chirality 6 (1994) 135–140.

[88] N. Sultana, M. Saeed Arayne, F. Hussain, A. Shakeel, J. Saudi Chem. Soc. 10 (2006) 141–148.

[89] M. Rambla-Alegre, J. Esteve-Romero, S. Garda-Broch, Anal. Chim Acta 633 (2009) 250–256.

[90] G.W. Mihaly, P.J. Prichard, R.A. Smallwood, N.D. Yeomans, W.J. Louis, J. Chromatogr. 278 (1983) 311–319.

[91] P.O. Lagerstrom, B.A. Persson, J. Chromatogr. 309 (1984) 347–356.

[92] B. Persson, S. Wendsjo, J. Chromatogr. 321 (1985) 375–384.

[93] I. Grundevik, G. Jerndal, K. Balmer, B.A. Persson, J. Pharm. Biomed. Anal. 4 (1986) 389–398.

[94] P. Erlandsson, R. Isaksson, P. Lorentzon, P. Lindberg, J. Chromatogr. 532 (1990) 305–319.

[95] T. Arvidsson, E. Collijn, A.M. Tivert, L. Rosen, J. Chromatogr. 586 (1991) 271–276.

[96] Z. He, Yaowu Fenxi Zazhi 11 (1991) 276–278.

[97] K. Kobayashi, K. Chiba, D.R. Sohn, Y. Kato, T. Ishizaki, J. Chromtogr. 579 (1992) 299–305.

[98] O. Gyllenhaal, J. Vessman, J. Chromatogr. 628 (1993) 275–281.

[99] T. Andersson, P.O. Lagerstrom, J.O. Miners, M.E. Veronese, L. Weidolf, D.J. Birkett, J. Chromatogr. 619 (1993) 291–297.

[100] S. Luo, T. Li, Yaowu Fenxi Zazhi 13 (1993) 263–264.

[101] K. Kobayashi, K. Chiba, M. Tani, Y. Kuroiwa, T. Ishizaki, J. Pharm. Biomed. Anal. 12 (1994) 839–844.

[102] A.M. Cairns, R.H.Y. Chiou, J.D. Rogers, J.L. Demertriades, J. Chromatogr. 666 (1995) 323–328.

[103] M. Tanaka, H. Yamazaki, H. Hakushi, Chirality 7 (1995) 612–615.

[104] X.P. Xu, C.Z. Dai, T. Li, Fenxi Ceshi Xuebao 16 (1997) 48–53.

[105] X.Y. Xu, J.H. Lu, M.L. Wang, C. Xu, R.L. Wang, L.Y. He, Yaowu Fenxi Zazhi 17 (1997) 169–171.

[106] J. Macek, P. Ptacek, J. Klima, J. Chromatogr. B. 689 (1997) 239–243.

[107] S. Gangadhar, G.S.R. Kumar, M.N.V.S. Rao, Indian Drugs 34 (1997) 99–101.

[108] P.K.F. Yeung, R. Little, Y. Jiang, S.J. Buckley, P.T. Pollak, H. Kapoor, et al., J. Pharm. Biomed. Anal. 17 (1998) 1393–1398.

[109] P.N.V. Tata, S.L. Bramer, Anal. Lett. 32 (1999) 2285–2295.

[110] L. Ding, J. Yang, H.L. Yan, Z.X. Zhang, D.K. An, Yaowu Fenxi Zazhi 19 (1999) 300–303.

[111] A. Ekpe, T. Jacobsen, Drug Dev. Ind. Pharm. 25 (1999) 1057–1065.

[112] Anonymous, Macherey-Nagel Applicaton Note A-1054, 2000, p. 2.

[113] K.H. Yuen, W.P. Choy, H.Y. Tan, J.W. Wong, S.P. Yap, J. Pharm. Biomed. Anal. 24 (2001) 715–719.

[114] Anonymous, ChromTech. Application Note, 2001, p. 1.

[115] D.S. Yim, J.E. Jeong, J.Y. Park, J. Chromatogr. B. 754 (2001) 487–493.

[116] E.J. Woolf, B.K. Matuszewski, J. Chromatogr. 828 (1998) 229–238.

[117] H. Stenhoff, A. Blomqvist, P.O. Lagerstrom, J. Chromatogr. B. 734 (1999) 191–201.

[118] H. Kanazawa, A. Okada, M. Higaki, H. Yokota, F. Mashige, K. Nakahara, J. Pharm. Biomed. Anal. 30 (2003) 1817–1824.

[119] J. Martens-Lopenhoffer, I. Reiche, U. Troger, K. Monkemuller, P. Malfertheiner, S.M. Bode-Boger, J. Chromatogr. B. 857 (2007) 301–307; Corrigendrum to this article, J. Chromatogr. B., 859(2007) 289.

[120] J. Macek, J. Klima, P. Ptacek, J. Chromatogr. B. 852 (2007) 282–287.

[121] L. Weidolf, T.R. Covey, Rapid Commun. Mass Spectrom. 6 (1992) 192–196.

[122] L. Weidolf, N. Castagnoli Jr., Rapid Commun. Mass Spectrom. 15 (2001) 283–290.

[123] W. Naidong, W.Z. Shou, T. Addison, S. Maleki, X. Jiang, Rapid Commun. Mass Spectrom. 16 (2002) 1965–1975.

[124] H. Kanazawa, A. Okada, Y. Matsushima, H. Yokota, S. Okubo, F. Mashige, et al., J. Chromatogr. 949 (2002) 1–9.

[125] B.P. Jensen, C. Smith, I.D. Wilson, L. Weidolf, Rapid Commun. Mass Spectrom. 18 (2004) 181–183.

[126] A. Tolonen, M. Turpeinen, J. Uusitalo, O. Pelkonen, Eur. J. Pharm. Sci. 25 (2005) 155–162.

[127] J. Wang, Y. Wang, J.P. Fawcett, Y. Wang, J. Gu, J. Pharm. Biomed. Anal. 39 (2005) 631–635.

[128] V.A. Frerichs, C. Zaranek, C.E. Haas, J. Chromatogr. B. 824 (2005) 71–80.

[129] Q. Song, W. Naidong, J. Chromatogr. B 830 (2006) 135–142.

[130] I. Hultman, H. Stenhoff, M. Liljeblad, J. Chromatogr. B 848 (2006) 317–322.

[131] U. Hofmann, M. Schwab, G. Treiber, U. Klotz, J. Chromatogr. B 831 (2006) 85–90.

[132] L. Toribio, C. Alonso, M.J. del Nozal, J.L. Bernal, M.T. Martin, J. Chromatogr. 1137 (2006) 30–35.

[133] G. McGrath, S. McClean, O. O'Kane, W.F. Smyth, F. Tagliaro, J. Chromatogr. 735 (1996) 237–247.

[134] D. Eberle, R.P. Hummel, R. Kuhn, J. Chromatogr. 759 (1997) 185–192.

[135] K.D. Altria, S.M. Bryant, T.A. Hadgett, J. Pharm. Biomed. Anal. 15 (1997) 1091–1101.

[136] P.S. Bonato, F.O. Paias, J. Braz. Chem. Soc. 15 (2004) 318–323.

[137] Y.H. Lin, S.M. Wu, LC.GC Europe 18 (2005) 164–167.

[138] J.J. Berzas Nevado, G. Castaneda Penalvo, R.M. Rodriguez Dorado, Anal. Chim. Acta 533 (2005) 127–133.

[139] J. Olsson, F. Stegander, N. Marlin, H. Wan, L.G. Blomberg, J. Chromatogr. 1129 (2006) 291–295.

[140] T. Perez-Ruiz, C. Martinez-Lozano, A. Sanz, E. Bravo, R. Galera, J. Pharm. Biomed. Anal. 42 (2006) 100–106.

[141] J. Olsson, L.G. Blomberg, J. Chromatogr. B 875 (2008) 329–332.

[142] C.G. Regardh, M. Gabrielsson, J.K. Hoffman, I. Lofberg, I. Skanberg, Scand. J. Gastroenterol. 108 (Suppl.) (1985) 79–94.

[143] C.G. Regardh, Scand. J. Gastroenterol. 118 (Suppl.) (1986) 99–104.

[144] J. Naesdal, T. Andersson, G. Bodemar, R. Larsson, C.G. Regardh, I. Skanberg, et al., Clin. Pharmacol. Ther. 40 (1986) 344–351.

[145] C. Cederberg, T. Andersson, I. Skanberg, Scan. J. Gastroenterol. 166 (Suppl.) (1989) 33–40.

[146] C.G. Regardh, T. Andersson, P.O. Lagerstrom, P. Lundborg, I. Skanberg, Ther. Drug Monit. 12 (1990) 163–172.

[147] T. Andersson, C. Cederberg, C.G. Regardh, I. Skanberg, Eur. J. Clin. Pharmacol. 39 (1990) 195–197.
[148] D.R. Sohn, K. Kobayashi, K. Chiba, K.H. Lee, S.G. Shin, T. Ishizaki, J. Pharmacol. Exp. Ther. 262 (1992) 1195–1202.
[149] S. Landahl, T. Andersson, M. Larsson, B. Lernfeldt, P. Lundborg, C.G. Regardh, et al., Clin. Pharmacokinet. 23 (1992) 469–476.
[150] T. Andersson, R. Olsson, C.G. Regardh, I. Skanberg, Clin. Pharmacokinet. 24 (1993) 71–78.
[151] A. Okada, H. Kanazawa, F. Mashige, S. Okubo, H. Yokota, Anal. Sci. 17 (Suppl.) (2001) i871.
[152] A. Abelo, T.B. Andersson, M. Antonsson, A.K. Naudot, I. Skanberg, L. Weidolf, Drug Metab. Dispos. 28 (2000) 966–972.
[153] J.M. Pique, F. Feu, G. de Prada, K. Rohss, G. Hasselgren, Clin. Pharmacokinet. 41 (2002) 999–1004.
[154] T. Kita, T. Sakaeda, N. Aoyama, T. Sakai, Y. Kawahara, M. Kasuga, et al., Biol. Pharm. Bull. 25 (2002) 923–927.
[155] M.J. Kendall, Aliment Pharmacol. Ther. 17 (Suppl. 1) (2003) 1–4.
[156] R. Kumar, Y.K. Chawla, S.K. Garg, P.K. Dixit, S.K. Satapathy, R.K. Dhiman, et al., Methods Find. Exp. Clin. Pharmacol. 25 (2003) 625–630.
[157] M. Hassan-Alin, T. Andersson, M. Niazi, K. Rohss, Eur. J. Clin. Pharmacol. 60 (2005) 779–784.
[158] H.F. Helander, C.H. Ramsay, C.G. Regardh, Scand. J. Gastroenterol. 108 (Suppl.) (1985) 95–104.
[159] K.J. Hoffmann, L. Renberg, S.G. Olovson, Drug Metab. Dispos. 14 (1986) 336–340.
[160] K.J. Hoffmann, Drug Metab. Dispos. 14 (1986) 341–348.
[161] L. Renberg, R. Simonsson, K.J. Hoffmann, Drug Metab. Dispos. 17 (1989) 69–76.
[162] L. Weidolf, K.E. Karlsson, I. Nilsson, Drug Metab. Dispos. 20 (1992) 262–267.
[163] K. Chiba, K. Kobayashi, K. Manabe, M. Tani, T. Kamataki, T. Ishizaki, J. Pharmacol. Exp. Ther. 266 (1993) 52–59.
[164] L. Zhao, Y.Q. Lou, Acta Pharm. Sin. 30 (1995) 248–253.
[165] U.A. Meyer, Eur. J. Gastroenterol. Hepatol. 1 (Suppl.) (1996) S21–S25.
[166] G. Tyrbing, Y. Bottiger, J. Widen, L. Bertilsson, Clin. Pharmacol. Ther. 62 (1997) 129–137.
[167] A. Tolonen, M. Turpeinen, J. Uusitalo, O. Pelkonen, Eur. J. Pharm. Sci. 25 (2005) 155–162.
[168] V.F. Roche, Am. J. Pharm. Educ. 70 (2006) 1–11.
[169] P. Lindberg, P. Nordberg, T. Alminger, A. Brandstrom, B. Wallmark, J. Med. Chem. 29 (1986) 1327–1329.
[170] I. Puscas, M. Coltau, M. Baican, G. Domuta, J. Pharmacol. Exp. Ther. 290 (1999) 530–534.
[171] R.A. Quercia, C. Fan, X. Liu, M.S. Chow, Am. J. Health Syst. Pharm. 54 (1997) 1833–1836.
[172] J.L. DiGiacinto, K.M. Olsen, K.L. Bergman, E.B. Hoie, Ann. Pharmacother. 34 (2000) 600–605.
[173] M. Palummo, A. Cingolani, L. Dall, M.G. Volonte, Bull. Chim. Farm. 139 (2000) 124–128.
[174] H.G. Choi, C.K. Kim, J. Contr. Rel. 68 (2000) 397–404.
[175] C.S. Yong, J.H. Jung, J.D. Rhee, C.K. Kim, H.G. Choi, Drug Dev. Ind. Pharm. 27 (2001) 447–455.
[176] A. Leitner, P. Zollner, Wein Med. Wochenschr 152 (2002) 568–573.
[177] J. Moschwitzer, G. Achleitner, H. Pomper, R.H. Muller, Eur. J. Pharm. Biopharm. 58 (2004) 615–619.

[178] A. Reidel, C.S. Leopold, Drug Dev. Ind. Pharm. 31 (2005) 151–160.
[179] J.E. Burnett, E.R. Balkin, Am. J. Health Syst. Pharm. 63 (2006) 2240–2247.
[180] M. Espinosa Bosch, A.J. Ruiz Sanchez, F. Sanchez Rojas, C. Bosch Ojeda, J. Pharm. Biomed Anal. 44 (2007) 831–844.

CHAPTER 5

Parbendazole

Badraddin M.H. Al-Hadiya

Contents

Department of Clinical Pharmacy, College of Pharmacy, King Saud University, Riyadh, Kingdom of Saudi Arabia

Profiles of Drug Substances, Excipients, and Related Methodology, Volume 35
ISSN 1871-5125, DOI: 10.1016/S1871-5125(10)35005-9

1. INTRODUCTION

The helminthic, or worm, diseases are caused by members of two phyla:

(a) The phylum *Platyhelminthes*, with classes
    (i) *Cestoda* (tapeworms), which infect fish, pork, beef, and dwarf tapeworms
    (ii) *Trematoda* (flukes), including lung, liver, and intestine flukes
(b) The phylum *Nemathelminthes*, with the important class *Nematoda* (roundworms) which are of higher organization than the flatworms, including hookworms, roundworm, whipworm, and pinworm. The counterparts of most human parasitic worms can be found in animals. The majority of helminth infections are acquired by contact with (i) infected animals, (ii) ground contaminated by human or animal excrement, (iii) water infected with cercariae, and (iv) ingestion of infected meat. Anthelmintics are therapeutic agents used to eradicate parasitic worms from the infected host. Most of the drugs used contain nitrogen and many are amines [1–7].

2. GENERAL INFORMATION

2.1. Nomenclature

2.1.1. Systematic chemical names

- (5-Butyl-1*H*-benzimidazol-2-yl)carbamic acid methyl ester.
- Methyl-5-butyl-2-benzimidazolecarbamate.
- 5-Butyl-2-(carbomethoxyamino)benzimidazole.
- Carbamic acid-(5-butyl-1*H*-benzimidazol-2-yl)-methyl ester.
- Methyl-*N*-(5-butyl-3*H*-benzoimidazol-2-yl)carbamate [1, 8].

2.1.2. Nonproprietary names

• Generic: Parbendazole BAN, DCF, USAN [1, 8].
• Synonyms: SKF-29044 [1].

2.1.3. Proprietary names
Helmatac (SK&F), Verminum (Squibb), Worm Guard (SK&F), Helatac [1].

2.2. Formulae
2.2.1. Emperical formula, molecular weight, and CAS number

Parbendazole	$C_{13}H_{17}N_3O_2$	247.30	14255-87-9

2.2.2. Structural formula

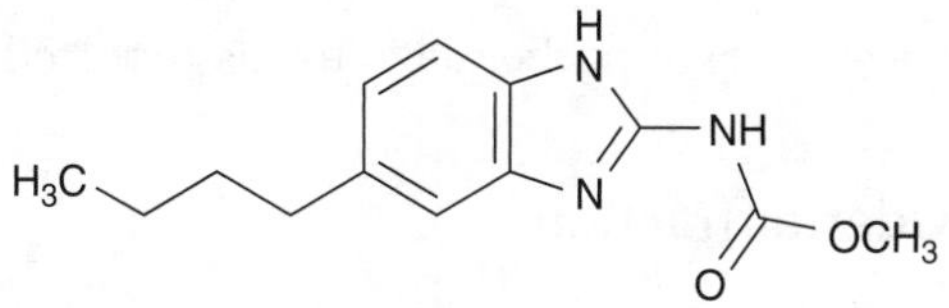

2.3. Elemental analysis

The theoretical elemental composition of parbendazole is as follows [1]:
C 63.14%, H 6.93%, N 16.99%, O 12.94%.

2.4. Appearance

Crystals or fine white crystalline powder [7].

3. METHODS OF PREPARATION

Parbendazole was prepared on the industrial scale by scientists in the SmithKline and French Laboratories [11, 12] by the reaction of 4-butyl-*o*-phenylene-diamine (1) with carbomethoxycyanamide (2) in boiling 2-propanol, to yield parbendazole (3):

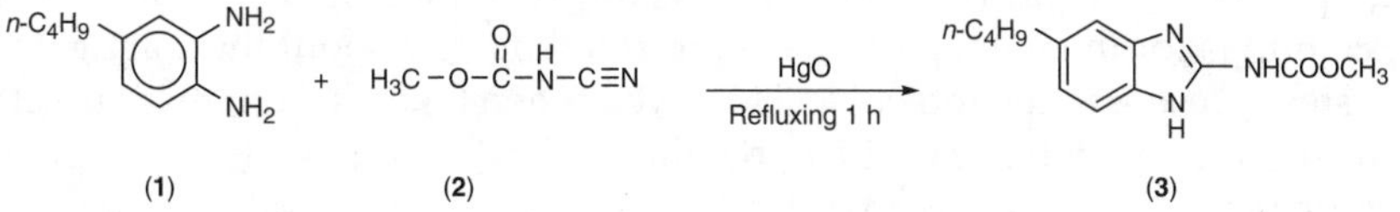

Another method for preparing (3) is by treating benzimidazole carbonyl chloride with trimethylsilylazide (Me_3SiN_3) and methanol [13].

4. PHYSICAL CHARACTERISTICS

4.1. Solubility characteristics

Parbendazole is practically insoluble in water, soluble to the extent of 1 part in 900 parts of ethanol, soluble to the extent of 1 part in 300 parts of chloroform. The substance is also very slightly soluble in ether, and soluble in dilute mineral acids [7].

4.2. Melting point

The melting range of parbendazole is between 224 and 227 °C, with decomposition [7].

4.3. Partition coefficient

The $\log P$ (octanol/water) of parbendazole was reported to be 3.6 [7].

4.4. X-Ray powder diffraction

The simplicity and advantages of the routine application of powder diffraction techniques for the chemical analysis and identification of polycrystalline materials were pointed out by Hull [9], who stated "that every crystalline substance gives a pattern; that the same substance always gives the same pattern," and by the pioneering work of Hanawalt *et al.* [10].

The X-ray powder diffraction pattern of parbendazole has been measured using a Philips PW-1050 diffractometer, equipped with a single-channel analyzer and using copper Kα radiation. The pattern obtained is shown in Fig. 5.1, and the data of scattering angle (° 2θ) and the relative intensities (I/I_{max}) are presented in Table 5.1 (as photographic and diffractometric pattern data) as an attempt to establish a reference chart for the purpose of identification of parbendazole.

4.5. Spectroscopic methods

4.5.1. UV–VIS spectroscopy

In practice, UV absorption spectroscopy (in the region from 200 to 380 nm) is for the most part yields information only about the conjugated system present in the molecule. However, when taken in conjunction with the wealth of detail provided by infrared (IR) and nuclear magnetic resonance (NMR) bands may lead to successful structural elucidations. The principal characteristics of an UV absorption band are its position (λ_{max}) and intensity (ε_{max} or $\log \varepsilon_{max}$).

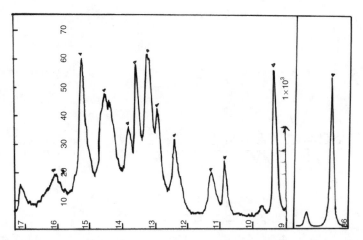

FIGURE 5.1   X-Ray powder diffraction pattern of parbendazole using goniometric diffractometer technique (Cu Kα radiation).

The UV absorption spectra of parbendazole were recorded in methanol, methanolic 0.1 N HCl, and methanolic 0.1 N NaOH to determine the relationship between structure and wavelength shifts in the UV region occurring on transfer to different pH media. The results are presented in Fig. 5.2, where the spectra were recorded using a Perkin-Elmer double beam model 550s UV–VIS spectrophotometer. The values of the log ε_{max} (log molar extinction coefficient) at their corresponding wavelength maximum in nanometers (λ_{max}) are shown in Table 5.2. The value of ε was calculated from the absorbance (A) using the equation:

$$A = \varepsilon c l$$

where l is the length of the absorbing solution (equal to 1 cm) and c is the molar concentration.

The bands show significant bathochromic shifts in base solution (Fig. 5.2C) relative to neutral (Fig. 5.2A) or acidic (Fig. 5.2B) solutions. This may be due to tautomeric [39] and resonance effects taking place in basic medium. Interactions occurred between the nonbonded electrons on the nitrogens and the π electrons of the fused rings as evidenced by the blurring of the bands. Furthermore, the enolic configuration

$$-N=\overset{\overset{\displaystyle O^-}{|}}{C}-OCH_3$$ in the carbamic acid methyl ester part of the molecule predominates in basic solution, causing extension for the chromophoric system in the molecule. The alkyl substitution at the 6-position of the benzimidazole may cause little bathochromic shifts due to hyperconjugation [14, 15].

TABLE 5.1 X-Ray diffraction data for parbendazole

Cu Kα d (Å)	(Diffractometer) I/I₁	Cu Kα d (Å)	(Camera) I/I₁	Co Kα d (Å)	(Camera) I/I₁	Cr Kα d (Å)	(Camera) I/I₁
13.1	100	12.9	100	12.9	100	13.0	100
9.77	11	9.72	51	9.69	63	9.67	36
8.63	1	8.63	4	8.55	6	8.52	3
6.51	4	6.50	8	6.49	10	6.49	5
6.01	3	5.93	8	5.93	11	5.91	6
5.00	2	–	–	–	–	–	–
4.87	5	4.84	18	4.84	28	4.85	13
4.48	7	4.48	25	4.48	32	4.47	15
4.29	11	4.27	32	4.28	41	4.27	18
4.10	10	4.08	31	4.08	40	4.08	17
3.95	6	3.94	19	3.93	26	3.94	10
3.69	7	–	–	3.69	27	–	–
3.67	6	–	–	–	–	–	–
3.61	8	3.62	25	–	–	3.63	5
3.58	7	–	–	3.57	29	–	–
3.34	10	3.34	27	3.34	33	3.34	15
3.15	1	–	–	–	–	–	–
3.11	1	–	–	–	–	–	–
3.08	1	–	–	–	–	–	–
3.06	1	–	–	–	–	–	–
2.76	1	2.76	6	2.76	6	2.75	4
–	–	2.47	4	2.471	5	2.46	3
2.38	1	2.41	4	2.41	3	2.41	3
2.36	1	–	–	–	–	–	–
2.34	1	2.26	2	2.26	4	–	–
2.03	1	–	–	2.08	2	–	–
2.03	1	–	–	2.01	2	–	–

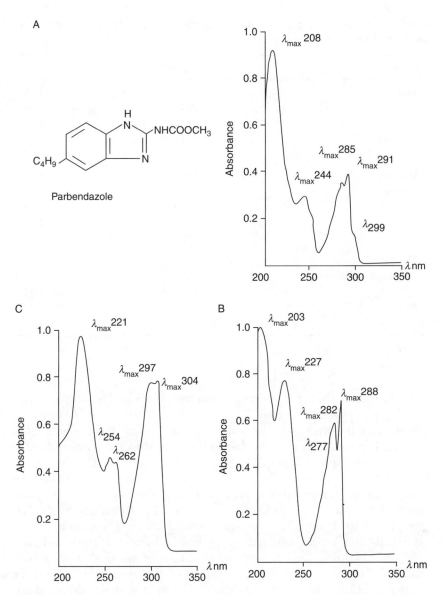

FIGURE 5.2   UV absorption spectrum of parbendazole in methanol (A), methanolic 0.1 M HCl (B), and methanolic 0.1 M NaOH (C).

4.5.2. Vibrational spectroscopy

The IR absorption spectrum of parbendazole is shown in Fig. 5.3, and was obtained in a KBr pellet using a Unicam SP200 IR spectrophotometer. The principal absorption bands were noted at 1220, 1390, 1445, 1510, 1585,

TABLE 5.2 Values of log ε_{max} at their corresponding wavelength maximum in the UV spectrum of parbendazole

Medium	λ_{max} (nm)	log ε
Methanol	208	4.58
	244	4.09
	285	4.17
	291	4.21
	299 (sh)	3.76
Methanolic N/10	203	4.43
hydrochloric acid	227	4.31
	277 (sh)	4.16
	282	4.18
	288	4.26
Methanolic N/10	221	4.41
sodium hydroxide	254	4.06
	262 (sh)	4.04
	297	4.30
	304	4.30

sh = shoulder.

1625, 2855, and 3360 cm^{-1} (summarized in Table 5.3). The single characteristic absorption band near 3360 cm^{-1} is due to the N–H stretching vibration for the open-chain secondary amide in solid state parbendazole. The bands near 2855 cm^{-1} are due to the C–H stretching absorptions in the saturated aliphatic groups; those near 1390 and 1445 cm^{-1} are due to C–H deformations of the CH$_3$ and CH$_2$ groups, respectively. Parbendazole contains the group R–NHCO–OR, and the carbonyl absorption within the amide-I band is likely to be lowered in frequency by the NH group, and is observed as the band at 1625 cm^{-1}. The amide-II band occurs at 1510 cm^{-1}, lowered by hydrogen bonding. The band near 1585 cm^{-1} is due to the vibrations of the characteristic skeletal stretching mode of the semiunsaturated C–C bonds of the benzene ring, and the band near 1220 cm^{-1} is due to the C–H in-plane deformation of the aromatic ring.

4.5.3. Nuclear magnetic resonance spectrometry

NMR spectroscopy concerns radio frequency induced transitions between quantized energy states of magnetic nuclei that have been oriented by magnetic fields. Of these nuclei proton (^{1}H) and carbon-13 (^{13}C)

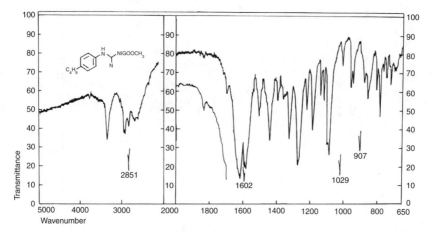

FIGURE 5.3   Infrared spectrum of parbendazole.

are the most important from the viewpoint of the organic and pharma-ceutical chemist. The technique permits the exploration of a molecule at the level of the individual atom, and affords information concerning the environment of that atom [16, 17].

The ^{1}H NMR spectra of parbendazole was recorded with a JEOL-PS 100 NMR spectrometer operating at a frequency of 100 MHz and a magnetic field strength of 2.349 T. Spectra were determined over the region 10.8–0.0 parts per million (ppm), with a sweep time of 250 s. Chemical shifts were recorded as δ (delta) ppm downfield from tetra-methylsilane (TMS). Proton noise and off-resonance decoupled ^{13}C NMR spectra were measured on a JEOL FX 90Q Fourier Transform NMR spectrometer operating at 90 MHR and spectral width of 5000 Hz (220 ppm). All measurements were obtained with the compound being dissolved in deuterated dimethyl sulfoxide (DMSO-d_6) for ^{1}H NMR and in deuterated trifluoroacetic acid (TFA-d_1) for ^{13}C NMR.

4.5.3.1. ^{1}H NMR spectrum   The full ^{1}H NMR spectrum of parbendazole is shown in Fig. 5.4. Assignments for the resonance bands are given in Table 5.4. From the table, the broad singlet band centered at $\delta = 12.64$ is integrated for the two amide protons (b) and (c). On deuteration this signal did not show up, confirming the assignment. The signal appeared at high δ value due mainly to intermolecular hydrogen bonding enhanced by the highly polar solvent, DMSO-d_6. This involved electron transfer from the hydrogen atoms to the neighboring electron negative atoms, with a net deshielding effect experienced by the protons. The broadness

TABLE 5.3 Band frequencies in the infrared absorption spectrum of parbendazole

Name of vibrational band	Position of absorption frequency (cm^{-1})	Name of vibrational band	Position of absorption frequency (cm^{-1})
C–H stretching vibrations		Aromatics: ν (=C–H) 3000 ± 50 cm^{-1}	2960
–CH$_3$ (2962 ± 10 cm^{-1})	2955 (sh)		
–CH$_2$ (2962 ± 10 cm^{-1})	2920		
C–H bending vibrations		Aromatics: overtones and combination bands of γ (C–H) 2000–1660	Two bands
Asym: 1450 ± 20 cm^{-1}	1445		
Sym: 1375 ± 20 cm^{-1}	1390		
Skeletal modes (CH$_2$)$_n$: 750–720 cm^{-1}	745, 720	Aromatics: ν (C---C) ring frequencies (skeletal modes): Near 1610, Near 1580, Near 1500	1602, 1585, 1502 (sh)
Amide bands			
Amide-I: 1680–1620 cm^{-1}	1625		
Amide-II: 1570–1515 cm^{-1}	1510		

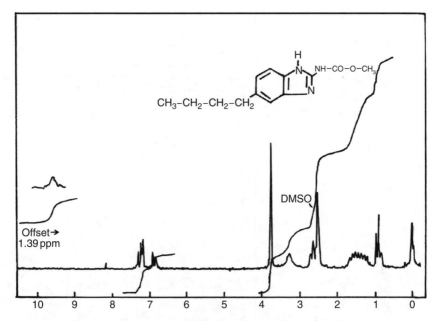

FIGURE 5.4 ^{1}H NMR spectrum of parbendazole in DMSO-d_6.

of the signal is due to the efficient spin-lattice relaxation for the protons attached to ^{14}N, possessing an electric quadrupole ($I = 1$) [18]. The methyl protons (1) and the methylenes (2, 3, 4) of the *n*-butyl chain substituted at position 4 in parbendazole showed virtual coupling, with the characteristic blurring of signals [18]. In the spectrum of parbendazole (Fig. 5.4), the three aromatic protons (d), (e), and (g) gave rise to complex bands ranging from about δ 6.8 to δ 7.3, showing an ABX pattern, with coupling constant $J_{AB} = 8$ Hz. Since *para*-coupling is mostly unobservable, $J_{AX} = 0$ and J_{BX} was found to be about 3 Hz. Due to the coupling between proton (e) and proton (g), with $J_{BX} = 3$ Hz, the low-field doublet (δ 7.24) splitting into four lines, representing a two-proton (e, g) multiplet, and the high-field (δ 6.86) represented a one-proton (d) doublet. This was confirmed by the integration values in Table 5.4.

4.5.3.2. ^{13}C NMR spectrum The ^{13}C NMR spectrum of parbendazole is shown in Fig. 5.5 (full ^{13}C NMR spectrum) and Fig. 5.6 (off-resonance spectrum). ^{13}C chemical shift data in TFA-d_1 are summarized in Table 5.5, which are consistent with the ^{13}C contents of parbendazole. The assignments for the observed bands are based on established chemical shift parameters, off-resonance spectra, and comparisons with the following

TABLE 5.4 ^{1}H NMR data and spectral assignments for parbendazole (solvent: DMSO-d_6)

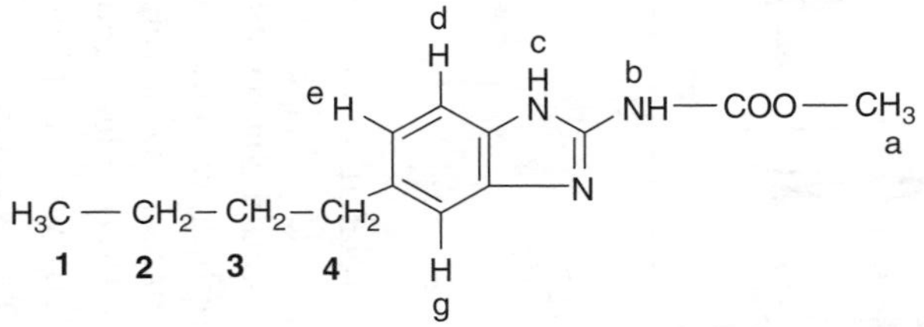

$$H_3C - CH_2 - CH_2 - CH_2$$
$$1 \quad\; 2 \quad\;\; 3 \quad\;\; 4$$

(benzimidazole ring with labels d, e, c, b, a, g; NH—COO—CH₃)

Chemical shift (ppm)	Integral trace in mm	Proton ratio	Multiplicity	Assignment	Remarks
0.91	30	3	Second-order triplet	1	Coupled to 2
1.40	47	4	Second-order multiplet	2, 3	
2.64	Less accurate	2	Second-order triplet	4	Coupled to 1
3.26	–				Peak due to H_2O
3.76	30	3	Singlet	a	
6.86	10	1	Doublet	d	Aromatic protons
7.26	20	2	Doublet	e, g	$J_{AB} = 8$ Hz; $J_{AX} = 0$ Hz; $J_{BX} = 3$ Hz
12.64	20	2	Broad singlet	b, c	Amide protons

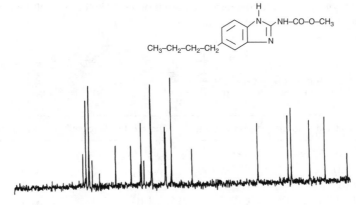

FIGURE 5.5 ^{13}C NMR spectrum of parbendazole in TFA-d_1.

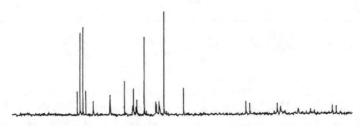

FIGURE 5.6   Off-resonance ^{13}C NMR spectrum (55.0 kHz) of parbendazole in TFA-d_1.

model compounds: benzimidazole (**1**) [19], benzimidazole-2-thiol (**2**) [20], carbamic acid methyl ester (**3**) [21], *n*-butyl-benzene (**4**) [22], benzophe-nones (**5**) [23–25], and thiazole (**6**) [26, 27]. The model compounds and their ^{13}C chemical shift data are displayed in Fig. 5.7. The resonance position of the benzimidazole carbon C-7, at δ 145.4 ppm is readily identified as the only peak having unit intensity in the spectrum. The bridgehead carbons C-1, -6 are readily identified, since a quaternary carbon does not exhibit the proton splitting pattern characteristic of a carbon atom with directly bonded protons. The chemical shifts of C-1, -6 have values of 127.7 and 129.9 ppm, respectively. In the off-resonance spectrum of parbenda-zole, Fig. 5.6, the triplets due to C-10 and C-11 overlapped.

4.5.4. Mass spectrometry
The role of functional groups in directing molecular fragmentation, and the value of mass spectrometry in elucidation of the structure of organic compounds was clearly demonstrated by Biemann [28], Beynon [29], and McLafferty [30].

TABLE 5.5 Assignments for the resonance bands observed in the ^{13}C NMR spectrum of parbendazole (solvent: TFA-d_1)

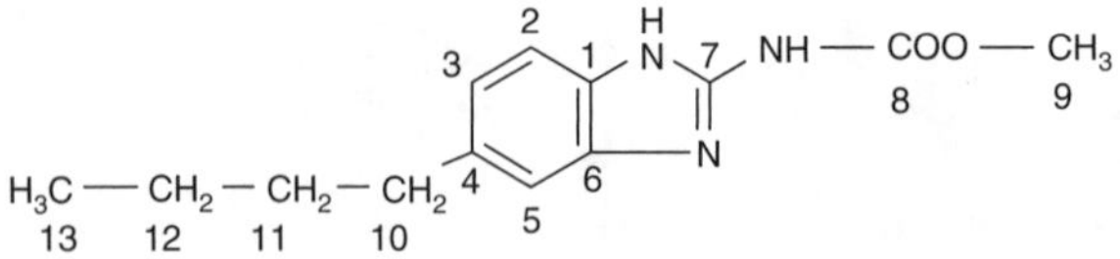

Chemical shift (ppm)[a]	Assignment	Chemical shift (ppm)[a]	Assignment
127.7 (s)	C-1	155.6 (s)	C-8
114.0 (d)	C-2	56.4 (q)	C-9
129.4 (d)	C-3	37.2 (t)	C-10
145.4 (s)	C-4	35.1 (t)	C-11
129.4 (d)	C-5	23.6 (t)	C-12
129.9 (s)	C-6	14.0 (q)	C-13
145.4 (s)	C-7		

(s), singlet; (d), doublet; (t), triplet; (q), quartet.
[a] Chemical shift frequencies (δ ppm from TMS).

The mass spectra of benzimidazoles and derivatives have been studied extensively [31–34]. The mass spectral fragmentation is shown below and the successive expulsion of two molecules of HCN affords the M-2HCN$^+$ ion at $m/z = 64$:

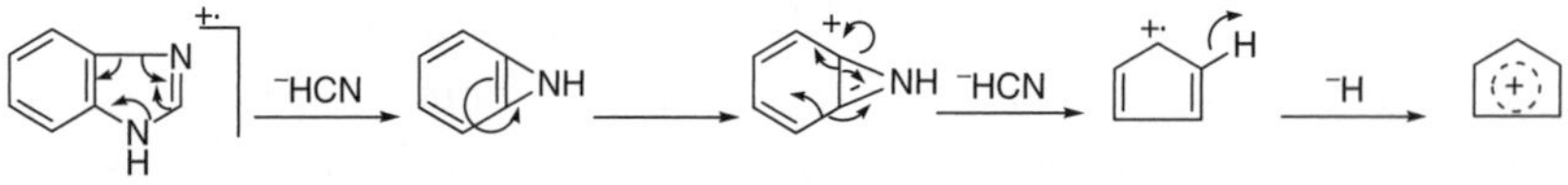

The $m/z = 63$ ion is a fragment ion observed in the mass spectra of aromatic compounds. This general decomposition pathway was also noted in the mass spectra of a series of substituted benzimidazoles by Lawesson *et al.* [34].

The mass spectrum of parbendazole was obtained utilizing an AEI single focusing mass spectrometer model MS12 with VG micromass 2S8, using an ion source temperature of 100–300 °C, an electron energy of 70 eV, and with a trap current of 100 μA.

The mass spectrum was made by plotting m/z against ion abundance. It is plotted in terms of relative abundance, with the most intense peak (the base-peak) being taken as 100%. All peaks above m/z 25 which are greater than 1% of intensity of the largest peak are shown in the spectrum. The ion corresponding to the parent compound is referred to as the molecular ion (M$^+$). The presence of a metastable peak, m^*, is an

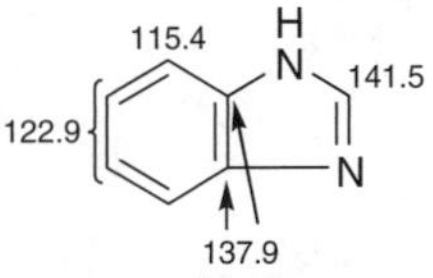

FIGURE 5.7 Model compounds and their ^{13}C chemical shift data.

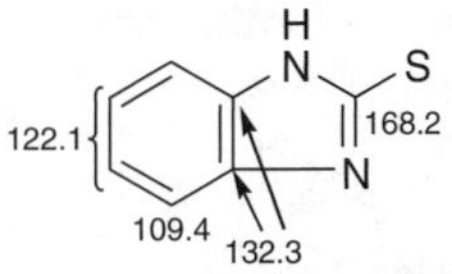

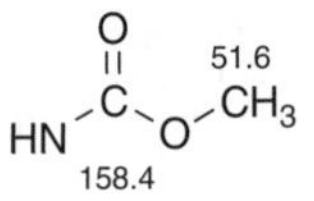

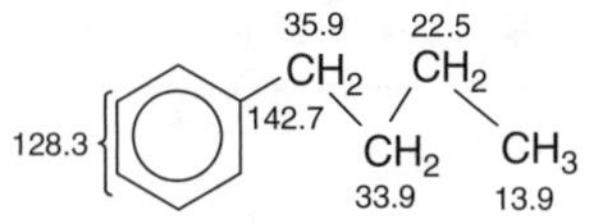

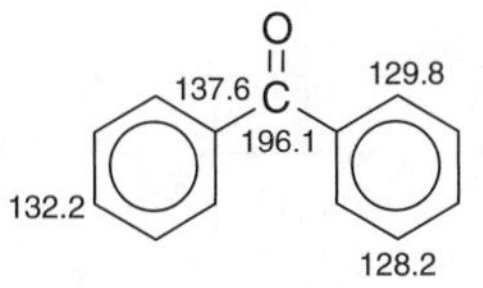

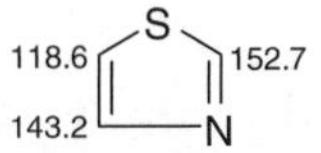

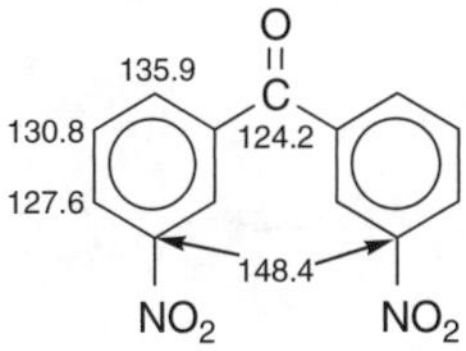

indication for the decomposition in one step of a parent ion, m_1, into a daughter ion, m_2 plus a neutral particle and they are related to each other as $m^* = (m_2)^2/m_1$.

Figure 5.8 shows the detected mass fragmentation pattern of parbendazole. The major peaks in the spectrum occur at m/z 247, 205, 204, 186, 173, 172, 160, 145, 131, 118. The mass spectrum and fragmentation pattern is typical of alkyl benzenes [35], carbamates [36], and benzimidazoles [31]. The most characteristic cleavage occurred β to the benzene ring along the *n*-butyl side-chain as this requires the least energy of dissociation. In the molecular ion region the relative intensities of $(M + 1)^+$ and $(M + 2)^+$ peaks nearly agree with those to be expected from the known isotopic ratios of C, N, and O. In Fig. 5.8, pathway (1) represents cleavage of the bond β to the benzene ring accompanied by hydrogen migration through a four-membered transition state:

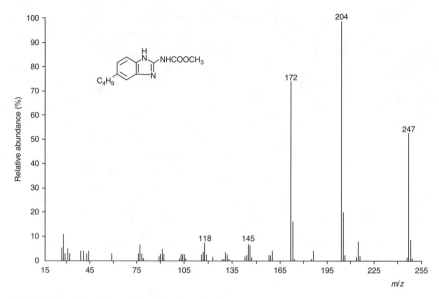

FIGURE 5.8   Mass spectrum of parbendazole.

Pathway (2) is also a cleavage of the bond β to the benzene ring accompanied by hydrogen migration via a six-membered transition state. This is analogous to the McLafferty rearrangement process [30]. Further loss of a hydrogen radical yields the most abundant ion, the base-peak, at m/z 204, with a metastable ion at m/z 168.5. The great stability of this ion is due to the formation of the bicyclic tropylium ion [37, 38]:

Further decomposition involves losses of CH_3OH, CO, and HCN as shown in Fig. 5.9. Hydrogen migration with α-bond breakage may proceed by a mechanism similar to that accompanying β-bond cleavage. In parbendazole β-bond cleavage predominates over α-bond cleavage [35]. The following table illustrates the metastable ions from parbendazole fragmentation.

m_2	m_1	$m^* = m_2^2/m_1$
204	247	168.5
172	204	145.02

5. METHODS OF ANALYSIS

5.1. Identification

Dissolve 5 mg in 5 ml of 0.1 M hydrochloric acid and add 3 mg of p-phenylenediamine dihydrochloride; shake to dissolve and add 0.1 g zinc powder. After mixing, allow to stand for 2 min and add 10 ml of ferric ammonium sulfate solution. A blue or violet-blue color develops [7].

5.2. Thin-layer chromatography

The thin-layer chromatography (TLC) system for screening of parbendazole consisted of [7]:

- *Stationary phase*: silica gel G, 250 μm thick, impregnated with 0.1 M potassium hydroxide in methanol, and dried.
- *Mobile phase*: methanol:strong ammonia solution (100:1.5); saturated chamber.
- *Location reagent*: acidified iodoplatinate solution oversprayed the plate, giving a positive color, with Rf value of 70.

5.3. Gas chromatography

The drug was not eluted when using general standard gas chromatography (GC) screening systems, involving the use of standard columns that are able to chromatograph a wide variety of drugs and chemicals [7].

5.4. High-performance liquid chromatography

The retention time of parbendazole was found to be 18.2 min using the following high-performance liquid chromatography (HPLC) conditions [7, 49]:

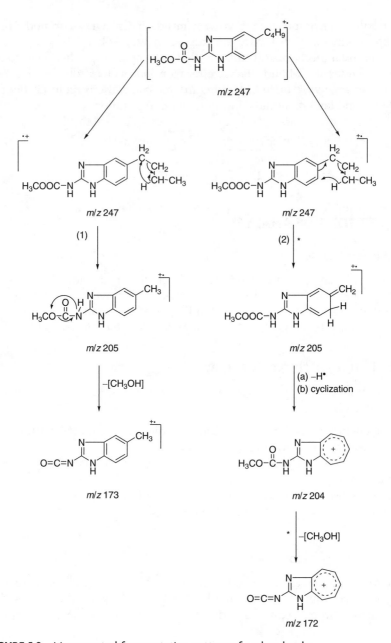

FIGURE 5.9   Mass spectral fragmentation pattern of parbendazole.

- *Column*: C$_8$ symmetry (250 × 4.6 mm i.d., 5 μm) with symmetry pre-column (20 mm).
- Column temperature: 30 °C.
- *Mobile phase*: (A:B) phosphate buffer (pH 3.8): acetonitrile.
- *Elution program*: (85:15) for 6.5 min to (65:35) until 25 min to (20:80) for 3 min and back to initial conditions for equilibration for 7 min.
- *Flow rate*: 1 ml/min for 6.5 min, then linear increase to 1.5 ml/min for 6.5–25 min and hold for 3 min (reequilibration is made at 1.5 ml/min).
- *Detection*: UV diode-array.

6. STABILITY

Parbendazole was found to be incompatible with strong oxidizing agents [7].

7. PHARMACOLOGICAL PROFILE

7.1. Uses

An extensive review of *in vivo* Anthelmintic activity testing has been given by Standen [40].

Benzimidazole derivatives act by interfering with metabolic pathways. Parbendazole has been shown to exhibit a broad-spectrum Anthelmintic activity. This compound is the most potent of a series of substituted 2-amino derivatives, is active against a wide variety of animal nematodes [41]. The drug showed marked activity against third stage larvae which were located in the muscular tissue. It inhibits monoamine oxidase in animal nematodes. The drug was used with good effects against the common parasites of cattles, pigs, and horses. It acts quickly reaching peak blood levels 6–8 h after administration. Its use in pregnant animals is contraindicated because of its teratogenicity; the defects are largely skeletal. Like other benzimidazoles it suffers from the problem of resistance developing in the resident worm population if it is used persistently [12, 42].

7.2. Mode of action

The mode of action is not clearly delineated but the following effects on the parasite have been reported [50, 51]:

- Degeneration of microtubules
- Inhibition of glucose transport and uptake

- Interference with energy production
- Inhibition of fumarate reductase
- *Ovicidal*: ova are inactivated early following treatment
- Cytoplasmic microtubules disappear in tegumentary and intestinal cells
- Secretory processes, for example, from Golgi apparatus and acetylcholine release, are impaired.

7.3. Pharmacokinetics

After i.v. administration, the plasma concentration of parbendazole drug at 15 min was 5.75 ± 0.22 μg/ml which decreased to 1.92 ± 0.29 μg/ml at 12 h. Pharmacokinetic data in Table 5.6 indicates that the distribution of drug was quite rapid (14.0 ± 1.38 min). The elimination half-life was found to be 720.0 ± 21.27 min.

The longer biological half-life appears to be due to low-body clearance value (1.26 ± 0.08 ml/kg/min). The cumulative urinary excretion of parbendazole for 5 days was found to be very low ($0.475 \pm 0.144\%$). Fecal excretion of the drug was slightly higher ($4.20 \pm 0.71\%$ of the total administered dose) and may be suggestive of biliary excretion [43].

Pharmacokinetics of radiolabeled parbendazole have been studied after oral administration in pig and sheep [44–46].

7.4. Toxicity investigations

The toxicity of parbendazole in different doses was investigated with group of animals. Side effects observed with the Anthelmintic were laxation (soft dung, diarrhea), anorexia, and listlessness. The seriousness of

TABLE 5.6 Pharmacokinetic data of parbendazole following single intravenous injection (5 mg/kg) in goats ($n = 5$)

Pharmacokinetic determinants	Mean $\pm$ SEM
C_0^p	8.30 ± 0.36 μg/ml
$t_{1/2\alpha}$	14.0 ± 1.38 min
$t_{1/2\beta}$	720 ± 21.27 min
K_{el}	0.002 ± 0.0001
K_{12}/K_{21}	1.14 ± 0.16
$AUC_{(0-\alpha)}$	4018.6 ± 268.2 μg/ml/min
$Vd_{(area)}$	1.32 ± 0.104 l/kg
Cl_B	1.26 ± 0.08 ml/kg/min

C_0^p plasma concentration at time zero; $t_{1/2\alpha}$, distribution half-life; $t_{1/2\beta}$, elimination half-life; K_{el}, elimination rate constant from central compartment; K_{12}/K_{21}, transfer rate constant between peripheral and central compartments; $AUC_{(0-\alpha)}$, total area under plasma drug concentration time curve; $Vd_{(area)}$, apparent volume of distribution; Cl_B, total body clearance.

these symptoms depended greatly on the dose used. The laxative side effect of the drug was greater during the performing of work (movement). During pregnancy, no unfavorable effect on the development of the fetus were observed. Neither clinical examination nor clinical chemical serum examination gave any indication of the occurrence of hyperlipaimia after treatment with parbendazole [47]. Overdosing of parbendazole could cause paralysis in lambs when given to ewes during the early stages of pregnancy [48].

ACKNOWLEDGMENT

The author thanks Mr. Tanvir A. Butt for typing this manuscript.

REFERENCES

[1] The Merck Index, nineth ed., Merck & Co., Inc., Whitehouse Station, NJ, USA, 1977.
[2] British Pharmacopoeia, The University Press, Cambridge, 1980.
[3] British Pharmaceutical Codex, The Pharmaceutical Press, London, 1973.
[4] The United States Pharmacopoeia XIX, United States Pharmacopeial Convention, New York, 1975.
[5] British Veterinary Codex, The Pharmaceutical Press, London, 1965 (Supplement 1970).
[6] Martindale, The Extra Pharmacopoeia, 27th ed., The Pharmaceutical Press, London, 1977.
[7] A.C. Moffat (Ed.), Clarke's Analysis of Drugs and Poisons in Pharmaceuticals, Body Fluids and Post-mortem Material, third ed., The Pharmaceutical Press, London, 2004, p. 1397.
[8] International Drug Directory, DirectorySwiss Pharmaceutical Society (Ed.), 17th ed., Medpharm Scientific Publishers, Stuttgart, Germany, 2004, p. 794.
[9] A.W. Hull, J. Am. Chem. Soc. 41 (1919) 1168.
[10] J.D. Hanawalt, H.W. Rinn, L.K. Frevel, Ind. Eng. Chem. Anal. Ed. 10 (1938) 457.
[11] France Patent Office, 2178385.
[12] P. Actor, E.L. Anderson, C.J. Dicuollo, R.J. Ferlauto, J.R.E. Hoover, J.F. Pagano, et al., Nature 215 (1967) 321.
[13] Fr. Demande Patent, 2290430.
[14] E.A. Steck, F.C. Nachod, G.W. Ewing, N.H. Gorman, J. Chem. Soc. 70 (1948) 3406.
[15] E.A. Steck, G.W. Ewing, J. Am. Chem. Soc. 70 (1948) 3397.
[16] H.H. Willard, L.L. Merritt, J.A. Dean, Instrumental Methods of Analysis, fifth ed., D. Van Nostrand Company, London, 1974.
[17] R.T. Parfitt, Nuclear magnetic resonance spectroscopy, in: A.H. Beckett, J.B. Stenlake (Eds.), Practical Pharmaceutical Chemistry, Part Two, The Athlone Press, London, 1970, p. 274.
[18] W. Kemp, Organic Spectroscopy, The Macmillan Press, London, 1975.
[19] R.J. Pugmire, D.M. Grant, J. Am. Chem. Soc. 93 (1971) 1880.
[20] W. Bremser, L. Ernst, W. Fachinger, R. Gerhards, A. Hardt, P.M.E. Lewis, Carbon-13 NMR Spectral Data, NR. CNMR 7300, Verlag Chemie, New York, 1979.
[21] W. Bremser, L. Ernst, W. Fachinger, R. Gerhards, A. Hardt, P.M.E. Lewis, Carbon-13 NMR Spectral Data, NR. CNMR 8067, Verlag Chemie, New York, 1979.

[22] L.F. Johnson, W.C. Jankowski, Carbon-13 NMR Spectra, NR. 380, Verlag Chemie, New York, 1979.
[23] P.E. Hansen, O.K. Poulsen, A. Berg, Org. Magn. Reson. 9 (1977) 649.
[24] G.A. Olah, P.W. Westerman, J. Am. Chem. Soc. 96 (1974) 3548.
[25] Bruker Carbon-13 Data Bank, vol. 1, Wiswesser Code WNR CVR CNW.
[26] W. Bremser, L. Ernst, W. Fachinger, R. Gerhards, A. Hardt, P.M.E. Lewis, Carbon-13 NMR Spectral Data, NR. CNMR 0445, Verlag Chemie, New York, 1979.
[27] H. Hartmann, R. Radeglia, J. Prakt. Chem. 317 (1975) 657.
[28] K. Biemann, Mass Spectrometry and Its Application to Organic Chemistry, McGraw-Hill, New York, 1962.
[29] J.H. Beynon, Mass Spectrometry and Its Application to Organic Chemistry, Elsevier Publishing Company, Amsterdam, 1960.
[30] F.W. McLafferty, Interpretation of Mass Spectra, second ed., W.A. Benjamin Inc., New York, 1973.
[31] S. Safe, O. Hutzinger, Mass Spectrometry of Pesticides and Pollutants, CRC Press, Cleveland, Canada, 1973.
[32] Y. Soeda, S. Kosaka, T. Noguchi, Agric. Biol. Chem. 36 (1972) 817.
[33] T. Nishiwaki, J. Chem. Soc. C (1968) 428.
[34] S.-O. Lawesson, G. Schroll, J.H. Bowie, R.G. Cooks, Tetrahedron 24 (1968) 1875.
[35] S. Meyerson, Appl. Spectrosc. 9 (1955) 120.
[36] C.P. Lewis, Anal. Chem. (1964) 176.
[37] H. Budzikiewicz, C. Djerassi, D.H. Williams, Mass Spectrometry of Organic Compounds, Holden-Day, Inc., London, 1967.
[38] H.M. Grubb, S. Meyerson, Mass Spectrometry of Organic Ions, in: F.W. McLafferty (Ed.), Academic Press, New York, 1963, p. 516.
[39] V.I. Isagulyants, R. Boeva, Z.D. Kustanovich, V.S. Markin, J. Appl. Chem. USSR 41 (1968) 1507.
[40] O.D. Standen, Experimental Chemotherapy, vol. 1, in: R.J. Schnitzer, F. Hawking (Eds.), Academic Press, New York, 1963, p. 70.
[41] A.S. Tomcufcik, E.M. Hardy, Textbook of Medicinal Chemistry, Part 1, in: A. Burger (Ed.), third ed., Wiley-InterScience, New York, 1970, p. 583.
[42] M. Sanchez Morino, J. Barrett, Parasitology 78 (1979) 1.
[43] L.C. Lahon, P.K. Gupta, Ind. J. Pharmacol. 26 (1994) 235.
[44] P. Actor, E.L. Anderson, C.J. Dicuollo, Nature 215 (1967) 321.
[45] C.J. Dicuollo, J.A. Miller, W.L. Mendelson, J.F. Pagano, J. Agric. Food Chem. 22 (1974) 948.
[46] L.M. Jones, H.N. Booth, L.E. McDonald, Veterinary Pharmacology and Therapeutics, fourth ed., Iowa State University Press, Ames, IA, 1977, p. 1002.
[47] L.R. Verberne, M.H. Mirck, Tijdschr Diergeneeskd 100 (21) (1975) 1143.
[48] L. Prozesky, J.P. Joubert, M.D. Ekron, J. Onderstepoort, Vet. Res. 48 (3) (1981) 159.
[49] Y. Gaillard, G. Pepin, J. Chromatogr. A 763 (1997) 149.
[50] L.C. Davidse, Ann. Rev. Phytopathol. 24 (1986) 43.
[51] J.C. Havercroft, R.A. Quinlan, K. Gull, J. Cell. Sci. 49 (1981) 195.

CHAPTER **6**

Rocuronium Bromide

Maria L.A.D. Lestari and Gunawan Indrayanto

Faculty of Pharmacy, Airlangga University, Dharmawangsa Dalam, Surabaya, Indonesia

Profiles of Drug Substances, Excipients, and Related Methodology, Volume 35
ISSN 1871-5125, DOI: 10.1016/S1871-5125(10)35006-0

1. INTRODUCTION

Rocuronium bromide is a nondepolarizing neuromuscular blocking agent which produces rapid muscle relaxation after being injected intravenously [1–3]. Furthermore, it is an aminosteroid neuromuscular blocking agent which has a propenyl ($-CH_2-CH=CH_2$) group on its active quaternary nitrogen atom. The compound is pharmacologically used to provide muscle relaxation in general anesthesia for surgical procedures and to aid in controlled ventilation [4]. In the market, rocuronium bromide is only available as a solution dosage form intended for intravenous or intramuscular injection [1, 2].

1.1. Chemical name

- 1-Allyl-1-(3α,17β-dihydroxy-2β-morpholino-5α-androstan-16β-yl)-pyrrolidinium bromide-17-acetate [1, 5].
- 1-(17β-Acetoxy-3α-hydroxy-2β-morpholino-5α-androstan-16β-yl)-1-(prop-2-enyl)pyrrolidinium bromide [6, 7].

1.2. Molecular formula and molecular weight

The molecular formula of rocuronium bromide is $C_{32}H_{53}BrN_2O_4$, and the substance is characterized by a molecular weight of 609.7 [1, 5–7].

1.3. Solubility characteristics and partition coefficient

Rocuronium bromide is freely soluble in either water or in anhydrous ethanol, and is characterized by a n-octanol/water partition coefficient of 0.5 at 20 °C [2, 8].

1.4. Melting point

The melting point of rocuronium bromide has been reported to be in the range of 161–169 °C [8, 9].

1.5. Solution characteristics

Rocuronium bromide has a specific optical rotation of +28.5° to +32°, when measured on an anhydrous and solvent-free basis at 20 °C. A 1% solution of the substance in water yields pH values ranging from 8.9 to 9.5 [5–7].

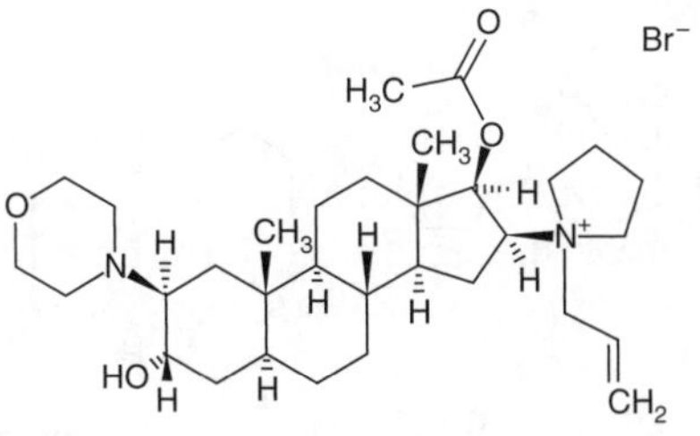

FIGURE 6.1 Chemical structure of rocuronium bromide.

1.6. Known impurities

Compendial reviews state that rocuronium bromide had been identified as having eight different impurities, which are labeled as impurities A, B, C, D, E, F, G, and H. Impurity A (i.e., related compound A) is 3α-hydroxy-2β (morpholin-4-yl)-16β-(pyrrolidin-1-yl)-5α-androstan-17β-yl acetate; impurity B (i.e., related compound B) is 1-[3α,17β-bis(acetyloxy)-2β-(morpholin-4-yl)-5α-androstan-16β-yl]-1-(prop-2-enyl)pyrrolidinium; impurity C (i.e., related compound C) is 1-[3α,17β-dihydroxy-2β-(morpholin-4-yl)-5α-androstan-16β-yl]-1-(prop-2-enyl)pyrrolidinium; and impurity D (i.e., related compound D) is 1-[3α-(acetoxy)-17β-hydroxy-2β-(morpholin-4-yl)-5α-androstan-16β-yl]-1-(prop-2-enyl)pyrrolidinium; impurity E (i.e., related compound E) is 1-[17β-(acetoxy)-3α-hydroxy-2β-(pyrrolidin-1-yl)-5α-androstan-16β-yl]-1-(prop-2-enyl)pyrrolidinium; impurity F (i.e., related compound F) is 1-[3α,17β-bis(acetiloxy)-2β-(pyrrolidin-1-yl)-5α-androstan-16β-yl]-1-(prop-2-enyl)pyrrolidinium; impurity G (i.e., related compound G) is 2β-(morpholin-4-yl)-16β-(pyrrolidin-1-yl)-5α-androstane-3α,17β-diol; and impurity H (i.e., related compound H) is 1-[17β-(acetyloxy)-2-(morpholin-4-yl)-3-oxo-5α-androst-1-en-16β-yl]-1-(prop-2-enyl)pyrrolidinium [5–7]. The molecular structure of rocuronium bromide is shown in Fig. 6.1, while the structures of the known impurities are illustrated in Fig. 6.2.

2. COMPENDIAL METHOD OF ANALYSIS

2.1. Identification

Identification methods of rocuronium bromide in compendias only specify an identification method for its bulk drug substance, and none of these mention identification of rocuronium bromide in any of its pharmaceutical dosage forms.

European Pharmacopoeia (EP) 6.0 [6] and British Pharmacopoeia (BP) 2009 [7] use infrared absorption spectroscopy and a solution test for bromide ion to identify rocuronium bromide. For the infrared absorption

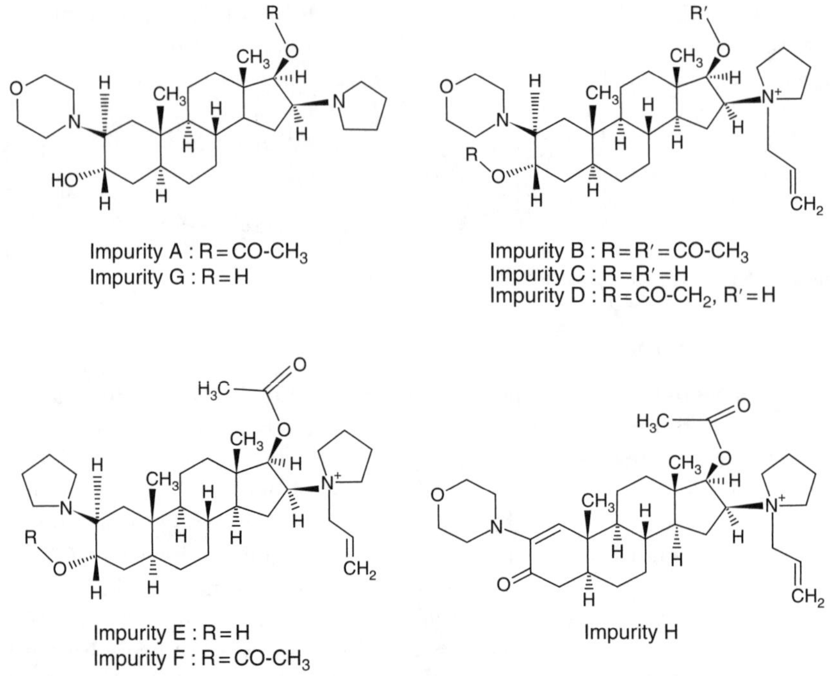

Impurity A : R=CO-CH₃
Impurity G : R=H

Impurity B : R=R'=CO-CH₃
Impurity C : R=R'=H
Impurity D : R=CO-CH₂, R'=H

Impurity E : R=H
Impurity F : R=CO-CH₃

Impurity H

FIGURE 6.2 Impurities of rocuronium bromide.

test, BP and EP recommend that the spectra are recorded in the region of 4000–650 cm^{-1} (2.5–15.4 μm), while United States Pharmacopoeia (USP) recommends that the spectra are recorded in the region of 3800–650 cm^{-1} (2.6–15 μm). The infrared absorption spectra obtained for the rocuronium bromide bulk drug substance sample must correspond with that of the reference standard (RS) for a successful outcome. For the sample preparation, USP details that the sample is finely ground and dispersed in mineral oil prior to infrared spectrophotometry examination, while BP and EP do not specify a sample preparation method.

BP and EP recommend that for the bromide test, a sample solution has to be first acidified with dilute nitric acid, and silver nitrate solution is added to produce a curdled, pale yellow precipitate. This precipitate will dissolve with difficulty when suspended in a solution consisting of 2 ml of water and 1.5 ml of ammonia. Unlike BP and EP, USP employs a silver nitrate test for identification of bromide. In this test, addition of silver nitrate test solution to the solution of rocuronium bromide (1 part of rocuronium bromide in 100 parts of water) will produce a white precipitate which is insoluble in nitric acid and slightly soluble in 6N ammonium hydroxide.

USP32 NF 27 [5] also adds a chromatographic method for identification purposes. Here, USP specifies that the retention time of the major peak obtained must be similar with the RS when the procedure is conducted as per the assay in Section 2.3. The liquid chromatographic method used for identification of rocuronium bromide is similar with the method used to quantify rocuronium bromide in bulk drug substance, as detailed in Section 2.3.1.

2.2. Impurity analysis

2.2.1. Chromatographic method

All compendias [5–7] specify use of a liquid chromatographic method to determine the eight known rocuronium bromide impurities. The chromatographic method corresponds to the chromatographic method used to determine rocuronium in bulk drug substance, as detailed in Section 2.3.1. For the purpose of analyzing impurities, the RS used is the RS of rocuronium that is intended for peak identification, where the identification standard contains impurities of rocuronium bromide and rocuronium bromide itself.

One milligram of this standard mixture is dissolved with solvent mixture (10:90, v/v, water/acetonitrile) and made up with suitable volume of solvent to obtain a final solution concentration of 1 mg/ml [5–7]. The relative retention time and percentage of each impurity allowed are shown in Table 6.1. To quantify the portion of each impurity in the bulk drug substance, USP suggests that the standard solution of rocuronium bromide in the solvent mixture is injected subsequently into the liquid chromatographic system, as well as the standard mixture of impurities and the test solution of bulk drug substance. The amounts of each impurity are then calculated using the following equation:

$$100 \times \left(\frac{C_S}{C_T}\right)\left(\frac{r_u}{r_s}\right)\left(\frac{1}{F}\right) \tag{1}$$

where C_S is the concentration of USP rocuronium bromide RS in the standard solution (in mg/ml), C_T is the concentration of rocuronium bromide in the test solution, r_u is the peak area for any impurity in the test solution, r_s is the peak area of rocuronium bromide obtained from the standard solution, and F is the relative response factor as detailed in Table 6.1 [5].

In contrast with the USP method, BP and EP [6, 7] employ a reference solution obtained from the test solution and use a correction factor to calculate the percentage of each impurity. In the preparation procedure, 0.100 g of the examined substance is dissolved, then diluted to 10.0 ml with the solvent mixture, and then used as the test solution. Furthermore,

TABLE 6.1 Relative retention time of each impurity of rocuronium bromide, its requirement, and limitation

Impurity	Relative retention time [5–7]	Correction factor [6, 7]	Limit (%) [5–7]	Limitation of peak area	Relative response factor (F) [5]
A	0.20	0.47	0.2	<2 times the area of the principal peak	2.1
B	0.80	n/a	0.3	<3 times the area of the principal peak	1.0
C	1.20	n/a	0.3	<3 times the area of the principal peak	1.0
D	0.90	n/a	0.1	Less than the area of the principal peak	1.0
E	1.53	n/a	0.1	Less than the area of the principal peak	1.0
F	0.75	1.26	0.1	Less than the area of the principal peak	0.79
G	0.44	0.43	0.1	Less than the area of the principal peak	2.3
H	0.95	0.35	0.1	Less than the area of the principal peak	2.9
Unspecified impurities	n/a	n/a	0.1	Less than the area of the principal peak	n/a
Rocuronium bromide	1.0	n/a	n/a	n/a	n/a

1.0 ml of the test solution is then diluted to 100.0 ml with solvent mixture, and 1.0 ml of this diluted solution is then diluted again to 10.0 ml with mobile phase to obtain the reference solution. Five microliters each of the test solution, the reference solution, and the standard mixture are injected subsequently.

Calculation of the levels of each impurity is performed by multiplying the peak areas of impurities A, F, G, and H with the corresponding correction factor of those impurities, as provided in Table 6.1. In addition, the requirement that the peak area of each impurity in the test solution corresponds with the principal peak of rocuronium bromide obtained from the RS is also detailed in Table 6.1. In addition, all compendias [5–7] suggest that any peak eluted prior to the elution of impurity A should be ignored due to the blank and to bromide ion that elutes just before impurity A. Furthermore, any peak with an area less than 0.5 times of the principal peak of rocuronium bromide should be disregarded.

2.3. Assay methods

2.3.1. Bulk drug substance

A liquid chromatographic method is utilized for quantifying rocuronium bromide in samples of bulk drug substance. USP employs a liquid chromatography (LC) column made from porous silica particles of 5–10 μm in diameter (i.e., the L3 column). On the other hand, BP and EP use silica gel for chromatography (5 μm diameter particle size) as the stationary phase, which is a commercial grade suitable for column chromatography. Both types of column used have a column length of 25 cm. All compendias specify use of the same mobile phase, which is a mixture of acetonitrile and 0.025 M tetramethylammonium hydroxide pentahydrate, where the pH is adjusted to 7.4 with 9:1 phosphoric acid. The flow rate is adjusted to 2.0 ml/min; the column temperature is set to 30 °C, and the detector at 210 nm. Both the rocuronium bromide standard and the test solution are dissolved with the solvent mixture (as used for the impurity analysis) to obtain a final concentration of 1 mg/ml. According to these compendial methods, the retention time of rocuronium bromide should be 9.0 min and the total running time will be around 20 min.

3. NUCLEAR MAGNETIC RESONANCE

A 400 MHz nuclear magnetic resonance (NMR) method for the identification of rocuronium bromide was reported by Fielding [10]. The ^{1}H spectra were referenced to an internal TMS standard, while the solvent signals at 77.0 ppm (CDCl$_3$) or 39.6 ppm (DMSO-d_6) were used as references for the ^{13}C data. Based on the NMR spectra, it was concluded that rocuronium bromide in the solution phase existed in a state of dynamic equilibrium between chair and twist-boat conformations of ring A. This twist-boat conformation was stabilized by the intramolecular hydrogen bonding within the *trans* amino-ol configuration. The 2*b*-morpholine group was able to rotate freely, and was noted to rapidly invert between the two chair conformations. The rotation of the morpholine group was associated with the nitrogen-ring inversion process for six-membered rings. Furthermore, the twist-boat conformation was favored by the non-polar solvents used (e.g., 85% twist-boat in $CDCl_3$), while the chair conformation was prevalent into the polar aprotic solvents used (e.g., 88% chair in DMSO-d_6).

Another NMR method was utilized by Cameron *et al.* [11] to determine rocuronium bromide in its complexed form with γ-cyclodextrin. In this method, the experiments were performed at 400 MHz on a Bruker DRX spectrometer at 303 K. The method entailed the use of ^{1}H NMR spectra of samples that had been dissolved in D_2O at pH 7.5. However, in this report, the authors did not specifically discuss the ^{1}H NMR result for rocuronium bromide as single compound.

4. LIQUID CHROMATOGRAPHY

Two LC methods have been reported for determining rocuronium bromide in pharmaceutical dosage forms [12, 13]. The first method reported was used to examine the stability of a propofol–rocuronium mixture, a mixture employed for the rapid induction of anesthesia. In this method, a μBondapak CN column (150 mm $\times$ 3.9 mm i.d.) was used as the stationary phase. A standard mixture of rocuronium bromide and propofol was prepared in acetonitrile, having 5:3 proportion of rocuronium bromide and propofol, respectively. The mobile phase consisted of 60:40 (v/v) acetonitrile/water, and the wavelength detection was set at 220 nm. It was found that in this mixture, rocuronium bromide was stable up to 48 h after mixing [13].

Błażewicz *et al.* [12] reported a liquid chromatographic method that used electrochemical detection to determine rocuronium bromide and its impurities, particularly impurities A and C. These latter impurities are also known to be two metabolites of rocuronium bromide, namely N-desallylrocuronium and 17-desacetylrocuronium (impurity A and impurity C, respectively). In this method, the mobile phase used was the same as mentioned in all compendias, and the column used was a Hypersil 100 Silica 5 μm (250 mm $\times$ 4.6 mm). Although the solvent and the composition of the mobile phase is the same as the compendial methods, and the flow rate was set up at a lower rate than the compendial methods (1.5 ml/min), the retention time of rocuronium bromide and its impurities were found to be lower than the compendial methods. The total running time was 600 s (10 min), which is seen to be twofold shorter than the compendial methods (20 min).

In this procedure, the amperometric detection mode was chosen since this is more specific and selective than UV detection, and therefore enables detection of very low concentration of impurities, as well as removal of the interferences caused by electroinactive substances [12, 14]. When using a glassy carbon electrode as the working electrode and silver–silver chloride (Ag/Ag Cl) as the reference electrode, the chromatograms were recorded at an amperometric detection potential of 0.9 V. The potential 0.9 V was chosen due to the fact that rocuronium was electrochemically active and underwent anodic oxidation at the potential of +0.65 and +0.95 V against the Ag/AgCl electrode. Furthermore, impurity B and impurity D underwent anodic oxidation at the potential close to the upper limit of available measuring (reported as >0.9 V), resulting in the lack of sensitivity for those impurities. Additionally, as the electrode potential increased, the noise of the electrochemical baseline also increased.

By using this method, the limit of detection (LOD) and limit of quantitation (LOQ) achieved were 7 and 25 ng/ml, respectively, for impurities A and G. The highest LOD and LOQ were 225 and 750 ng/ml for impurity B. Recoveries of rocuronium bromide and impurity C ranged between 98.91% and 104.20%. Moreover, when stored at 6 ± 2 °C, stock solutions of rocuronium and its impurities were stable in the solvent system for 2 weeks. After 2 months of storage, the content of impurity A in the mixture decreased while the levels of impurities C and G increased.

5. DETERMINATION OF ROCURONIUM BROMIDE IN BIOLOGICAL SAMPLES

5.1. Liquid chromatography

In the human biological system, rocuronium bromide is eliminated unchanged by the biliary and urinary routes [15], and has two metabolites as mentioned in Section 4. As detailed in Table 6.2, determination of rocuronium bromide in biological samples was mostly carried out using LC–MS detection [16–23], although some workers made use of electrochemical detector [24–26] and only one report used a fluorimetric detector [27].

The main advantage of using a fluorimetric detector is its inherently high sensitivity for pharmaceutical analysis, but very often application of this method to nonfluorescent drug substances requires derivatization with a strong fluorophore. The derivatization process can be performed either precolumn or postcolumn to increase the detection sensitivity [14].

TABLE 6.2 Summary of HPLC methods used to analyze rocuronium bromide and its metabolites in biological sample

Analyte(s)	HPLC conditions	Sample	Solvent preparation of standard, sample extraction, and cleanup method	LOD, LOQ, and accuracy (Rec)	References
Rocuronium bromide 17-Desacetyl derivative of rocuronium bromide (Org 9943) N-Desallyl derivative of rocuronium bromide (Org 20860) 3,17-Dihydroxy derivative of vecuronium (i.s.)	Column: Lichrospher 100-RP 18 (150 × 39 mm i.d.) and μBondapack C_{18} precolumn (4 × 6 mm i.d.) at room temperature Mobile phase: water: dioxane = 84:16 (v/v). Water contained 0.1 M NaH_2PO_4 0.11 mM DAS and 0.11 mM 1-heptane-sulfonic acid, adjusted to pH 3.0 with orthophosporic acid Organic extractant: dichloroethane After each series of analysis, HPLC column was flushed with 15 ml water and 75 ml MeOH Detector: fluorimetric detector operated at 385 nm (excitation) and 452 nm (emission)	Human plasma, urine, bile, tissue homogenates, stoma fluid	Standard: 0.1 M NaH_2PO_4, pH 3.0 Sample: LLE: sample acidified with NaH_2PO_4. Acidified sample was then made to volume with water up to 2.0 ml, added with 1.0 ml KI-glycine buffer and extracted with dichloromethane. Organic layer was evaporated at 37 °C at air dry stream then dissolved with mobile phase	*Rocuronium bromide* In plasma: LOD: 3 ng LOQ: 10 ng/ml Rec: 91.5–107.1% In urine and bile: LOD: 4 ng LOQ: 100 ng/ml (bile) Rec: 88.0–109.4% In tissue homogenates: LOD: 5 ng LOQ: 100 ng/ml Rec: 89.8–102.8% *Org 9943* In plasma: LOD: 5 ng LOQ: 20 ng/ml Rec: 91.5–105.6% In urine and bile: LOD: 8 ng LOQ: 250 ng/ml (bile) Rec: 89.7–114.0% In tissue homogenates: LOD: 10 ng LOQ: 100 ng/ml Rec: 93.4–106.2%	[27]

| Rocuronium bromide Pancuronium (i.s.) | Human plasma | Standard: n/a
Sample:
LLE: sample acidified with sodium hydrogen phosphate then frozen. Before extraction, sample was added with 1 M sodium hydrogen phosphate and i.s. Afterward, 1 ml of 6 M potassium iodide was added to 0.5 ml of plasma and extracted with 5 ml of toluene, mixed and centrifuged. Organic layer was then separated and dried by evaporated. Dried residue was dissolved in mobile phase | Column: Nucleosil C_{18} HD column (12.5 cm × 4 mm) protected with Nucleosil C_{18} HD guard column
Mobile phase: 50 mmol/1 ammonium formate buffer (pH 3): MeOH = 73:27 (v/v)
Detector: MS/MS operated in electrospray ionization (ESI)
Ion monitoring:
Rocuronium bromide: m/z 529.4 → 487.4
Pancuronium: m/z 286.4 → 236.7 | *Org 20860*
In plasma:
LOD: 15 ng
LOQ: 20 ng/ml
Rec: 92.6–111.1%
In urine and bile:
LOD: 10 ng
LOQ: 250 ng/ml (bile)
Rec: 88.6–186.4%
In tissue homogenates:
LOD: 25 ng
LOQ: 250 ng/ml
Rec: 85.0–105.2%
LOD: n/a
LOQ: 5 μg/ml
Rec: 90% | [20] |

(continued)

TABLE 6.2 (*continued*)

Analyte(s)	Sample	HPLC conditions	Solvent preparation of standard, sample extraction, and cleanup method	LOD, LOQ, and accuracy (Rec)	References
Rocuronium bromide Verapamil (i.s.)	Human plasma	Column: Symmetry Shield RP18 cartridge (50 × 2.1 mm) Detector: ESI-MS Mobile phase: linear gradient from 10% to 90% of ACN in water containing 0.1% TFA, applied in 15 min. Column was then washed for 3 min at final gradient condition and set back to initial condition in 1 min and equilibrated for 5 min Single ion recording: Rocuronium bromide: m/z 265 and m/z 529 Verapamil: m/z 455	Standard: Rocuronium bromide: 0.1 M phosphate buffer, pH 3.0 Verapamil: MeOH: distilled water (1:100) Both rocuronium bromide and verapamil were diluted with ACN Sample: 0.5 ml sample was added with 200 μl 1 M phosphate buffer, 50 μl i.s., 500 μl saturated aqueous potassium iodide solution. Extraction was done by adding 5 ml of dichloromethane. The dichloromethane phase was evaporated under nitrogen stream at 30 °C. Dried residue was then dissolved in ACN	LOD: n/a LOQ: 25 ng/ml Rec: 94–106%	[19]

Rocuronium bromide Org 25289 (i.s.)	Guinea pig plasma and urine	Column: Jupiter C4 (50 × 4.6 mm) from Phenomenex. Mobile phase: 20 mM ammonium acetate (A) and ACN (B). Gradient mode with 90% A for the first 1.0 min dropping to 10% by 2.5 min and maintained for the next 1.0 min before being equilibrated for the next 0.5 min at its initial composition. Detector: MS. Ionization source was the turbo-ion spray (T-SP) operated in positive ion mode. MRM monitoring: Rocuronium bromide: m/z 529 → 487. Org 24748: m/z 310 → 91	Standard: ACN: water = 50:50. Sample: Plasma: plasma sample added with 300 μl ACN then vortexed and centrifuged. Supernatant was then concentrated down using sample concentrator, then added with 5 μl of 10 μg/ml i.s. and 10 μl TFA to form precipitation and centrifuged. The resultant supernatant was then injected. Urine: similar with the plasma	LOD: n/a. LOQ: 25 ng/ml (plasma and urine). Rec: 92–107% (plasma) 88–113% (urine). [18]
Rocuronium bromide and other quaternary nitrogen muscle relaxants	Human urine, bile, liver, serum, whole blood	Column: Inertsil ODS2 column (Chrompack). Mobile phase: 50 mM ammonium acetate buffer (pH 5.0) and	Standard: standards in the form of injection fluids were diluted with MeOH and kept at −18 °C when not in use	LOD: n/a. LOQ: n/a. Rec: 97%. [21]

(continued)

TABLE 6.2 (continued)

Analyte(s)	HPLC conditions	Sample	Solvent preparation of standard, sample extraction, and cleanup method	LOD, LOQ, and accuracy (Rec)	References
	ACN in the gradient mode. 95% buffer was set for 5 min then changed to 70% buffer. This condition maintained for the next 10 min then downgraded to 65% at 25th min at the 30th min until 33rd min, percentage of A remained constant at 30% and brought back to the initial condition (95%) in the next 0.5 min. This condition remained constant for the next 1.5 min and back again to the initial condition Detector: ESI-MS m/z 529.33		Sample: Preparation: 1.0 ml sample was added with 2.0 ml ammonium carbonate buffer (pH 9.3, 0.01 M), then vortexed and centrifuged. Supernatant was then used. For tissue samples, the tissue was minced then added with ammonium carbonate buffer with doubled volume SPE: BondElut C18 HF cartridge was rinsed with 2 ml MeOH and 2 ml ammonium carbonate buffer. Sample was then introduced to the column and rinsed twice with 1.5 ml ammonium carbonate. Cartridge was then vacuum dried, added with 50 μl hexane, and vacuum dried again. Retained drugs were		

Analyte	Chromatographic conditions	Sample preparation	Performance	Ref.
Rocuronium bromide and other quaternary nitrogen muscle relaxants	Column: Nucleosil C$_{18}$ LC Packings (150 × 1.0 mm) protected by a 0.5 μm frit Mobile phase: MeOH: 50 mM NH$_4$COOH (pH 3.0), using linear gradient method (30% MeOH to 90% in 6 min) Detector: ESI-MS m/z 487 (all MS data were recorded in the full scan mode with m/z 200 to 720)	eluted with 1 ml MeOH containing 0.1 M acetic acid and eluted under gravity force. The elutes were dried in a vacuum then reconstituted in 100 μl of a solution of ammonium acetate: ACN (7:3) and centrifuged for 5 min at 4000 rpm Standard: n/a (standard was in the form of injection fluid) Sample: 1 ml sample mixed with 1 ml saturated KI solution and 1 ml of 0.8 M phosphate buffer (NaH$_2$PO$_4$), pH 5.4, then extracted with 5 ml methylene chloride. The organic phase was then transferred and evaporated. Dry extract was reconstituted in 2 mM NH$_4$COOH, pH 3.0	LOD: 0.1 mg/ml LOQ: n/a Rec: 85.3%	[17]

(continued)

TABLE 6.2 (*continued*)

Analyte(s)	HPLC conditions	Sample	Solvent preparation of standard, sample extraction, and cleanup method	LOD, LOQ, and accuracy (Rec)	References
Rocuronium bromide	Similar with the method developed by Gutteck	Human cerebrospinal fluid, plasma		CSF: LOD: 0.5 ng/ml LOQ: n/a Rec: n/a Plasma: LOD: 5 ng/ml LOQ: n/a Rec: n/a	[37]
Rocuronium bromide and other nondepolarizing neuromuscular blocking agents Ambenonium (i.s.)	Column: X-Terra MS C18 (150 × 1 mm) Mobile phase: (A) 2 mM NH₄COOH (pH 3); (B) ACN/2 mM NH₄COOH (pH 3) (90:10, *v*/*v*). Gradient elution started from 15% B for 2 min then increased to 43% for 8 min and followed by 1 min reequilibration at 15% B Detector: atmospheric pressure ionization (API) MS with an electrospray type. *m/z* 529.4 (quantitation ion) and *m/z* 358.4 (confirmation ion)	Human serum, plasma, whole blood, urine, and gastric content	Standard: n/a Sample: n/a Standard: stock solutions were prepared in MeOH. Working solutions were prepared by diluting stock solutions with 0.625 mM sulfuric acid Sample: sample was acidified with 0.5 M sulfuric acid, vortexed and preserved at −20 °C. Sample was then added with i.s. in MeOH and 1 ml ACN then vortexed and centrifuged. Supernatant was drawn and evaporated under nitrogen stream to 100 or 150 μl volume	LOD: 2.5 μg/ml LOQ: 5 μg/ml Rec: n/a	[22]

| Rocuronium bromide and other quartenary ammonium drugs Benzyldimethylphenylammonium chloride (i.s.) | Equine urine | Column: Supelcosil LC-8-DB column (10 cm × 2.1 mm) Mobile phase: (A) 10 mM ammonium formate (pH 3.0); (B) ACN. Gradient mode initiated from 100% A and decreased to 0% in 12 min. This condition was hold for 4 min and brought back to 100% A in the next 2 min. This condition was kept stable for 4 min Detector: MS–MS with API source was operated in the positive ESI mode m/z: 529 → 487 | Standard: n/a Sample: 1 ml diluted with 0.01 M ammonium carbonate and adjusted to pH 9.3 SPE: ISOLUTE® CBA SPE column, conditioned with 5 ml MeOH, 5 ml water, and 5 ml 0.01 M ammonium carbonate, pH 9.3. Sample then passed through the SPE and washed with 0.01 M ammonium carbonate (pH 9.3), MeOH, chloroform, and MeOH. Each was given in 5 ml volume. Cartridge was then eluted with MeOH containing 1% formic acid. Eluate was evaporated under nitrogen and the residue was reconstituted in 10 mM ammonium formate (pH 3.0):ACN (8:2, v/v) | LOD: n/a LOQ: n/a Rec: 60% | [23] |

(continued)

TABLE 6.2 (continued)

Analyte(s)	HPLC conditions	Sample	Solvent preparation of standard, sample extraction, and cleanup method	LOD, LOQ, and accuracy (Rec)	References
Rocuronium Vecuronium	Column: CN spherisorb column (15 cm × 4.6 mm). Column temperature maintained at 35 °C Mobile phase: water: ACN:MeOH = 55:35:10 (adjusted to pH 4.75) Detector: electrochemical detector (Coulochem II, ESA) set at 750 mV	Venous plasma of dog	Standard: stock solution was made in 2 M H_2SO_4 and working solution was made by diluting with Modified Krebs Ringer bicarbonate buffer Sample: SPE on Bond Elut C1 cartridge	LOD: n/a LOQ: 5 ng/ml Rec: n/a	[24, 25]
Rocuronium	Column: X-Terra RP-C18 column (20 × 4.6 mm) Detector: amperometric electrochemical detection with electrode potential 1.1 V Mobile phase: n/a	Human plasma	Standard: n/a Sample: LLE using dichloroethane and extraction liquid ACN: NH_3 = 96:4 (v/v)	LOD: n/a LOQ: 20 ng/ml Rec: n/a	[26]

302

Analyte	Chromatographic conditions	Matrix	Sample preparation	Performance	Ref.
Rocuronium bromide, other quaternary ammonium drugs and herbicides	Column: Atlantis® dC18 (100 mm × 2.1 mm) Mobile phase: (A) 15 mM HFBA–20 mM ammonium formate buffer adjusted to pH 3.30 by formic acid (B) MeOH 100%. Linear gradient run from 5% to 90% of solvent B within 18 min. Detector: MS/MS using ESI in the positive mode m/z 529 → 487.2	Human whole blood	Standard: rocuronium bromide in the injection fluid diluted with MeOH Sample: 1 ml blood sample was diluted with 4 ml ammonium acetate solution, pH 8.0. SPE: BondElut® LRC-CBA, conditioned with 3 ml MeOH followed by 3 ml phosphate buffer, pH 6.0. Sample passed through was washed with 3 ml phosphate buffer (pH 6.0) followed by 3 ml MeOH. Cartridge was then eluted with 1 ml of 1.0 M HCl:MeOH (70:30, v/v) and evaporated to dryness under nitrogen. Residue dissolved with initial HPLC mobile phase	LOD: 4.9 ng/ml LOQ: 16.2 ng/ml Rec: 97.2–102.4%	[16]

In the method developed by Kleef *et al.* [27], the fluorescent anion 9,10-dimethoxyanthracene-2-sulfonate and postcolumn ion pair extraction into dichloromethane with a second pump device was used to increase the fluorescence of rocuronium bromide.

It is widely known that electrochemical detection can be used as an alternative for the analysis of drug substances that have no chromophores, but do have electroactive sites such as quaternary ammonium groups. Rocuronium bromide is an *N*-substituted piperazine, which has an electroactive center and therefore enables the use of electrochemical detection [26, 28]. Moroever, this detection method leads to the observation acceptable sensitivity and selectivity in compound analysis, and can be used without prior derivatization [14].

Compared to other LC detection methods, LC–MS provides better analysis for biological samples, particularly due to its excellent selectivity against interferences in the sample [29]. As reported by Kleef *et al.* [27], the desallyl derivative (one of the metabolites of rocuronium) could be detected in stomach fluid, but could not be quantitated by fluorescence detection due the existence of fluorescent endogenous substances in the sample. In addition, LC–MS requires no prior sample derivatization as do other detection methods [27, 29]. Detection of analytes in biological samples using LC–MS with electrospray ionization is suitable for the ionization of polar compounds such as rocuronium bromide [30].

As noted by Farenc *et al.* [19], due to the presence of the ammonium moiety, the absence of any organic acid modifier would lead to peak tailing on a reversed-phase column during the analysis of rocuronium bromide. When using formic acid or acetic acid, rocuronium bromide was eluted in the solvent front through the use of trifluoroacetic acid, enabling good separation between rocuronium, and the internal standard verapamil. In contrast, Yiu *et al.* [23] found that when ammonium formate was present in the eluting solvent system, the system became more sensitive by 2–4 times in detecting quaternary ammonium groups (such as rocuronium bromide), and when compared to formic acid, led to the observation of good peak shapes.

Although compendial methods [5–7] suggest the use of polar columns to analyze rocuronium bromide, almost all of the methods found in the literature for determining rocuronium bromide in biological samples used the nonpolar C18 column [16, 17, 19–22, 26, 27]. Other methods reported successfully utilized C8, C4, and CN columns for analyzing rocuronium bromide in biological samples, where these column types are more polar than C18 columns [18, 23–25, 31].

It is widely accepted that the extraction of quaternary ammonium drugs is difficult due to the presence of positively charged ionic compounds, and this applies to rocuronium bromide as well [23]. As a consequence, some of the extraction methods reported made use of ion pair formation through iodide or iodide–glycine complexes [19, 23, 27]. For the extraction procedure, Table 6.2 illustrates that most of the methods applied liquid–liquid extraction (LLE), although a few also employed solid phase extraction. With respect to the solid phase extraction (SPE) method, elution of the compound from the cartridge is mostly done under acidic conditions when the silanol groups are protonated. This is due to the condition that rocuronium bromide as a quaternary nitrogen has both a very polar group and a large apolar site. Under alkaline conditions, residual silanol groups on the solid phase sorbent will be negatively charged, causing interaction with the positively charged quaternary nitrogen groups. At the same time, the apolar site of the molecule interacts with the C18 chains of the sorbent [21].

The most common SPE types used in extracting rocuronium bromide from biological samples are the C18, C1, and LRC-CBA cartridges [16, 21, 24, 25]. Compared to the other SPE types, the LRC-CBA cartridge is specified for weak cation exchange. Although the C18 and C1 cartridges are not reported to have cation exchange capacity, these cartridge types are suitable for polar compounds due to the predominant effect of the long hydrocarbon chain of C18 and the endcapped sorbent of the C1 cartridge [32–34].

In order to avoid degradation of rocuronium in biological samples, the most common method employed requires first acidifying the sample using NaH_2PO_4 or sulfuric acid [20, 22, 27]. However, as shown by Farenc et al. [19], rocuronium bromide was also stable in plasma samples at 20 and 4 °C for 4 h when stored at physiological pH, and could be recovered with average percent of 99%. It was also found that the frozen plasma sample containing rocuronium bromide at -30 °C could maintain drug stability for at least 75 days. A similar result was also found by Kleef et al. [27], who reported that rocuronium bromide in plasma stored at physiological pH was stable after 9.5 months storage at -20 °C [27]. Therefore, it can be concluded that the acidification process can be avoided when the plasma sample is frozen immediately following the sampling process.

5.2. Thin-layer chromatography

A qualitative thin-layer chromatography method has been described by Kleef et al. [27] for the detection of rocuronium bromide and its metabolites in biological samples. This method was developed to confirm the identity of rocuronium bromide and its metabolites prior to their determination using HPLC coupled with fluorescence detection. In this method, the dried residue from the extraction process was dissolved in 0.05 ml of 0.01 M HCl. The stationary phase used was silicagel plates, that were developed in a mobile phase consisting of 2% solution of NaI in 2-propanol. The elution process was run for 4 h, and after the elution process was complete, the plates were dried and colored by spraying with iodoplatinate reagent (which is specific for nitrogen-containing organic compounds). The eluted fractions appeared as small brown spots on a light yellow background, having R_F values 31.8 for rocuronium, 62.0 for Org 9943, and 52.3 for Org 20860 [27].

5.3. Gas chromatography

Two gas chromatographic methods have been reported for the determination of rocuronium bromide and its metabolites in biological samples [35, 36]. The first method published for analyzing rocuronium bromide and 17-desacetylrocuronium required prior derivatization of the plasma sample using N-methyl-N-(tert-butyldimethylsilyl)-trifluoroacetamide at 70 °C overnight after ion pairing with saturated aqueous KI and LLE with dichloromethane through an Extrelut cartridge. Through the use of this particular LLE cartridge, shorter extraction times could be achieved and the water-free separated organic phase avoided the longer separation process of dichloromethane from the aqueous phase.

In this extraction process, the upper liquid phase after extraction was evaporated at 60 °C and the residue dissolved in acetonitrile before being derivatized. Prior to ion pairing and extraction, the plasma sample was mixed with 0.8 M phosphate buffer (pH 5.4) to prevent degradation of rocuronium bromide. The gas chromatography (GC) system made use of a nitrogen-specific detector or thermionic sensitive detector (TSD), which was modified by applying a rubidium dotted glass bead and a DB1 column with a 0.25-μm film thickness and 0.32 mm i.d. The purpose of using rubidium dotted glass bead was to prolong the life span of the TSD bead, since this type of TSD bead is extremely sensitive to moisture, acetone, and silicone compounds. The carrier gas used was helium, and the system ran with a temperature program ranging from 120 to 300 °C. This GC method was characterized by detection limits of 10 ng/ml of rocuronium bromide and 50 ng/ml for 17-desacetylrocuronium. Quantitation limits of 50 ng/ml for rocuronium bromide and 80 ng/ml for 17-desacetylrocuronium were reported as well. Recovery values for rocuronium bromide ranged from 88% to 103%, and from 108% to 121% for 17-desacetylrocuronium [36].

In contrast with the previous method, Gao et al. [35] reported a GC–MS method which did not require prior derivatization for the analysis of rocuronium bromide and 17-desacetylrocuronium in biological samples. To prevent hydrolysis of rocuronium, solutions containing 1 M NaH_2PO_4 was added to plasma samples, which were then stored at -20 °C prior to analysis. Similar with the previously described method, the internal standard was 3-desacetylvecuronium. The extraction process employed here also involved liquid–liquid ion pair extraction with saturated

KI followed by elution with dichloromethane. The organic phase was subsequently evaporated under a gentle stream of nitrogen at room temperature, and the residue dissolved in acetone prior to its injection onto the column.

Analysis of rocuronium bromide and its metabolite was conducted in a DB-5 chemical bonded capillary column and the system comprised helium as the carrier gas with temperature program ranging from 120 to 300 °C with an electron impact mass spectrometer as the detector [35]. Selected ion monitoring technique was applied to quantify rocuronium bromide, 17-desacetylrocuronium, and the internal standard at m/z values of 413 (for rocuronium bromide), 236 and 447 (for metabolites of rocuronium), and 425 (for the internal standard). The extraction efficiency obtained here ranged from 65% to 92% for rocuronium bromide, and from 41% to 65% for 17-desacetylrocuronium. The LOQ achieved for rocuronium bromide was 26 ng/ml, and was 870 ng/ml for 17-desacetylrocuronium. This method used a mixture of sodium chloride with potassium iodide as the ion pair buffer to optimize partitioning of drug–iodide ion pairs into the organic phase. In addition, the volume of KI added per milliliter of plasma was lower compared to the previous method, with more dichloromethane being used to extract the analytes. It is thought that these factors might contribute to the higher extraction efficiency acquired using this latter GC–MS method. However, the lower degree of extraction efficiency obtained for 17-desacetylrocuronium might be due to the more polar nature of this metabolite.

REFERENCES

[1] Martindale: The Complete Drug Reference, 35th ed., CD ROM, The Pharmaceutical Press, London, 2007.
[2] Zemuron Product Information, www.spfiles.com/pizemuron.pdf (July 20, 2009).
[3] E. Adar, D. Sondack, O. Friedman, I. Manascu, T. Fizitzki, B. Freger, O. Arad, A. Weisman, J. Kaspi, US Patent 2005/0159398 A1, 2005.
[4] C. Lee, Br. J. Anaesth. 87 (2001) 755.
[5] United States Pharmacopoeia 32 National Formulary 27, The United States Pharmacopeial Convention, Rockville, MD, 2009.
[6] European Pharmacopoeia 6.0, Council of Europe, London, 2006.
[7] British Pharmacopoeia 2009, The Stationery Office, London, 2009.
[8] A.C. Moffat, M.D. Osselton, B. Widdop (Eds.), Clarke's Analysis of Drugs and Poisons, Pharmaceutical Press, London, 2004, p. 1541.
[9] M.J. O'Neil, The Merck Index, Merck & Co. Inc., New Jersey, 2001, p. 1481.
[10] L. Fielding, Magn. Reson. Chem. 36 (1998) 387.
[11] K.S. Cameron, J.K. Clark, A. Cooper, L. Fielding, R. Palin, S.J. Rutherford, M.-Q. Zhang, Org. Lett. 4 (2002) 3403.
[12] A. Błażewicz, Z. Fijalek, M. Warowna-Grześkiewicz, M. Boruta, J. Chromatogr. A 1149 (2007) 66.
[13] S. Karaca, Z. Salihoglu, A. Duran, S. Rollas, Can. J. Anaesth. 50 (2003) 314.

[14] C. Wang, J. Xu, G. Zhou, Q. Qu, G. Yang, X. Hu, Comb. Chem. High Throughput Screen. 10 (2007) 547.

[15] J.H. Proost, L.I. Eriksson, R.K. Mirakhur, G. Roest, J.M. Wierda, Br. J. Anaesth. 85 (2000) 717.

[16] M.M. Ariffin, R.A. Anderson, J. Chromatogr. B 842 (2006) 91.

[17] V. Cirimele, M. Villain, G. Pépin, B. Ludes, P. Kintz, J. Chromatogr. B 789 (2003) 107.

[18] O. Epemolu, I. Mayer, F. Hope, P. Scullion, P. Desmond, Rapid Commun. Mass Spectrom. 2002 (1946) 16.

[19] C. Farenc, C. Enjalbal, P. Sanchez, F. Bressolle, M. Audran, J. Martinez, J.-L. Aubagnac, J. Chromatogr. A 910 (2001) 61.

[20] U. Gutteck-Amsler, K.M. Rentsch, Clin. Chem. 46 (2000) 1413.

[21] C.H.M. Kerskes, K.J. Lusthof, P.G.M. Zweipfenning, J.P. Franke, J. Anal. Toxicol. 26 (2002) 29.

[22] H. Sayer, O. Quintela, P. Marquet, J.-L. Dupuy, J.-M. Gaulier, J. Anal. Toxicol. (2004) 28.

[23] K.C.H. Yiu, E.N.M. Ho, T.S.M. Wan, Chromatographia 59 (2004) S45.

[24] S. Ezzine, F. Varin, Br. J. Anaesth. 94 (2005) 49.

[25] S. Ezzine, N. Yamaguchi, F. Varin, J. Pharmacol. Toxicol. Methods 49 (2004) 121.

[26] B. Woloszczuk-Gebicka, E. Wyska, T. Grabowski, A. Świerczewska, R. Sawicka, Paediatr. Anaesth. 16 (2006) 761.

[27] U.W. Kleef, J.H. Proost, J. Roggeveld, J.M.K.H. Wierda, J. Chromatogr. 621 (1993) 65.

[28] J. Ducharme, F. Varin, D.R. Bevan, F. Donati, Y. Theoret, J. Chromatogr. 573 (1992) 79.

[29] W.M.A. Niessen, J. Chromatogr. A 856 (1999) 179.

[30] S.J. Gaskell, J. Mass Spectrom. 32 (1997) 677.

[31] M.C. McMaster, HPLC, a Practical User's Guide, John Wiley & Sons, New Jersey, 2007.

[32] Varian Bond Elut—Non Polar SPE Columns, www.chromtech.com/2001catalog/separatePgs/118.pdf (August 20, 2009).

[33] Varian Bond Elut—Cation Exchange SPE Columns, www.chromtech.com/2001catalog/separatePgS/125.pdf (August 20, 2009).

[34] Bond Elut C1, www.chromtech.com/2001catalog/SeparatePgs/120.pdf (August 20, 2009).

[35] L. Gao, I. Ramzan, B. Baker, J. Chromatogr. B 757 (2001) 207.

[36] R. Probst, M. Blobner, P. Luppa, D. Neumeier, J. Chromatogr. B 702 (1997) 111.

[37] T. Fuchs-Buder, M. Strowitzki, K. Rentsch, J.U. Schreiber, S. Philipp-Osterman, S. Kleinschmidt, Br. J. Anaesth. 92 (2004) 419.

CHAPTER **7**

Vigabatrin

Abdulrahman Al-Majed

Contents

Department of Pharmaceutical Chemistry, College of Pharmacy, King Saud University, Riyadh, Saudi Arabia

Profiles of Drug Substances, Excipients, and Related Methodology, Volume 35
ISSN 1871-5125, DOI: 10.1016/S1871-5125(10)35007-2

1. DESCRIPTION

1.1. Nomenclature

1.1.1. Chemical names [1–4]
4-Aamino-5-hexenoic acid
4-Aminohex-5-enoic acid
γ-Vinyl-γ-aminobutyric acid
γ-Vinyl-GABA
(*RS*)-4-Aminohex-5-enoic acid

1.1.2. Nonproprietary names [1]
Vigabatrin

1.1.3. Preparation [1]
Vigabatrin oral powder, vigabatrin tablets.

1.1.4. Proprietary names
Sabril, Sabrilex, Sabrilan.

1.1.5. Definition [3]
Vigabatrin is (*RS*)-4-aminohex-5-enoic acid. It is a structural analog of γ-aminobuyric acid (GABA) with a vinyl appendage. It contains not less than 98.0% and not more than 102.0% of $C_6H_{11}NO_2$, calculated with reference to the anhydrous, ethanol-free substance. The compound exists as a racemic mixture of *R*(−) and *S*(+) isomers.

1.2. Formulae [1]

1.2.1. Empirical formula, molecular weight, and CAS number

Vigabatrin	$C_6H_{11}NO_2$	129.16	[60643-86-9]

1.2.2. Structure

1.3. Elemental composition [1]

The theoretical elemental composition of vigabatrin is as follows:

Carbon	55.80%
Hydrogen	8.85%
Nitrogen	10.84%
Oxygen	24.77%

1.4. Appearance [2]

Vigabatrin is a white to off-white crystalline solid.

2. METHODS OF PREPARATION

1. Wei and Knaus [5, 6] reported the synthesis of (*S*)-vigabatrin. Selective protection of L-glutamic acid **1** using concentrated sulfuric acid in EtOH, and subsequent reaction with methyl chloroformate afforded

the glutamate ethyl ester derivative **2** in 93% yield. Treatment of compound **2** with $ClCO_2i$-Bu in the presence of triethylamine followed by reduction with $NaBH_4$ led to alcohol **3** in 72% yield. Swren oxidation of compound **3** followed by olefination under Wittig reaction conditions using Ph_3=CH_2 produced the vinyl derivative **4** in 64% yield. Cleavage of N-methoxycarbonyl protective group in compound **4** with TMSI and subsequent hydrolysis of the ester group afforded the enantiometrically pure (S)-vigabatrin in 89% yield (Scheme 7.1).

2. Wei and Knaus [6, 7] also reported the synthesis of (S)-vigabatrin using (R)-methionine **1** as the starting material via a one-pot reduction–homologation procedure. In this context, esterification of (R)-methionine with thionyl chloride chloroformate in methanol followed by treatment with benzyl chloromate gave (R)-NCbz-α-amino-carboxylic methyl ester **2** in 82% yield. Methyl ester **2** was transformed into (R)-N-benzyloxycarbonyl-γ-amino-α,β-unsaturated carboxylate **3** in 68% yield. Hydrogenation of the double bond in compound **3** using magnesium-methanol gave γ-lactam **4**. Oxidation of the sulfide function into the corresponding sulfoxide followed by a thermal elimination reaction afforded the (S)-5-vinyl-γ-lactam **5** in 56% yield. Basic hydrolysis of γ-lactam **5** led to (S)-vigabatrin (Scheme 7.2).

3. Dangoneau *et al.* [6, 8] described the preparation of (S)-vigabatrin by nucleophilic addition of the lithiated anion generated from propiolate methyl ester **1** and n-BuLi, to enantiometrically pure (S)-nitrone **2** at −78 °C in THF afforded the N-hydroxylamino derivative *syn*-**3** as a single diastereoisomer in 94% yield. Catalytic hydrogenation of *syn*-**3** in the presence of Raney-nickel and (Boc)$_2$O gave the N-Boc-γ-amino-methyl ester derivative **4**, in 76% yield. Cleavage of the isopropylidine protective group in compound **4** with p-toluenesulfonic acid (PTSA) gave the corresponding diol **5** in 71% yield, which by reductive elimination of the two hydroxyl groups using Ph_3P and I_2 gave the

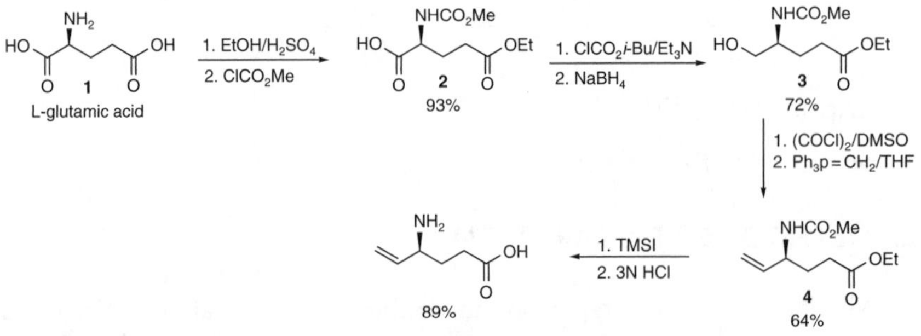

SCHEME 7.1

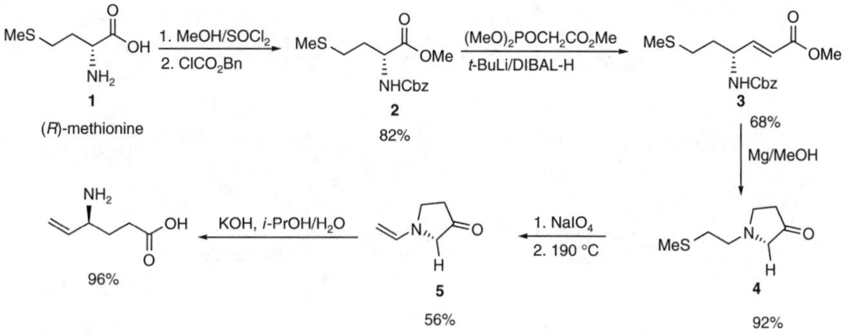

SCHEME 7.2

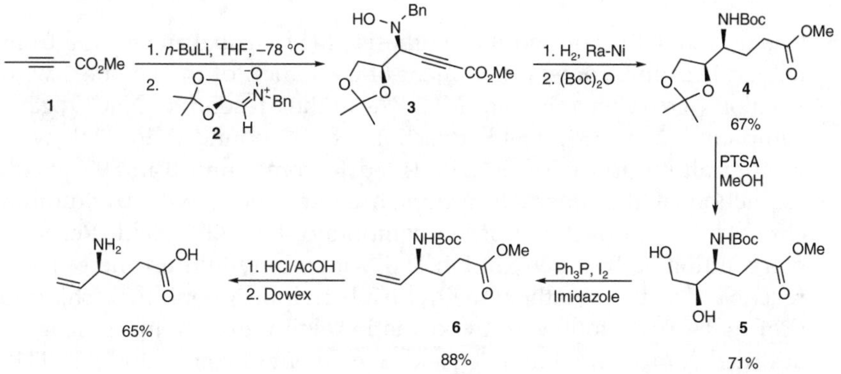

SCHEME 7.3

N-Boc-vigabatrin methyl ester **6** in 88% yield. Finally, acidic hydrolysis of compound **6** led to (S)-vigabatrin in 65% yield (Scheme 7.3). In a same manner, enantiometrically pure (R)-vigabatrin was obtained using (R)-nitrone **2** as the starting material.

4. Gheorghe *et al.* [9] described the synthesis of (S)-vigabatrin using the (R)-1,3-didehydro-pyrohomoglutamate **2** obtained from pyrrole **1** as the starting material. In this context, conjugate reduction of compound **2** with NaBH$_4$ in the presence of NiCl$_2$·6H$_2$O followed by N-Boc deprotection using AlCl$_3$ gave methyl ester derivative **3** in 41% yield and 99% after crystallization. Reduction of the methyl ester group in compound **3** with LiBH$_4$ afforded the corresponding alcohol **4** in 93% yield. Treatment of compound **4** with PBr$_3$ and subsequent dehydobromination with KOt-Bu produced vinylpyrrolidinone **5**, which by basic hydrolysis led to (S)-vigabatrin (Scheme 7.4).

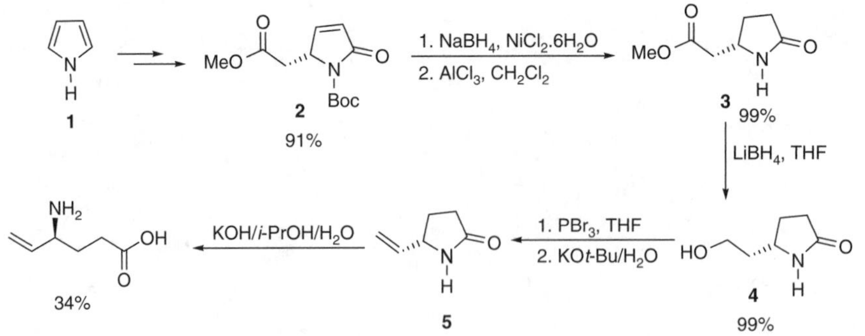

SCHEME 7.4

5. Chang *et al.* [10] reported the synthesis of (S)-vigabatrin starting from ketone **1**. In this context, treatment of a solution of the ketone **1** with *m*-chloroperoxybenzoic acid (MCPBA) in the presence of $NaCO_3$ gave compound **2** in 86% yield. Reaction of compound **2** in THF with lithium aluminum hydride (LAH) led to compound **3** in 89% yield. Reduction of the aldehyde group in compound **3** with pyridinium chlorochromate (PCC) afforded compound **4** in 82% yield. Reaction of a solution of the aldehyde **4** in THF with *n*-butyllithium and sequent reaction with methoxymethyl-triphenylphosphonium chloride (PhP=CHOMe) followed by oxidation with excess Jones reagent afforded compound **5** in 68% yield. Treatment of compound **5** in THF with sodium naphthalenide in the presence of HCl led to (S)-vigabatrin in 71% yield (Scheme 7.5).

6. Recently Raj and Sudalai [11] described asymmetric synthesis of (S)-vigabatrin started via cocatalyzed hydrolysis kinetic resolution of the racemic epoxide **1** which led to the chiral diol **2** in 51% yield and the chiral epoxide **2** in 47% yield. Regiospecific ring opening of the epoxide **2** with dimethylsulfonic methylide (formed by the treatment of $(CH_3)_3S^+I^-$ with *n*-BuLi at $-10\ ^\circ C$) gave the allyl alcohol **3** in 89% yield. Direct nucleophilic displacement of the alcohol in compound **4** with azide anion gave azide **5**, which was smoothly reduced to the amine (Ph_3P, THF, H_2O) and subsequently protected (Ac_2O, Py) as *N*-acetate **6** in 97% yield. The PMB group was selectively deprotected with DDQ to give the corresponding alcohol **7**, which was further subjected to oxidations (IBX and $NaClO_2$) to give carboxylic acid **7** in 77% yield. Finally, the *N*-acetyl moiety was deprotected on reduction with hydrazine hydrate to afford the (S)-vigabatrin in 87% (Scheme 7.6).

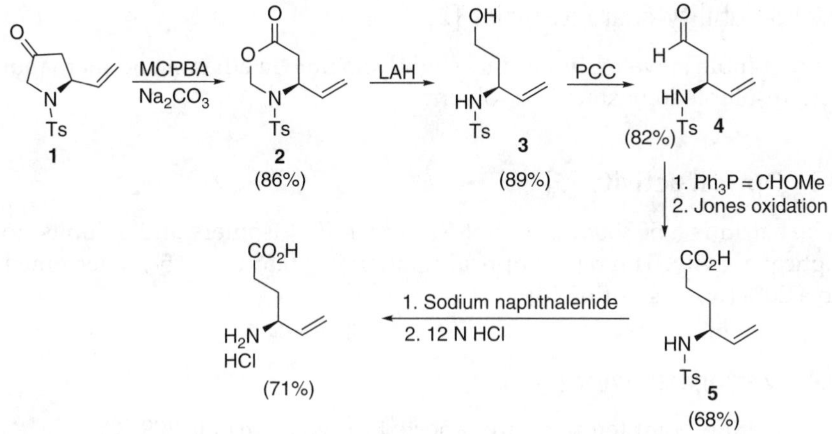

SCHEME 7.5

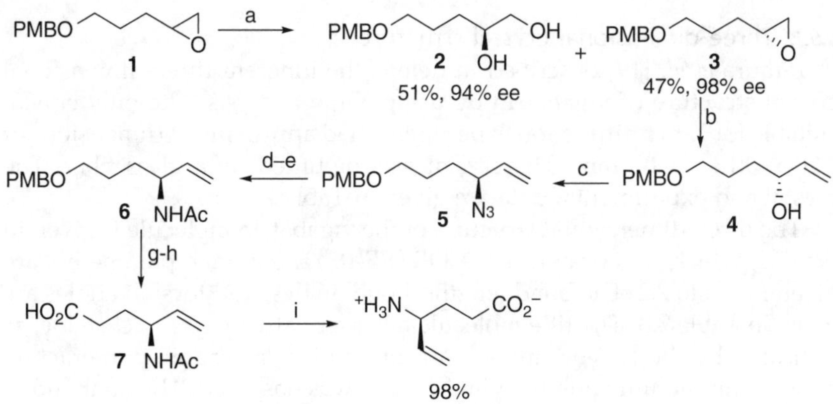

SCHEME 7.6 Reagents and conditions: (a) (R,R)-Co(III)-salen-OAc (0.5 mol%), H_2O (0.55 equiv.), 25 °C; (b) $Me_3S^+I^-$, n-BuLi, THF, -10 °C, 89%; (c) NaN_3, PPH_3, DMF, CCl_4, 60 °C, 79%; (d) PPH_3, THF, H_2O, 89%; (e) Ac_2O, CH_2Cl_2, pyridine, 97%; (f) DDQ, CH_2Cl_2, 25 °C, 93%; (g) IBX, DMSO, 25 °C; (h) $NaClO_2$, NaH_2PO_4, DMSO, 77%; (i) $N_2H_4 \cdot H_2O$, THF, MeOH, 87%.

Several other methods have been published for the preparation of vigabatrin [12–18].

3. PHYSICAL CHARACTERISTICS

3.1. Ionization constant [4]

Vigabatrin has two ionizable groups with pK_a values of 4.02 (carboxylic moiety) and 9.72 (γ-amine moiety).

3.2. Solubility characteristics [2]

Very soluble in water, but is only slightly soluble in ethanol and methanol and insoluble in hexane and toluene

3.3. Optical activity [3]

Vigabatrin is a 50/50 mixture of $R(-)$ and $S(+)$ isomers and exhibits no optical activity. The report optical rotation is $+0.5°$ to $-0.5°$, determined in a 20% (w/v) solution [3].

3.4. Meting behavior [1]

The melting point temperature reported for vigabatrin is 209 °C.

3.5. Crystal structure

3.5.1. Three-dimensional crystal structure

Haramura *et al.* [19] described in details the inherent three-dimensional crystal structure of vigabatrin by using X-ray analysis. The only crystal suitable for X-ray diffraction experiments had approximate dimensions of $0.20 \times 0.18 \times 0.07$ mm. The crystal was mounted on a glass fiber. The crystal and experimental data are given in Table 7.1.

The three-dimensional structure of the vigabatrin molecule is given in Fig. 7.1, which was drawn using ORTEP-III. The atomic parameters are given in Table 7.2. The bond lengths, bond angles, and torsion angles are given in Table 7.3. The title molecule takes a relatively compact shape, as indicated by the torsion angles. An intramolecular cation–π interaction between ammonium and the vinyl groups was observed. The nonbonded intramolecular N$\cdots$CS and N$\cdots$C6 distances are 2.482(7) and 2.880(7) Å, respectively, and the N–C–C=C torsion angle is $-12.6(7)°$. These geometrical parameters indicate that the vinyl group points toward the ammonium nitrogen atom. In addition, the C=C bond length of the vinyl group is relatively short. This intramolecular cation–π interaction plays an important role to restrain the conformation of vigabatrin, and which should help this molecule to more preferentially bind to GABA-T. There were four intermolecular hydrogen bonds between the carboxylate and ammonium groups. The geometrical parameters are as follows: N(1)$\cdots$O (1)^i = 2.777(6) Å, N(1)–H = 1.04(5) Å, H$\cdots$O(1)^i = 1.74(5) Å, $\angle$N(1)–H$\cdots$O (1)^i = 172(5)°;   N(1)$\cdots$O(2)^{ii} = 2.784(5) Å,   N(1)–H = 0.96(4) Å,   H$\cdots$O (2)^{ii} = 1.86(4) Å, $\angle$N(1)–H$\cdots$O(2)^{ii} = 161(4)°;   N(1)$\cdots$O(1)^{iii} = 2.843(8) Å, N(1)–H = 0.92(5) Å, H$\cdots$O(1)^{iii} = 1.93(5) Å, $\angle$N(1)–H$\cdots$O(1)^{iii} = 173(5)°;

TABLE 7.1   Crystal and experimental data of vigabatrin

Formula: $C_6H_{11}NO_2$
Formula weight: 129.16
Crystal system: orthorhombic
Space group: $Fdd2$
$Z = 16$
$a = 16.471(2)$ Å
$b = 22.717(3)$ Å
$c = 7.2027(7)$ Å
$V = 2695.1(5)$ Å^3
$D_x = 1.273$ g/cm^3
No. of observations $(I > 2.00\sigma(I)) = 395$
$\theta_{max} = 68.14°$ with Cu Kα
Residuals: $R(I > 2.00\sigma(I)) = 0.036$
$(\Delta/\sigma)_{max} = 0.005$
$(\Delta/\rho)_{max} = 0.14$ e/Å^3
Measurement: Rigaku RAXIS-RAPID
Program system: crystal structure [20]
Structure determination: SIR92[2]
Refinement: full-matrix

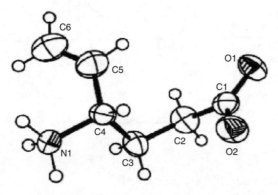

FIGURE 7.1   ORTEP-III drawing of vigabatrin. Thermal ellipsoids of non-H atoms are drawn at the 50% probability level.

$N(1)\cdots O(2)^{iii} = 3.223(8)$ Å, $N(1)–H = 0.92(5)$ Å, $h\cdots O(2)^{iii} = 2.55(6)$ Å, $\angle N(1)–H\cdots O(1)^{iii} = 130(4)°$ [symmetry   operations: (i) $-1/4 + x$, $1/4 - y, 3/4 + z$; (ii) $1 - x, 1/2 - y, 1/2 + z$; (iii) $x, y, 1 + z$]. The fourth hydrogen bond is marginal.

TABLE 7.2 Atomic coordinates and equivalent isotropic thermal parameters (B_{eq})

Atom	x	y	z	B_{eq}
O(1)	0.6179(2)	0.1477(1)	0.138(1)	4.13(8)
O(2)	0.5380(2)	0.2143(1)	0.264(1)	4.36(8)
N(1)	0.5067(3)	0.1703(2	0.8446(7)	3.2(1)
C(1)	0.5991(3)	0.1815(2)	0.2658(8)	3.3(1)
C(2)	0.6539(3)	0.1832(3)	0.4361(7)	4.3(1)
C(3)	0.6114(3)	0.1962(2)	0.6191(7)	3.7(1)
C(4)	0.5433(3)	0.1547(2)	0.6600(7)	3.3(1)
C(5)	0.5675(3)	0.0914(2)	0.651(1)	5.3(2)
C(6)	0.5732(3)	0.0548(2)	0.781(1)	5.9(2)

$B_{eq} = 8/3\ \pi^2(U_{11}(aa^*)^2 + U_{22}(bb^*)^2 + U_{33}(cc^*)^2 + 2U_{12}(aa^*bb^*)\cos\gamma + 2U_{13}(aa^*cc^*)\cos\beta + 2U_{23}(bb^*cc^*)\cos\alpha)$.

TABLE 7.3 Bond lengths (Å), bond angles (°), and torsion angles (°)

O(1)–C(1)	1.236(8)	O(2)–C(1)	1.253(6)
N(1)–C(4)	1.502(7)	C(1)–C(2)	1.523(7)
C(2)–C(3)	1.521(7)	C(3)–C(4)	1.494(7)
C(4)–C(5)	1.494(7)	C(5)–C(6)	1.256(9)
C(2)–C(1)–O1)	117.8(5)	C(2)–C(1)–O(2)	118.0(5)
O(1)–C(1)–O(2)	124.2(6)	C(3)–C(2)–C(1)	115.5(4)
C(4)–C(3)–C(2)	113.2(4)	C(5)–C(4)–N(1)	111.8(4)
C(5)–C(4)–C(3)	113.4(4)	N(1)–C(4)–C(3)	109.1(4)
C(6)–C(5)–C(4)	128.8(6)		
O(1)–C(1)–C(2)–C(3)	−148.3(5)	O(2)–C(1)–C(2)–C(3)	31.7(7)
C(1)–C(2)–C(3)–C(4)	55.4(6)	C(2)–C(3)–C(4)–N(1)	178.7(4)
C(2)–C(3)–C(4)–C(5)	53.4(6)	N(1)–C(4)–C(5)–C(6)	−12.6(7)
C(3)–C(4)–C(5)–C(6)			

3.5.2. X-Ray powder diffraction pattern

The X-ray powder diffraction pattern of vigabatrin was performed at College of Science, King Saud University using a Simons XRD-5000 diffractometer. Table 7.4 shows the values of the scattering angles (° 2θ), the interplanar d-spacings (Å), and the relative intensities (%) for vigabatrin, which were automatically obtained on a digital printer. Figure 7.2 shows the X-ray powder diffraction pattern of the pure sample of the drug.

TABLE 7.4 Crystallographic data from the X-ray powder diffraction pattern for vagabatrin

Scattering angle (° 2θ)	d-Spacing (Å)	Relative intensity (%)	Scattering angle (° 2θ)	d-Spacing (Å)	Relative intensity (%)
12.920	6.8464	4	37.460	2.3988	8
13.160	6.7220	17	37.900	2.3720	5
15.220	5.8165	5	39.160	2.2985	4
15.360	5.7638	11	39.340	2.2884	9
17.480	5.0693	5	40.120	2.2457	6
18.300	4.8439	6	40.440	2.2287	8
18.400	4.8178	9	40.760	2.2119	80
18.520	4.7869	17	41.040	2.1974	11
18.660	4.7513	23	41.340	2.1822	7
18.840	4.7063	50	41.520	2.1731	16
20.240	4.3838	5	41.720	2.1632	23
20.340	4.3625	13	42.120	2.1436	4
20.500	4.3288	48	42.180	2.1407	6
21.160	4.1952	9	42.380	2.1310	6
21.300	4.1680	5	42.580	2.1215	4
22.500	3.9483	19	43.660	2.0715	16
22.700	3.9140	100	44.460	2.0360	6
22.920	3.8769	21	44.640	2.0282	7
23.060	3.8537	69	44.800	2.0214	14
23.300	3.8145	21	45.240	2.0027	4
23.480	3.7857	9	45.500	1.9919	4
25.360	3.5092	4	45.600	1.9877	4
25.440	3.4983	5	45.680	1.9844	5
26.520	3.3582	10	45.740	1.9820	4
26.470	3.3311	14	46.220	1.9625	4
27.500	3.2408	5	46.360	1.9569	9
27.660	3.2224	50	46.420	1.9545	9
27.900	3.1952	74	46.560	1.9490	7
30.160	2.9607	8	46.800	1.9395	5
30.920	2.8896	17	46.900	1.9356	6
31.180	2.8661	7	47.420	1.9156	4
31.600	2.8290	44	47.620	1.9080	6
31.720	2.8186	20	47.860	1.8990	6
32.980	2.7137	5	48.080	1.8908	6
33.260	2.6915	5	52.300	1.7478	5
33.460	2.6759	12	52.360	1.7459	6
33.480	2.6743	13	54.600	1.6795	4

(*continued*)

TABLE 7.4   *(continued)*

Scattering angle (° 2θ)	*d*-Spacing (Å)	Relative intensity (%)	Scattering angle (° 2θ)	*d*-Spacing (Å)	Relative intensity (%)
33.540	2.6697	18	54.880	1.6715	5
33.720	2.6558	26	54.940	1.6699	4
33.860	2.6452	11	58.740	1.5706	4
33.960	2.6376	5	60.540	1.5281	5
35.340	2.5377	13	60.660	1.5254	5
35.540	2.5239	9	60.780	1.5226	6
35.600	2.5157	7	60.900	1.5199	4
35.740	2.5102	6	61.040	1.5168	5
35.960	2.4954	4	61.940	1.4969	8
36.160	2.4820	10	66.620	1.4026	4

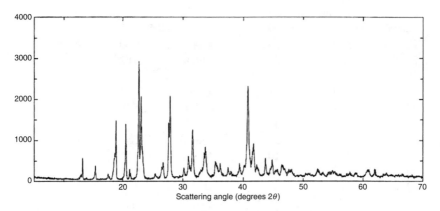

FIGURE 7.2   X-Ray powder diffraction pattern of vigabatrin.

3.6. Spectroscopy

3.6.1. UV–VIS spectrophotometry
The UV absorption spectrum of vigabatrin was performed at College of Pharmacy, King Saud University using a Shimadzu UV–VIS model 1601 PC Spectrophotometer. The drug exhibited two maxima at 213 nm (*A* 1%, 1 cm = 28) and at 226 nm (*A* 1%, 1 cm = 57).

3.6.2. Vibrational spectroscopy

The IR absorption spectrum of vigabatrin was performed at College of Pharmacy, King Saud University using a Perkin-Elmer model Spectrum BX FT-IR apparatus. The spectrum shown in Fig. 7.3 was obtained with the compound being compressed in a KBr pellet. The assignments for the major IR absorption bands are shown in Table 7.5.

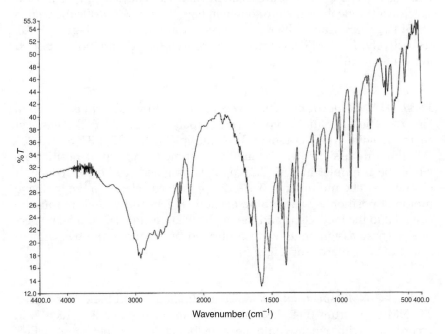

FIGURE 7.3 The infrared absorption spectrum of vigabatrin (KBr pellet).

TABLE 7.5 Assignment for the main infra-red absorption bands of vigabatrin

Assignments	Frequency (cm^{-1})
COOH	2450–3100
NH$_2$	3300–3400
O–H stretching	2600–3200
C–H olefinic stretching	(3030w)
C–H stretching	(2924w)
C=O stretching	1646
C=C stretching	1580

3.6.3. Nuclear magnetic resonance spectroscopy

Nuclear magnetic resonance (NMR) spectra were performed at College of Pharmacy, King Saud University using a Bruker 500 MHz apparatus. Both ^{1}H and ^{13}C NMR spectra of vigabatrin (racemic mixture and the pure S-isomer) were performed in D_2O using TMS as an internal standard (IS). Summaries of the observed ^{1}H and ^{13}C resonance bands, together with the peak assignments, are listed in Table 7.6. The detailed ^{1}H and ^{13}C spectra of (S)-vigabatrin were shown in Figs. 7.4 and 7.5, while the results of DEPT, COSY, and HSQC experiments are shown in Figs. 7.6–7.8, respectively. The NMR spectral data were accordance with those reported in the literature [5, 9, 10, 21].

3.6.3.1. ^{1}H NMR spectrum   The ^{1}H spectrum of (S)-vigabatrin is shown in Fig. 7.4. The aliphatic protons were initially assigned on the basis of chemical shift considerations. The multiplets at 1.75–1.83 and 1.90–1.95 ppm are assigned to the methylene function group at position 3-, while the multiplet at 2.11–2.23 ppm is assigned to the methylene at position 2-. The multiplet at 3.67–3.71 ppm is correlated to the methane proton at position 4-. The multiplets at 5.31–5.35 and 5.69–5.76 ppm are assigned to the methylene and the methane at positions 6- and 5-, respectively. These assignments were confirmed by means of the two dimensional COSY experiment (Fig. 7.5).

3.6.3.2. ^{13}C NMR spectrum   A noise-modulated broad-band decoupled ^{13}C NMR spectrum (Fig. 7.6) showed six resonance bands, which is in accordance with what would be anticipated for vigabatrin. A DEPT experiment (Fig. 7.7) permitted the identification of two methines at 53.9 and 132.9 ppm in addition to three methylenes at 28.9, 33.3, and 121.0 ppm. One carbonyl absorption was assigned at 181.3 ppm. These assignments were all conformed through the performance of HSQC experiment (Fig. 7.8).

3.6.4. Mass spectrometry

The mass spectrum of vigabatrin was obtained utilizing the instrument Agilent 6410 QQQ LC/MSMS, Agilent Technologies. It was made by ESI-VE electron spray ionization. Figure 7.9 shows the detected mass fragmentation pattern of vigabatrin. The major peaks in the spectrum occur at m/z 128.1, 110, 82.1, 71, and 53.9. Table 7.7 shows the detailed mass fragmentation pattern of the drug. Clarke reported the following principal peaks at m/z 65, 84, 111, 69, 82, 54, 67, and 45 [4].

TABLE 7.6 ^1H and ^{13}C resonance assignments for vigabatrin racemic mixture and (Δ) isomer

	Racemic vigabatrin			(S)-Vigabatrin				Assignment
$\Delta\,^1H$ (ppm)	No. of H	M^a	$\Delta\,^{13}C$ (ppm)	$\Delta\,^1H$ (ppm)	No. of H	M^a	$\Delta\,^{13}C$ (ppm)	
1.72–1.80	1 H	m	–	1.75–1.83	1 H	m	–	
1.86–1.93	1 H	m	28.8	1.90–1.95	1 H	m	28.9	3—CH_2
2.07–2.19	2 H	m	33.3	2.11–2.23	2 H	m	33.3	2—CH_2
2.64–3.68	1 H	m	53.9	3.67–3.71	1 H	m	53.9	4—CH
5.28–5.33	2 H	m	121.0	5.31–5.35	2 H	m	121.0	6—CH_2
5.66–5.71	1 H	m	132.9	5.69–5.76	1 H	m	132.9	5—CH
–	–	–	181.3	–	–	–	181.3	CO

^{a} M = multiplicity.

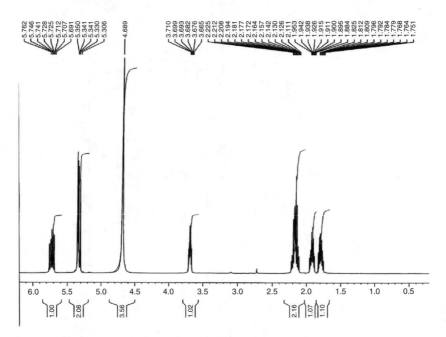

FIGURE 7.4 ¹H NMR spectrum of (S)-vigabatrin.

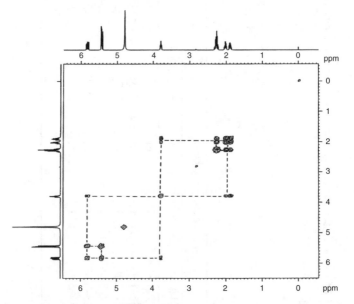

FIGURE 7.5 COSY experiment of (S)-vigabatrin.

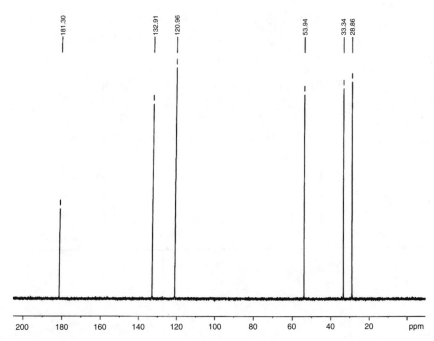

FIGURE 7.6 ^{13}C NMR spectrum of (S)-vigabatrin.

4. METHODS OF ANALYSIS

4.1. British Pharmacopoeia compendial methods [3]

4.1.1. Identification
The following identifications are listed in the compendia:

A. The *infrared absorption spectrum*, Appendix II A, is concordant with the *reference spectrum* of vigabatrin *(RS 360)*.
B. In the assay, the retention time of the principal peak in the chromatogram obtained with solution (1) is the same as that of the principal peak in the chromatogram obtained with solution (2).

4.1.2. Impurity analysis
The sum of the impurities determined by methods A and B below is not more than 0.5%.

A. Not more than 0.1% of any single impurity determined by the following method. Carry out the method for *liquid chromatography*, Appendix III D, using the following solutions in the mobile phase. Solution (1)

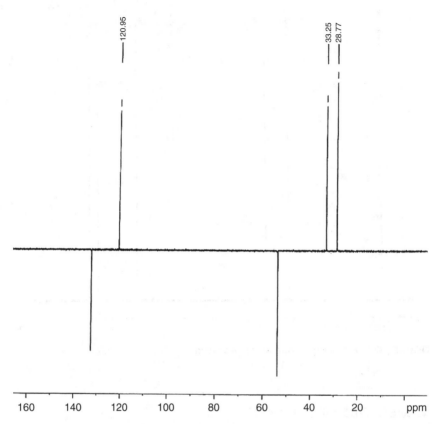

FIGURE 7.7 DEPT experiment of (S)-vigabatrin.

contains 0.4% (w/v) of the substance being examined. Solution (2) contains 0.0004% (w/v) of *3-aminopent-4-ene-1, 1-dicarboxylic acid BPCRS*. Solution (3) contains 0.0004% (w/v) of *5-vinyl-2-pyrrolidone BPCRS*. Solution (5) contains 0.002% (w/v) of *5-vinyl-2-pyrrolidone BPCRS* and 0.4% (w/v) of *vigabatrin BPCRS*.

The chromatographic procedure may be carried out using (a) two stainless steel columns in series; the first (25 cm × 4.6 mm) packed with particles of silica, the surface of which has been modified with chemically bonded hexysilyl groups (5 μm) (Spherisorb C6 is suitable) and the second (25 cm × 4.6 mm) packed with cation exchange resin (10 μm) (Partisil-10 SCX is suitable); (b) as the mobile phase at a flow rate of 1.0 ml/min, a mixture of 25 volumes of *acetonitrile*, 25 volumes of a phosphate buffer solution prepared by dissolving 58.5 g of *sodium dihydrogen orthophosphate*

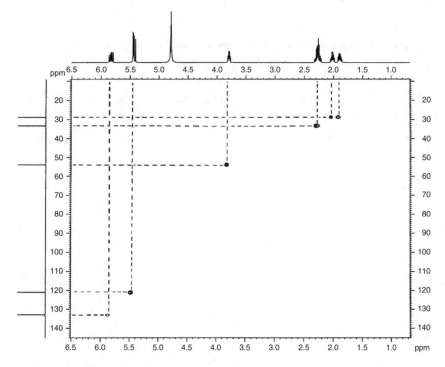

FIGURE 7.8 HSQC experiment of (S)-vigabatrin.

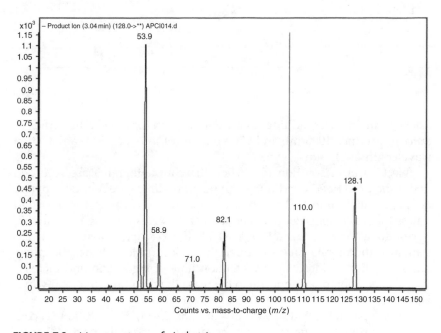

FIGURE 7.9 Mass spectrum of vigabatrin.

TABLE 7.7 Mass spectral fragmentation pattern of vigabatrin

m/z^a	Relative intensity (%)	Formula	Structure
128.1	41	$C_6H_{11}NO_2$	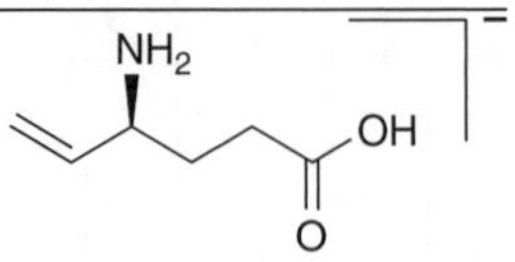
110.0	30	C_6H_9NO	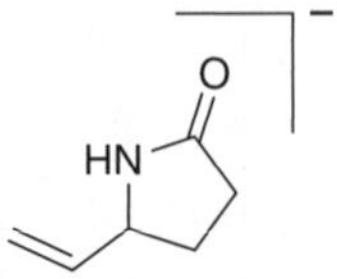
82.1	24	C_5H_9N	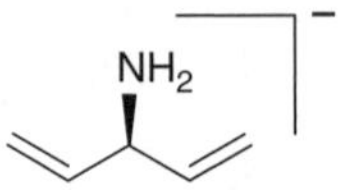
71.0	8	$C_3H_4O_2$	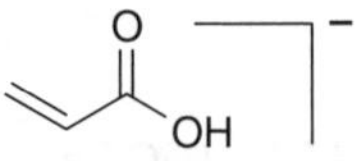
53.9	100	C_3H_5N	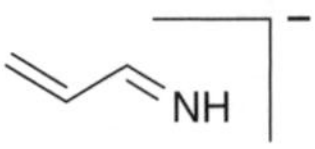

[a] All are negative ions (mass − 1).

monohydrate in *water*, adding 25 ml of *orthophosphoric acid* and sufficient *water* to produce 1000 ml, and 950 volumes of *water*; and (c) a detection wavelength of 210 nm.

Inject 20 µl of solution (5). When the chromatograms are recorded under the prescribed conditions, the retention times are 5-vinyl-2-pyrrolidone, about 18 min and vigabatrin, about 21 min. The test is not valid unless the *resolution factor* between the peaks corresponding to 5-vinyl-2-pyrrolidone and vigabatrin is at least 1.5; if necessary, adjust the concentration of the mobile phase (reduce the concentration of the phosphate buffer solution to increase the retention time of vigabatrin and increase the concentration of acetonitrile to decrease the retention time of 5-vinyl-2-pyrrolidone).

Inject separately 20 µl of each of solutions (1)–(4). For solution (1) continue the chromatography for twice the retention time of the principal peak.

In the chromatogram obtained with solution (1), the areas of any peaks corresponding to 5-vinyl-2-pyrrolidone and (*E*)-4-amino-2-ethylidenebutyric acid are not greater than the areas of the peaks in the chromatograms obtained with solutions (3) and (4), respectively (0.1% of each). The area of any other *secondary peak* is not greater than the area of the peak in the chromatogram obtained with solution (2) (0.1%). Calculate the percentage content of 5-vinyl-2-pyrrolidone and (*E*)-4-amino-2-ethylidenebutyric acid using the areas of the peaks in the chromatograms obtained with solutions (3) and (4), respectively, and of any other impurity from the peak in the chromatogram obtained with solution (2) taking, for the purposes of calculation, that this peak is equivalent to 0.1%, and hence determine the sum of the contents of the impurities.

B. Not more than 0.2% of 4-aminobutyric acid when determined by the following method. Carry out the method for *liquid chromatography*, Appendix III D, using the following solutions. For solution (1) add to 1 ml of a 0.20% (w/v) solution of the substance being examined 2 ml of a solution prepared by dissolving 7.7 g of *boric acid* in water, adjusting to pH 7.7 with a 50% (w/v) solution of *sodium hydroxide* and diluting to 250 ml with water and mix. Add 3 ml of a 0.16% (w/v) solution of *(9-fluorenyl)methyl chloroformate* in acetone, mix and allow to stand for 5 min. Add 3 ml of *ethyl acetate*, shake vigorously for a few seconds, and allow to separate; use the lower layer within 8 h of preparation. Prepare solution (2) in the same manner using 1 ml of a solution containing 0.20% (w/v) of the substance being examined and 0.002% (w/v) of *4-aminobutyric acid* in place of the solution of the substance being examined.

The chromatographic procedure may be carried out using (a) a stainless steel column (15 cm × 4.6 mm) packed with particles of silica the surface of which has been modified by chemically bonded phenyl groups (5 μm) (vydac phenyl is suitable); (b) as the mobile phase at a flow rate of 1.0 ml/min, a mixture of 25 volumes of *acetonitrile* and 75 volumes of a solution prepared by dissolving 8.2 g of *anhydrous sodium acetate* in *water*, adjusting the pH to 4.2 with *glacial acetic acid* and diluting to 2000 ml with *water*; and (c) a detection wavelength of 263 nm.

Inject 25 μl of solution (2). When the chromatogram is recorded under the prescribed conditions, the retention times for the derivatives of the following are (9-fluorenyl)methanol, about 6 min; 4-aminobutyric acid, about 9 min; and vigabatrin, about 14 min. The test is not valid unless the *resolution factor* between the peaks corresponding to the derivatives of 4-aminobutyric acid and (9-fluorenyl)-1-methanol is at least 2; if necessary, adjust the proportion of the components of the mobile phase.

Inject separately 25 μl of each of solutions (1) and (2) and calculate the content of 4-aminobutyric acid in the substance being examined from the chromatograms obtained.

4.1.3. Other tests
4.1.3.1. Heavy metals   Two grams complies with *limit test C for heavy metals,* Appendix VII. Use 2 ml of *lead standard solution (10 ppm Pb)* to prepare the standard (10 ppm).

4.1.3.2. Ethanol   Not more than 0.6% (w/w) of ethanol when determined by the following method. Carry out the method for headspace *gas chromatography* (GC), Appendix III B. Solution (1) contains 2 g of the substance being examined in 10 ml of a solution containing 0.0025% (w/v) of *1,2-dichloroethane* (IS) in *water* in a 20-ml headspace vial. Heat the vial at 60 °C for 30 min. Solution (2) contains 0.025% (w/v) of *absolute ethanol* and 0.0025% (w/v) of *1,2-dichloro-ethane* in water. Use *water* as the blank.

The chromatographic procedure may be carried out using a fused-silica column (60 m × 0.32 mm) coated with a 1.0-μm film of bonded methylsilicone (SPB-1 is suitable) at an initial temperature of 35 °C for 12 min, increasing to 175 °C at a constant rate of 10 °C/min, maintaining the injector at 150 °C and the detector at 250 °C. Use *helium* as the carrier gas. Calculate the content of ethanol in the substance being examined from the areas of the peaks in the chromatograms obtained with solutions (1) and (2).

4.1.3.3. Water   Not more than 0.5% (w/w), Appendix IX C. Use 0.3 g dissolved in 50 ml of *anhydrous methanol.*

4.1.3.4. Sulphated ash   Not more than 0.1%, Appendix IX A.

4.1.4. Assay method
Carry out the method for *liquid chromatography,* Appendix III D, using the following solutions in the mobile phase. Solution (1) contains 0.2% (w/v) of the substance being examined. Solution (2) contains 0.2% (w/v) of *vigabatrin BPCRS.* Solution (3) contains 0.002% (w/v) of *5-vinyl-2-pyrrolidone BPCRS* and 0.2% (w/v) of *vigabatrin BPCRS.* The chromatographic procedure may be carried out using (a) a stainless steel column (25 cm × 4.6 mm) packed with cation exchange resin (10 μm) (Whatman Partisil 10 SCX is suitable); (b) as the mobile phase with a flow rate of 1.5 ml/min a mixture of 4 volumes of *acetonitrile,* 40 volumes of *methanol,* and 1000 volumes of a 0.34% (w/v) solution of *potassium dihydrogen orthophosphate,* the mixture adjusted to pH 2.8 with *orthophosphoric acid;* and (c) a detection wavelength of 210 nm.

Inject 20 μl of solution (3). When the chromatogram is recorded under the prescribed conditions, the retention times are 5-vinyl-2-pyrrolidone, about 5 min and vigabatrin, about 8 min. The test is not valid unless the *resolution factor* between the peaks corresponding to 5-vinyl-2-pyrrolidone and vigabatrin is at least 1.5.

Inject separately 20 μl of solutions (1) and (2) and calculate the percentage content of vigabatrin from the areas of the peaks using the declared content of $C_6H_{11}NO_2$ in *vigabatrin BPCRS*.

4.1.5. Impurities

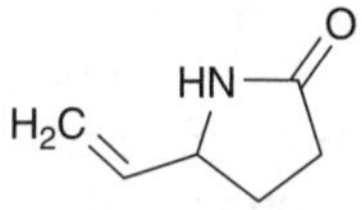

A. 5-Vinyl-2-pyrrolidone

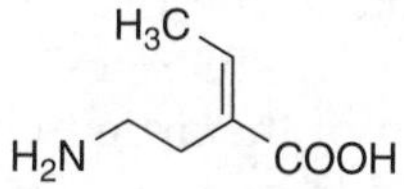

B. (*E*)-4-Amino-2-ethylidenebutyric acid

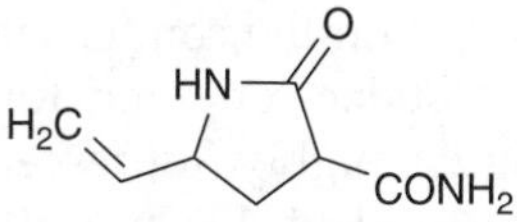

C. 2-oxo-5-Vinylpyrrolidine-3-carboxamide

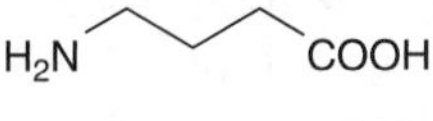

D. 4-Aminobutyric acid

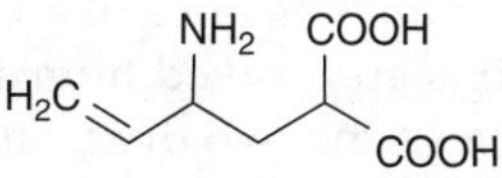

E. 3-Aminopent-4-ene-1,1-dicarboxylic acid

4.2. Spectroscopic methods

4.2.1. Spectrofluorotometry

Grove *et al.* [22] described a sensitive method for determination vigabatrin in plasma and urine. The method was based on coupling the drug with o-phthaladehyde and measuring the resulting fluorescence. Plasma and acidified urine was deproteinized with triacetic acid in the presence of Li citrate buffer (pH 2.2) for direct analysis on DC-6 resin in a liquimat II amino acid analyzer with 0.668 M Li citrate buffer (pH 4.6). The drug was determined by fluorimetry after online derivatization with o-phthaladehyde. The fluorescence intensity was a linear function of the concentration of vigabatrin of the range of 0.2–5.0 nmol. Recovery of the drug was 93.8–99%. The method was used to study the pharmacokinetics of the drug after a 1-g oral dose.

Hassan *et al.* [23] described a highly sensitive method for determination vigabatrin and gabapentin in their dosage forms and spiked human plasma. The method was based on coupling the drugs with 4-chloro-7-nitrbenzo-2-oxa-1,3-diazole (NBD-Cl) in borate buffer and measuring the resulting fluorescence at 532 nm after excitation at 465 nm. The fluorescence intensity was a linear function of the concentration of vigabatrin of the range of 1.3–6.5 μg/ml. The proposed method was successfully applied to the determination of the drug in its dosage form, and the percent recovery ± SD was 104.53 ± 1–2. The method was further applied to the determination of the drug in spiked plasma samples. No interference was encountered from coadministrated drugs.

Al-Zehouri *et al.* [24] developed a selective and sensitive method for determination of vigabatrin and gabapetin in dosage forms and biological fluids. The method was based on the condensation of the drugs through their amino groups with acetylacetone and formaldehyde according to Hantzsch reaction yielding the highly fluorescent derivatives. The resulting fluorescence was measured at 406 nm after excitation at 466 nm. The color was measured spectrophotometrically at 410 nm for vigabatrin and the absorbance–concentration plots were rectilinear over the range 10–70 μg/ml. The fluorescence intensity was a linear function of the concentration of vigabatrin of the range of 0.5–10 μg/ml. No interference was encountered from coadministrated drugs, except for cimetidine.

Belal *et al.* [25] described a stability-indicating selective spectrofluorimetric method for the determination of vigabatrin and gabapentin in their pharmaceutical formulations and spiked human urine. The method was based on the reaction between the two drugs and fluorscamine in borate buffer to give highly fluorescent derivatives that are measured at 472 nm using an excitation at 390 nm. The endogenous amino acids and coadministrated drugs did not interfere with the assay. The fluorescence intensity was a linear function of the concentration of vigabatrin of the range of 0.2–4.0 μg/ml.

Olgun *et al.* [26] used two sensitive and selective spectrofluorimetric and spectrophotometric methods for the determination of vigabatrin in tablets. The methods were based on derivatization with NBD-Cl. The resulting fluorescence was measured at 520 nm after excitation at

460 nm in ethyl acetate. The reaction obeys Beer's law over the ranges of 2–10 and 0.05–1.00 μg/ml for the spectrophtometric and spectrofluorimetric methods, respectively. Excipients presented in the tablet do not interfere with the analysis.

4.2.2. IR spectrophotometry

Torano and Van Hattum [27] used Fourier transform mid-infrared (FTMIR) spectroscopy with an attenuated total reflection (RTR) module for determination of vigabatrin. The drug was extracted from the capsule content after addition of sodium thiosulfate IS solution. The extract was concentrated and applied to the FTMIR-ART module. The ratio of the area of the drug peak to that of the IS was linear over the concentration range of 90–110 mg/ml. The accuracy of the method in this range was 99.7–100.5% with a variability of 0.4–1.3%.

4.2.3. Mass spectrometry

Matar and Abdel-Hamid [28] described a liquid chromatographic–mass spectrometric method for quantification of vigabatrin in human plasma using 4-phenyl-4-aminobutanoic acid as an IS. The method involved deproteinization of plasma sample with acetonitrile followed by detection of analytes by tandem mass spectrometry. The drug was chromatographed using XTerra C18 column and a mobile phase consisting of acetonitrile:water (50:50) and 0.025% formic acid at a flow rate of 0.1 ml/ min. The effluents were detected in multiple reaction monitoring (MRM) mode using transitions m/z 129.75 > 70.99 and m/z 179.7 > 116.92 for vigabatrin and IS, respectively. The method was linear over the vigabatrin concentration range of 0.5–10 μg/ml with a limit of detection of 0.05 μg/ml. The intraday coefficient of variation (CV) ranged from 3.84% to 6.53% and the interday CV range was 2.78%. Mean recovery percentage of the drug in spiked human plasma was 99.59%.

Oertel et al. [29] reported an automated method to quantify the anticonvulsants vigabatrin, gabapentin, and pregabalin simultaneously using hydrophilic interaction liquid chromatography–mass spectrometric method. The hydrophilic interaction chromatography (HILIC) with a mobile phase gradient was used to divide off ions of the matrix and for separation of the analytes. Four different HILIC columns and two different column temperatures were used. The anticonvulsants including vigabatrin were detected in the MRM mode with ESI-MS–MS method. The lower limit of quantifications was 312 ng/ml. The described HILIC-MS–MS method was suitable for therapeutic drug monitoring and for clinical and pharmacokinetic investigations of the anticonvulsant drugs.

4.3. Chromatography

4.3.1. High-performance liquid chromatography

Smithers *et al.* [30] described a reversed-phase high-performance liquid chromatography (HPLC) method for determination of vigabatrin in plasma and urine utilizing dansyl chloride as a derivatizing reagent and fluorimetric detection at 418 nm with excitation at 345 nm. A column (25 cm × 4.6 mm) of Zorbax C8 (6 μm) with acetonitrile–dioxan–0.5 M H_3PO_4 (60:30:10) as a mobile phase (flow rate 1.0 ml/min) and 4-amino-4-phenylbutyric acid as an IS was selected. The calibration graphs for the drug in plasma and urine were rectilinear for 5–200 μg/ml (limit of detection $\sim$ 0.5 μg/ml) and 200–4000 μg/ml (limit of detection $\sim$ 10 μg/ml), respectively. The CV in plasma was 3.2–9.2%.

Chen and Contario [31] reported an HPLC resolution of enantiomers of vigabatrin. The samples were treated with N-t-butoxycarbonyl-L-leucine N-hdroxysuccinimide esters in the presence of sodium carbonate and then with trifluoroacetic acid. The resulting diastereoisomers were separated on a column (25 cm × 4 mm) of Lichrosorb RP-8 (10 μm) with acetonitrile–dioxan–0.5 M H_3PO_4 (pH 7)–acetonitrile (24:1) as mobile phase with a flow rate of 2 ml/min and detection at 210 nm. Good recoveries of the inactive R-(−)-form (0.5–2%) added to the active S-(+)-form of the aminobutyrate aminotransferase inhibitor were obtained. The results on a sample containing 0.24% of the R-(−)-form agreed with those obtained by GC.

Chen and Fike [32] reported an HPLC method for determination of vigabatrin and its primary degradation product (5-vinylpyrrolidine-2-one) in pharmaceutical tablet formulation. An extract of tablets containing vigabatrin with vinylpyroolidinone and other excipients was analyzed on a column (25 cm × 4.6 mm) of Partisil SCX (10 mm) with potassium phosphate (pH 2.8, 25 mM)–methanol–acetonitrile (89:10:1) as a mobile phase (flow rate 1.5 ml/min) and detection at 210 nm. The calibration graph for vigabatrin was rectilinear from 20% to 120% of the normal amount injected (2 mg/ml). The mean recoveries of the drug and its degradation product were 99.8% and 99–102%, respectively.

Tsanaclis *et al.* [33] described a reversed-phase HPLC method for determination of vigabatrin in plasma after derivatization with phthalaldehyde. Serum was mixed with γ-amino-γ-phenylbutyric acid and methanol. Derivatization was carried out with phthaldehyde in borate buffer (pH 9.5) containing 2-mercaptoethanol. The resulting mixture was separated on a column (15 cm × 4.6 mm) of C18 Microsorb (5 μm) with 10 M H_3PO_4–acetonitrile–methanol (6:3:1) as mobile phase with a flow rate of 2 ml/min and fluorimetric detection between 418 and 700 nm (excitation at 370 nm). The detection limit was 0.08 μg/ml of vigabatrin. The CVs were 9%, 5%, and 5% at 2.14, 20.1, and 83.61 μg/ml/min, respectively. Common anticonvulsant did not interfere.

Juergens *et al.* [34] utilized *o*-phthalaldehyde as a derivatizing agent for simultaneous HPLC determination of vigabatrin and gabapentin in serum with automated preinjection derivatization. The serum mixture containing the two drugs were separated on a column (25 cm × 4 mm) of C18 ASMT BANsil (5 μm) with gradient elution (1 ml/min) with 0.1% H_3PO_4 (pH 2)–acetonitrile–methanol (8:1:1; solvent A) and acetonitrile–methanol (1:1:solvent B) [A:B, 9:1] and fluorimetric detection at 435 nm (excitation at 235 nm). Calibration graphs were linear from 0.1 to 300 μg/ml for both drugs. The detection limits were 0.5 and 1 μg/ml and within and between run relative standard deviation (RSD) were 0.82% and 1.15% and 1.10% and 2.5% for vigabatrin and gabapentin, respectively.

Wad and Kramer [35] described an HPLC method for simultaneous determination of vigabatrin and gabapentin in serum and urine after precolumn derivatization with *o*-phthaldialdehyde and fluorimetric detection at 455 nm with excitation at 230 nm. A column (12.5 cm × 3.0 mm) of Superspher 60 RP-Selected B (5 μm) with acetonitrile in 20 mM 0.5 M KH_2PO_4 as a mobile phase (flow rate 0.7 ml/min). The day-to-day CV of vigabatrin in a pooled serum was 3.1%. The lower limit of detection was 0.5 μmol/l and the calibration graph was rectilinear from 1300 μmol/l.

Vermeij and Edelbroek [36] reported HPLC analysis of vigabatrin enantiomers in human serum by precolumn derivatization with *o*-phthaldialdehyde-*N*-acetyl-ʟ-cystein and fluorescence detector. Separation was achieved on a Spherisorb 3ODS2 column using a gradient solvent program. Within-day precisions were 2.8% and 1.1%, respectively, for the (*R*)-(−)- and (*S*)-(+)-enantiomer in serum containing 15.4 mg/l (*RS*)-vigabatrin. The method was linear in the 0–45 mg/l range for both enantiomers and the minimum quantitation limit was 0.20 mg/l for (*R*)-(−)-vigabatrin and 0.14 mg/l for (*S*)-(+)-vigabatrin. No interferences were found from commonly coadministered antiepileptic drugs and from endogenous amino acids.

Ratnaraj and Patsalos [37] described an HPLC method for simultaneous determination of vigabatrin and gabapentin in human serum after precolumn derivatization with *o*-phthaldialdehyde in the presence of mercaptoethanol and fluorimetric detection at 440 nm with excitation at 340 nm. Separation was achieved on a Hypersil BDS C18 (3 μm) column (12.5 cm × 3 mm) with a mixture of 250 mM phosphate buffer, acetonitrile, water, and methanol as a gradient mobile phase (flow rate 0.45 ml/min). The method was linear over the concentration range of 25–400 μg/ml for vigabatrin. The lower limit of detection was 1 μM for both analytes. Within- and between-batch RSD were 2–4% and 3–4%, respectively. No interference from other commonly prescribed antiepileptic drugs was observed.

George *et al.* [38] described an HPLC method for determination of vigabatrin concentrations in plasma or serum. A column (10 cm × 4.6 mm) of a Spherisorb ODS2 (5 μm) with acetonitrile/H_2O/40 mM acetate buffer (pH 3.5) (52:35:13) as a mobile phase a (flow rate 0.7 ml/min) with detection at 340 nm was used. The assay was linear over a concentration range of 1–50 mg/l and the limit of detection was 1 mg/l. The RSD at 50 mg/l was 2.7% and the between-batch RSD at 20 and 40 mg/l were 4.4% and 3.3%, respectively. There was no evidence of interferences in the assay from other commonly prescribed antiepileptic drugs.

Chollet *et al.* [39] developed an isocratic HPLC method for simultaneous determination of vigabatrin and gabapentin in human serum after precolumn derivatization with *o*-phthaldialdehyde and fluorimetric detection at 435 nm with excitation at 235 nm. A column (25 cm × 3.0 mm) of Nucleosil C_{18} (5 μm) with a mixture of 0.022 M phosphoric acid (pH 2)–acetonitrile (45:55) as a mobile phase (flow rate 0.6 ml/min) was used. The calibration graph was rectilinear from 2.0 to 40.0 μg/ml for vigabatrin. The limit of detection and the limit of quantification for vigabatrin were 0.6 and 2.0 μg/ml, respectively. No interference from endogenous compounds or commonly prescribed antiepileptic drugs and their metabolites was observed.

Erturk *et al.* [40] determined vigabatrin in human plasma and urine by HPLC after derivatization with 4-chloro-7-nitrobenzofurazan with fluorescence detection at 520 nm with excitation at 460 nm. A column (20 cm × 3.9 mm) of Shim-Pack C_{18} (5 μm) with a mixture of 10 mM phosphoric acid–acetonitrile (60:40) as a mobile phase (flow rate 1.0 ml/min) was used. The assay was rectilinear over the concentration range of 2.0–20.0 μg/ml for plasma and 1.0–15.0 μg/ml for urine. The lower limit of detection and the lower limit of quantification for vigabatrin were 0.1 and 0.2 μg/ml, respectively, in plasma and urine. Both the within-day and day-to-day reproducibilities and accuracies were less than 5.46% and 1.6%, respectively.

Cetin and Atmaca [41] reported HPLC analysis of vigabatrin in tablets by precolumn derivatization with 1,2-naphthoquinone-4-sulfonic acid sodium salt and detected at 451 nm. Separation was achieved on a Shim-Pack CLS-ODS (5 μm) column (25 cm × 4 mm) with a mixture of 10 mM phosphoric acid–acetonitrile (75:25) as a mobile phase (flow rate 1.0 ml/min). The method was linear over the concentration range of 1.15–43.2 μg/ml. The lower limit of detection and the lower limit of quantification for vigabatrin were 0.86 and 0.23 μg/ml, respectively. The results were compared statistically with those obtained by HPLC method reported previously using *t*- and *F*-tests.

Cetin and Atmaca [42] also reported HPLC analysis of vigabatrin in human plasma and urine after precolumn derivatization with 1,2-naphthoquinone-4-sulfonic acid sodium salt using UV–VIS detection at 451 nm. Separation was achieved on a Shim-Pack CLS-ODS (5 μm) column (25 cm $\times$ 4.6 mm) with a mobile phase consisting of 10 mM phosphoric acid and acetonitrile gradient elution. The method was linear over the concentration range of 0.8–30 μg/ml. The lower limit of detection and the lower limit of quantification for vigabatrin were 0.8 and 0.5 μg/ml, respectively, in plasma and urine.

Vermeij and Edelbroek [43] described an HPLC method for simultaneous determination of vigabatrin, gabapentin, and pregabalin in human serum after precolumn derivatization with o-phthaldialdehyde and fluorimetric detection at 450 nm with excitation at 330 nm. Separation was performed on a homemade reversed column (15 cm $\times$ 0.46 mm) packed with Alltima 3C18. The mobile phase consisted of a mixture of methanol–acetonitrile–20 mM phosphate buffer (pH 7) (8:17.5:74.5) at a flow rate of 0.7 ml/min was used. The calibration graph for vigabatrin was linear up to 62 mg/ml. The lower limit of quantification for vigabatrin was 0.06 mg/ml. No interferences were found from commonly coadministered antiepileptic drugs and from endogenous amino acids.

Al-Majed [44] reported HPLC analysis of vigabatrin and gabapentin in dosage forms and biological fluids by precolumn derivatization with fluorescamine and fluorescence detection at 472 nm with excitation at 390 nm. Separation was achieved on an ASMT BANsil CN (5 μm) column (30 cm $\times$ 3.9 mm) with a mixture of acetonitrile–TBAH (20:80) as a mobile phase. The method was linear over the concentration range of 0.2–1.0 μg/ml for vigabatrin. The lower limit of detection for vigabatrin was 0.03 μg/ml. No interference was encountered from the excipients, coadministered drugs, and the possible degradation products of vigabatrin.

Franco et al. [45] described an HPLC method for simultaneous determination of the R-(−) and (S)-(+)-enantiomers of vigabatrin in human serum after precolumn derivatization with 2,4,6-trinitrobenzene sulfonic acid (TNBSA) and detection at 340 nm. Separation was achieved on a reversed phase chiral column (Chiralcel-ODR, 25 cm $\times$ 4.6 mm) using 0.05 M potassium hexafluorophosphate (pH 4.5):acetonitrile:ethanol (50:40:10) as a mobile phase at a flow rate of 0.9 ml/min. The calibration graphs for each enantiomer were linear over the concentration range of 0.5–40 μg/ml with a limit of quantification of 0.5 μg/ml. No interferences were found from commonly coadministered antiepileptic drugs.

Hsieh et al. [46] described an HPLC method for simultaneous determination of the R-(−) and (S)-(+)-enantiomers of vigabatrin in human serum after precolumn derivatization with a fluorescent chiral reagent (naproxen acyl chloride) and detection at 350 nm with excitation at 230 nm. Separation was achieved on an Agilent Eclipse XDB-C8 (5 μm) column (15 cm $\times$ 4.6 mm) using a mixed solvent of acetate buffer (pH 7)–methanol (60:40) as a mobile phase at a flow rate of 1.2 ml/min. The calibration graphs for each enantiomer were linear over the concentration range of 0.025–2.00 μg/ml. The lower limit of detection and the lower limit of quantification for each enantiomer were 2.5 and 25 nM, respectively. No interferences were found from commonly coadministered antiepileptic drugs and from the structure related γ-aminobutyric acid.

Lee et al. [47] reported an HPLC method for resolution of vigabatrin and analog γ-amino acids on chiral stationary phases (CSPs) based on (+)-(18-crown-6)-2,3,11,12-tetracarboxylic acid. Between the two CSPs which contain three methylene-unit or 11 methyene-unit spacer group, the latter was found to be greater than the former in the resolution of vigabatrin and its analog γ-amino acids. The separation, α, and the resolution factor, Rs, for the resolution of the drug on the latter being 1.91 and 4.57, respectively. The chromatographic behaviors for the resolution of the drug its analog γ-amino acids on the two CSPs were found to be independent on the type and the content of organic and acidic modifiers in aqueous mobile phase.

4.3.2. Gas chromatography

Schramm et al. [48] described a gas chromatographic method for determination of vigabatrin enantiomers in plasma. The method used a double derivatization step on a megabore Chirasil-Val capillary column and thermionic specific detection. The calibration graphs for the R-($-$) and (S)-(+)-enantiomers were linear over the concentration range of 1.0–200 and 0.5–100 μg/ml, respectively. The assay was suitable for pharmacokinetic studies and routine therapeutic drug monitoring in humans.

Borrey et al. [49] described a simple gas chromatography–mass spectrometry (GC–MS) method for the simultaneous quantitative determination of vigabatrin and gabapentin in human serum. The chromatographic system consisted of a Varian VF-5MS capillary column (30 mm $\times$ 0.25 mm, 0.25 μm film thickness). The compounds were derivatized by methylation and analyzed on a polyimethylsiloxane column using splitless. The assay for vigabatrin was rectilinear over the concentration range of 5–80 μg/ml. The within-day and day-to-day standard deviations at different concentration levels was < 10%. The lower limit of detection was 2 μg/ml.

4.3.3. Electrophoresis

Chang and Lin [50] determined vigabatrin in human plasma by capillary electrophoresis and laser-induced fluorescence after precolumn derivatization with 5-carboxytetramethylrhodamine succinimidyl ester. Optimal separation and detection were obtained with an electrophoric buffer of 50 mM sodium borate (pH 9.5) containing 10 mM sodium dodecyl sulfate and a green He–Ne laser with fluorescence detection at 589 nm and excitation at 543 nm. The assay was rectilinear over the concentration range of 1.5–200 μM and the lower limit of detection was 0.13 μM. Both the within-day and day-to-day reproducibilities and accuracies were less than 14.3% and 4.9%, respectively.

Benturquia *et al.* [51] described simultaneous determination of vigabatrin and amino acid neurotransmitters in brain microdialysates by capillary electrophoresis and laser-induced fluorescence after precolumn derivatization with naphthalene-2,3-dicaroxaldehyde (NDA). Optimal separation and detection were obtained with a sodium borate buffer (pH 9.2) containing 60 mM sodium dodecyl sulfate and 5 mM hydroxypropyl-β-cyclodextrin with fluorescence detection at 589 nm and excitation at 543 nm. The assay was rectilinear over the concentration range of 1.5–200 μM and the lower limit of detection was 0.13 μM. Both the within-day and day-to-day reproducibilities and accuracies were less than 14.3% and 4.9%, respectively.

Lin *et al.* [52] reported simultaneous determination of vigabatrin and gabapentin in pharmaceutical preparations by simple capillary zone electrophoresis after derivatization with ofloxacin acyl chloride (OAC) with UV detection at 300 nm. The assay was rectilinear for both drugs over the concentration range of 50–500 μM and the lower limit of detection was 5 μM. Both the within-day and day-to-day reproducibilities and accuracies are below 3.1% and 4.8%, respectively.

Shafaati and Lucy [53] determined vigabatrin in its dosage forms by capillary zone electrophoresis with indirect UV detection at 214 nm. Optimal separation and quantification of vigabatrin was obtained using 5 mM sodium phosphate buffer (pH 2.2) containing 5 mM benzyl triethyl ammonium hydroxide. The assay was rectilinear over the concentration range of 5–150 μg/ml and the lower limit of quantification was 5 μg/ml. The RSD of migration time for 10 consecutive injections of a standard solution of the drug was 0.19%.

Zhao *et al.* [54] presented an enantioslective separation and determination of vigabatrin enantiomers in spiked human plasma by capillary electrophoresis and UV–VIS detection at 202 nm after precolumn derivatization with dehydroabietylisthiocynate. Optimal separation and detection were obtained with an electrophoric buffer of 50 mM sodium phosphate (pH 9.0) containing 17 mM sodium dodecyl sulfate and 25% actonitrile. The assay was rectilinear over the concentration range of 0.3–6.0 μg/ml and the lower limit of detection was 0.15 μg/ml. The results obtained indicate that the RSD was less than 5%. No interferences were found from endogenous amino acids.

Musenga *et al.* [55] described a capillary electrophoresis method for determination of vigabatrin in human plasma after precolumn derivatization with 6-carboxyfluorescein-N-succinimididyl ester. Optimal separation and detection were obtained with 50 mM borate buffer (pH 9.0) containing 100 mM N-methylglucamine with laser-induced fluorescence detector (λ_{exc} = 488 nm). The assay was rectilinear over the concentration range of 10–120 μg/ml. The lower limit of detection and the lower limit of quantification for vigabatrin were 2.0 and 5.0 μg/ml, respectively. Both the RSD and accuracy were 6.7% and between 97.0% and 101.6%, respectively.

5. STABILITY

Yang *et al.* [56] described the stability of powder vigabatrin package in vinyl film. The stability studies were performed into light shield package and without shield envelope. Powdered vigabatrin contents were not changed for 11 weeks. At high temperature (37 °C), room temperature, and cool temperature, the content of vigabatrin at 12 weeks were 88.7%, 85.9%, and 86.4%, respectively.

6. USES AND ADMINISTRATION

Vigabatrin is used as an adjunctive antiepileptic in patients with resistant partial epilepsy with or without secondary generalization, unresponsive to other therapy [2]. Nowadays, vigabatrin is rarely used in the treatment of partial seizures due to several irreversible visual field constrictions associated with its chronic use [57–62]. It is regarded by many authorities as a drug of choice in infants with west syndrome (infantile spasms), particularly in cases associated with tuberous sclerosis [62].

The recommended initial dose of vigabatrin as adjunctive therapy in adults is 1.0 g daily by mouth, increased if necessary in increments of 0.5 g at weekly intervals to a maximum of 3.0 g daily. A recommended initial dose in children is 40 mg/kg body-weight daily. For infantile spasms the dose is from 50 mg to 150 mg/kg daily [2].

7. PHARMACOKINETICS

Table 7.8 shows the pharmacokinetic characteristics of vigabatrin in a single-dose kinetic study for both enantiomers [63]. Haegele and Schechter [64] demonstrated that peak plasma concentrations for both enantiomers were achieved within 0.5 and 2 h after a 1500-mg dose. The concentration of the two enantiomers did not differ after 24 h. The AUC for the (+) enantiomer is less than the $R(-)$, this discrepancy is due to that only the $S(+)$ enantiomer is used by GABA-T as a substrate, whereas the $R(-)$ enantiomer is inactive and is not recognized by GABA-T.

TABLE 7.8 Key pharmacokinetic characteristics of vigabatrin [63]

T_{max}	1–4 h (unaffected by food)
$t_{1/2}$	5–7 h
Protein binding	95% free
Elimination	70% renal
Active metabolites	No
Drug interaction	Minimal ($\downarrow$ phenytion levels)
Blood levels	Not related to efficacy
Liver enzyme induction	None
Accumulation	None

The pharmacokinetics of the $S(+)$ enantiomer were not influenced by the $R(-)$ enantiomer. Furthermore, no $R(-)$ enantiomer was detected after the administration of pure $S(+)$ vigabatrin, thus demonstrating that no chiral inversion occurs in humans [63].

Tong *et al.* [65] studied the pharmacokinetics of vigabatrin in rat blood and cerebrospinal fluid (CSF), and the major findings of this study are that: (i) the pharmacokinetics of vigabatrin in serum are linear and dose-dependent, while in CSF are dose-independent; (ii) vigabatrin is not protein bound in serum; (iii) the elimination of vigabatrin from serum is rapid; and (vi) vigabatrin is rapidly penetrated the blood–brain barrier (BBB).

7.1. Absorption

Vigabatrin is rapidly and completely absorbed after oral administration and about 80% of the dose is recovered in the urine [4]. In healthy volunteers, its absorption was rapid and beak plasma concentrations occurred within 2 h [65]. The effect of food on the bioavailability of vigabatrin tablets has been studied [4, 66]. The AUC for fasted and fed volunteers was not significantly different. This indicates that food does not alter the extent of absorption. The pharmacokinetics profiles of vigabatrin in tablet form and in solution were strikingly similar [66].

7.2. Distribution

Because vigabatrin is not protein-bound and is a highly water-soluble compound, a wide distribution in the body is observed [63], and readily crosses the BBB. CSF concentrations are about 10–15% of those in plasma [62, 64]. It has a volume of distribution of about 0.8 l/kg [65–68]. It distributes into red blood cells and saliva at a concentration of 30–80% and 10% of that in the plasma, respectively [69]. The transfer of the drug from maternal to fetal blood across the placenta was low and was comparable to other acidic α-amino acids [70, 71]. The drug is also distributed into breast milk [67].

Vigabatrin distribution in the brain is region specific, with frontal cortex concentrations substantially greater than those seen in the hippocampus. Elevation of GABA concentrations does not reflect the concentration profile of vigabatrin but reflect its regional distribution [68].

7.3. Metabolism

Vigabatrin is not appearing to be extensively metabolized by the liver, nor influence hepatic metabolism [65]. Durham et al. [73] studied the metabolism of vigabatrin following a single oral dose of vigabatrin in healthy male volunteer. Two minor metabolites (<5% of total dose) of vigabatrin were detected using liquid scintillation counting of HPLC eluant fractions. One urinary metabolite was identified by thermospray-LC/MS as the lactam metabolite of vigabatrin.

7.4. Elimination

The major route of elimination for vigabatrin is renal excretion of the unchanged drug. The elimination half-life is 5–8 h in young adults and the total clearance is approximately 1.7–1.9 ml/min/kg, with renal clearance accounting for 70% of the total clearance. The elimination of the drug is affected by age, being slower in the elderly because of the lower creatinine clearance [63, 73]. Elimination is not influenced by dosage or by duration of treatment [72].

8. PHARMACOLOGICAL EFFECTS

8.1. Mechanism of action

Vigabatrin is a selective, irreversible inhibitor of GABA-T, which catalyzed the inactivation of GABA. Competitive inhibition of the enzyme leads to an increase in GABA concentration in presynaptic terminals within CNS [74]. The drug is also selective in its antagonism of GABA degradation. GABA is an important inhibitory neurotransmitter: increases in its concentration are associated with anticonvulsant activity in animal models [75, 76]. The $S(+)$-enantiomer potently inhibits GABA-T, whereas the $R(-)$-enantiomer has minimal activity [77].

The pharmacologic activity and the toxic effects of vigabatrin is associated only with $S(+)$ enantiomer. The $R(-)$ enantiomer appears to be entirely inactive [78]. Because the target enzyme, GABA transaminase (GABA-T), has a longer half-life than the drug itself [79, 80]. The pharmacologic effects are determined by the half-life of the enzyme rather than by the drug or the $S(+)$ enantiomer.

Vigabatrin is accepted as a substrate of GABA-T by forming a Schiff base with pyridoxal phosphate in the active site of the enzyme, which abstracts a proton from the Schiff base. The resulting charge stabilization by the pyridine ring induces the aldimine to ketimine tautomerism that occurs in the normal transamination process. The reactive unsaturated ketimine forms a stable bond with a nucleophilic residue of GABA-Ts active site, resulting in irreversible inhibition of the enzyme and eliminating its ability to transaminate new substrate [81].

Significant elevations of homocarnosine (dipeptide of GABA) have also been reported. Elevations of these markers of GAB-ergic transmission are associated with a significant reduction in seizure frequency in patients with refractory seizures [68, 79].

8.2. Toxicity and adverse reactions

There has been significant concern with respect to a potentially serious toxicity of vigabatrin in humans as a result of the microvacuolization of white matter demonstrated in rodents and dogs. In all animal studies the microvacuolization was not dose-dependent and was completely reversible with discontinuation of the drug [73].

Another major concern for toxicity of vigabatrin is sever persistent visual field constriction associated with retinal cone system dysfunction [63, 82]. The effect did not appear to be reversible upon discontinuation of the drug. Vigabatrin also causes GABA to accumulate in retinal glial cells in rats, suggesting a mechanism for the toxic effect [83].

The most commonly reported adverse effects are sedation, headaches, and fatigue, although in children excitation and agitation occur more frequently. Depression, confusion, and other behavior abnormalities account for about 5% of the reported adverse effects, and frequency of these unwanted effects can be reduced by gradual increase in dosage. Weight gain has occurred in 40% of epileptic patients treated for prolonged periods with add-on vigabatrin therapy [1, 53, 60].

8.3. Drug interactions

Because vigabatrin is not protein bound, nor is it metabolized by the hepatic microsomal system, it would not expect clinically significant interactions with other drugs. However, vigabatrin was shown to decrease levels by 20–30% in clinical trials. No mechanism of action could account for the decreased level [73, 84]. A study found that vigabatrin causes a statistically significant increase in plasma clearance of carbamazepine [85].

REFERENCES

[1] S. Budavari (Ed.), The Merck Index, 12th ed., Merck and Co., N. J., 1996, p. 10114.
[2] S. Sweetman (Ed.), Martindale, the Complete Drug Reference, 33rd ed., Pharmaceutical Press, London, Chicago, 2002, pp. 370–371.
[3] British Pharmacopoeia, vol. II, The Stationary Office, London, UK, 2005, pp. 2057–2059.
[4] A.C. Moffat (Ed.), Clarke's Analysis of Drugs and Poisons, third ed., Pharmaceutical Press, London, 2004, p. 1697.
[5] Z.-Y. Wei, E.E. Knaus, J. Org. Chem. 58 (1993) 1586.
[6] M. Ordonez, C. Cativiela, Tetrahedron: Asymmetry 18 (2007) 3.
[7] Z.-Y. Wei, E.E. Knaus, Tetrahedron 50 (1994) 5569.
[8] C. Dangoneau, A. Tomassini, J.-N. Denis, Y. Vallee, Synthesis (2001) 150.
[9] A. Gheorghe, O. Schulte, O. Reiser, J. Org. Chem. 71 (2006) 2173.
[10] M.-Y. Chang, C.-Y. Lin, C.-W. Ong, Heterocycles 68 (2006) 2031.
[11] V.P. Raj, A. Sudalai, Tetrahedron Lett. 49 (2008) 2646.
[12] T.W. Kwon, P.F. Keusenkothen, M.B. Smith, J. Org. Chem. 57 (1992) 6169.
[13] B.M. Trost, R.C. Lemoine, Tetrahedron Lett. 37 (1996) 9161.
[14] M. Alcon, M. Poch, A. Moyano, M.A. Pericas, A. Riera, Tetrahedron: Asymmetry 8 (1997) 2967.
[15] H. McAlonan, P.J. Stevenson, Tetrahedron: Asymmetry 6 (1995) 239.
[16] S. Chandrasekhar, S. Mohapatra, Tetrahedron Lett. 39 (1998) 6415.
[17] C.E. Anderson, L.E. Overman, J. Am. Chem. Soc. 125 (2003) 12412.
[18] M.Y. Chang, C.-W. Ong, C.-Y. Lin, Heterocycles 68 (2006) 2031.
[19] M. Haramura, A. Tanaka, T. Akimoto, N. Hirayama, Anal. Sci. 20 (2004) x9.
[20] A. Alltomare, G. Cascarano, C. Giacovazzo, A. Guagliardi, M. Burla, G. Polidori, M. Camalli, J. Appl. Cryst. 27 (1994) 435.
[21] C. Gnamm, G. Franck, G. Miller, N. Stork, T. Brodner, K. Hel, Synthesis-Stuttgart 20 (2008) 3331.
[22] J. Grove, J. Alken, P.J. Schechter, Chromatogr. Biomed. Appl. 31 (1984) 383.
[23] E.M. Hassan, F. Belal, O.A. Al-Deeb, N.Y. Khalil, J. AOAC Int. 84 (2001) 1017.
[24] J. Al-Zehouri, S. A-Madi, F. Belal, Arzneim. -Forech. Drug Res. 51 (2001) 97.
[25] F. Belal, H. Abdine, A. Al-Majed, N.Y. Khalil, J. Pharm. Biomed. Anal. (2001).
[26] N. Olgun, S. Erturk, S. Atmaca, J. Pharm. Biomed. Anal. 29 (2002) 1.
[27] J.S. Torano, S.H. Van Hattum, Anal. Biomed. Chem. 371 (2001) 532.
[28] K.M. Matar, M.E. Abdel-Hamid, J. Liq. Chromatogr. Relat. Tech. 28 (2005) 395.
[29] R. Oertel, N. Arenz, J. Pietsch, W. Kirch, J. Sep. Sci. (2008) 15.
[30] J.A. Smithers, J.F. Lang, R.A. Okerholm, J. Chromatogr. Biomed. Appl. 42 (1985) 232.
[31] T.M. Chen, J.J. Contario, J. Chromatogr. 314 (1984) 495.
[32] T.M. Chen, R.R. Fike, Chromatogram 11 (1990) 4.
[33] L.M. Tsanaclis, J. Wicks, J. Williams, A. Richens, Ther. Drug Monit. 13 (1991) 251.
[34] U.H. Juergens, T.W. May, B. Rambeck, J. Liq. Chromatogr. Relat. Technol. 19 (1996) 1459.
[35] N. Wad, G. Kramer, J. Chromatogr. B 705 (1998) 154.
[36] T.A. Vermeij, P.M. Edelbroek, J. Chromatogr. B 716 (1998) 233.
[37] N. Ratnaraj, P.N. Patsalos, Ther. Drug Monit. 20 (1998) 430.
[38] S. George, L. Gill, R.A. Braithwaite, Ann. Clin. Biochem. 37 (2000) 338.
[39] D.F. Chollet, L. Goumaz, C. Juliano, G. Anderegg, J. Chromatogr. B 746 (2000) 31.
[40] S. Erturk, E.S. Aktas, S. Atmaca, J. Chromatogr. B 760 (2001) 207.
[41] S.M. Cetin, S. Atmaca, Acta Pharmaceutica Turcica 44 (2002) 57.
[42] S.M. Cetin, S. Atmaca, J. Chromatogr. A 1031 (2004) 237.
[43] T.A.C. Vermeij, P.M. Edelbroek, J. Chromatogr. B 810 (2004) 297.
[44] A.A. Al-Majed, J. Liq. Chromatogr. Relat. Tech. 28 (2005) 3119.

[45] V. Franco, I. Mazzucchelli, C. Fattore, R. Marchiselli, G. Gatti, E. Perucca, J. Chromatogr. B 864 (2007) 63.

[46] C.-Y. Hsieh, S.-Y. Wang, A.-L. Kwan, H.-L. Wu, J. Chromatogr. A 1178 (2008) 166.

[47] S.J. Lee, H.S. Cho, H.J. Choi, M.H. Hyun, J. Chromatogr. A 1188 (2008) 318.

[48] T.M. Schramm, G.E. McKinnon, M.J. Eadie, J. Chromatogr. 616 (1993) 39.

[49] D.C. Borrey, K.O. Godderis, V.I. Engelrelst, D.R. Bernard, M.R. Langlois, Clin. Chim. Acta 354 (2005) 147.

[50] S.Y. Chang, W.-C. Lin, J. Chromatogr. B 794 (2003) 17.

[51] N. Benturquia, S. Parrot, V. Sauvinet, B. Renaud, L. Denoroy, J. Chromatogr. B 806 (2004) 237.

[52] F.-M. Lin, H.-S. Kou, S.-M. Wu, S.-H. Chen, H.-L. Wu, Anal. Chim. Acta 9 (2004) 523.

[53] A. Shafaati, C. Lucy, J. Pharm. Pharmac. Sci. 8 (2005) 190.

[54] S. Zhao, R. Zhang, H. Wang, L. Tang, Y. Pan, J. Chromatogr. B 833 (2006) 186.

[55] A. Musenga, R. Mandrioli, I. Comin, E. Kenndler, M.A. Raggi, Electrophoresis 28 (2007) 3535.

[56] E.Y. Yang, K.J. Park, N.C.C. Cho, J. Kor. Soc. Hosp. Pharm. 16 (1999) 22.

[57] T. Eke, J.F. Talbot, M.C. Lawden, Br. Med. J. 314 (1997) 180.

[58] R. Kalviane, I. Nousianen, CNS Drugs 15 (2001) 217.

[59] K. Malmgren, E. Ben-Manachem, L. Frisen, Epilepia 42 (2001) 609.

[60] S.J. You, H. Ahn, T.-S. Ko, J. Korean, Med. Sci. 21 (2006) 728.

[61] R. Riikonen, Curr. Opin. Neurol. 18 (2005) 91.

[62] P. Curatolo, R. Bombardieri, C. Cerminara, Curr. Opin. Neurol. (2006) 19.

[63] E. Ben-Menachem, Epilepsia 36 (Suppl. 2) (1995) S95.

[64] K.D. Haegele, P.J. Schechter, Clin. Pharmacol. Ther. 40 (1986) 581.

[65] X. Tong, N. Ratnaraj, P.N. Patsalos, Seizure 16 (2007) 43.

[66] Drug Facts and Comparison, A Wolters Kluwer Company, St. Louis, 2001, p. ku-47.

[67] J.F. Hoke, E.M. Chi, K. Antony, Epilepsy Res. 23 (Suppl. 3) (1991) 7.

[68] E. Rey, G. Pons, G. Olive, Clin. Pharmacokinet. 23 (1992) 267.

[69] S.M. Grant, R.C. Heel, Drugs 4 (1991) 889.

[70] A. Tran, T.O. Mahoney, E. Rey, J. Mail, J.P. Mumford, G. Olive, Br. J. Clin. Pharmacol. 45 (1998) 409.

[71] E. Rey, G. Pons, M.O. Richard, Br. J. Clin. Pharmacol. 30 (1990) 253.

[72] M.J. Jung, B. Lippert, B.W. Metcalf, Neurochemistry 29 (1997) 797.

[73] S.L. Durham, J.F. Hoke, T.M. Chen, Drug Metab. Dispos. 21 (1993) 480.

[74] J.C. Challier, T. Binten, G. Olive, E. Rey, Br. J. Clin. Pharmacol. 34 (1992) 139.

[75] X. Tong, N. Ratnaraj, P.N. Patsalos, Epilepsia (2008) 1.

[76] R. Sankar, A.T. Derdiarian, CNS Drug Rev. 4 (1998) 260.

[77] O.A. Petroff, D.L. Rothman, K.L. Behar, R.H. Mattson, Neurology 46 (1996) 1459.

[78] L.R. Macdonald, K.M. Kelly, Epilepsia 36 (Suppl. 2) (1995) S2.

[79] K. Gale, Adv. Neurol. 44 (1986) 343.

[80] O.M. Larsson, L. Gram, I. Schousboe, Neuropharmacology 25 (1986) 617.

[81] O.C. Petroff, R.H. Mattson, K.L. Behar, Ann. Neurol. 44 (1998) 948.

[82] G.L. Krauss, M.A. Johnson, N.R. Miller, Neurology 50 (1998) 614.

[83] M.J. Neal, J.R. Cunningham, M.A. Shahm, S. Yazulla, Neurosci. Lett. 98 (1989) 29.

[84] E.M. Rimmer, A. Richens, Br. J. Clin. Pharmacol. 27 (1998) 27S.

[85] S. Alcaraz, M.A. Quintana, E. Lopez, I. Rodriguez, P. Liopis, J. Clin. Pharmacol. Ther. 27 (2002) 427.

CHAPTER **8**

Zaleplon

Nagwa H. Foda and **Rana B. Bakhaidar**

Department of Pharmaceutics, College of Pharmacy, King Abdulaziz University, Jeddah, Kingdom of
Saudi Arabia

Profiles of Drug Substances, Excipients, and Related Methodology, Volume 35
ISSN 1871-5125, DOI: 10.1016/S1871-5125(10)35008-4

347

1. PHYSICAL PROFILES OF DRUG SUBSTANCES AND EXCIPIENTS

1.1. General information [1]

1.1.1. Nomenclature
1.1.1.1. Systematic chemical name   3′-(3-Cyanopyrazolo[1,5-*a*]pyrimidin-7-yl)-*N*-ethylacetanilide.

1.1.1.2. Nonproprietary names   Zaleplon.

1.1.1.3. Proprietary names   Sonata, Starnoc.

1.1.2. Formulae
1.1.2.1. Empirical formula, molecular weight, and CAS registry number

Zaleplon	$C_{17}H_{15}N_5O$	305.34	151319-34-5

1.1.2.2. Structural formula [2]

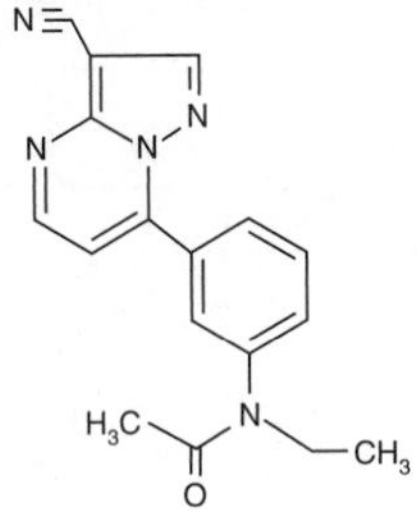

1.1.3. Elemental analysis
The calculated elemental content of zaleplon is given in Table 8.1.

1.1.4. Appearance [2]
Zaleplon is a white to off-white powder.

1.2. Physical characteristics [2]

1.2.1. Solubility characteristics
Pure zaleplon has very low solubility in water as well as low solubility in alcohol and propylene glycol.

1.2.2. Partition coefficient
Its partition coefficient (PC) in octanol–water is constant (log PC = 1.23) over the pH range from 1 to 7.

1.2.3. Crystallographic properties [3]
1.2.3.1. X-Ray powder diffraction pattern   The X-ray powder diffraction pattern of zaleplon powder sample was obtained using a Philips diffractometer system (model PW1710). The pattern was obtained using nickel-filtered copper radiation (=1.5405 Å), and is shown in Fig. 8.1. A full data summary is provided in Table 8.2.

TABLE 8.1 Elemental analysis

Element	Zaleplon (%, w/w)
C	66.87
H	4.95
N	22.94
O	5.24

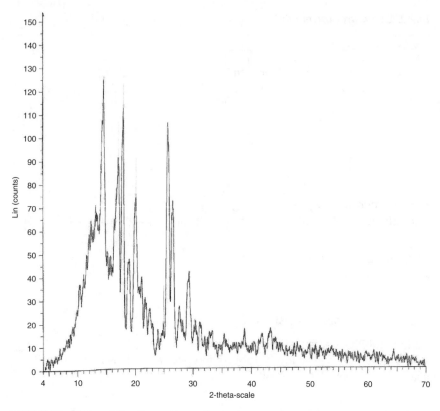

FIGURE 8.1   X-Ray diffraction pattern of zaleplon.

1.2.4. Thermal methods of analysis

1.2.4.1. Melting behavior   The melting range of zaleplon has been reported by a number of workers and summarized as follows:

157–159 °C	[1]
187 °C	[4]
186 °C	[3]

1.2.4.2. Differential scanning calorimetry (DSC) [3]   The DSC thermogram for zaleplon is shown in Figs. 8.2 and 8.3, and was obtained using Shimadzu DSC-50 thermal analyzer at scan rate of 10 and 5 °C/min under a nitrogen atmosphere over a temperature interval of 20–300 °C.

A sharp endothermic peak was observed having an onset of 181.35 and 181.14 °C and a maximum at 187.49 and 186.6 °C for 10 and 5 °C/min, respectively.

TABLE 8.2 X-Ray powder diffraction pattern

Scattering angle (2θ)	d-Spacing (Å)	Relative intensity (%)
14.406	6.14328	100
16.357	5.41476	51.9
17.030	5.20237	71.8
17.903	4.95060	90.2
18.922	4.68614	37.9
20.043	4.42662	59.2
21.098	4.20760	31.6
22.512	3.94632	20.7
23.927	3.71607	12.9
24.874	3.57674	14.9
25.631	3.47275	84.5
26.526	3.35761	57.6
27.663	3.22209	21.4
28.253	3.15611	16.4
29.335	3.04217	32.2
30.214	2.95566	15.5
30.416	2.93642	16.4
31.227	2.86197	14.9
33.255	2.69195	13.4
35.350	2.53704	11.2
38.798	2.31919	13.1
41.772	2.16068	10.9
43.259	2.08979	14.0

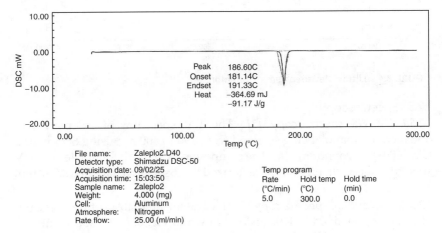

FIGURE 8.2 Differential scanning calorimetry thermogram of zaleplon, obtained at a heating rate of 5 °C/min.

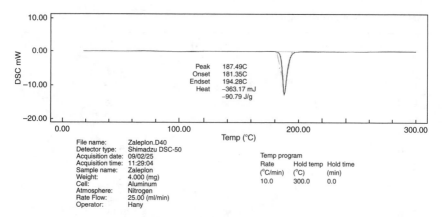

File name: Zaleplon.D40
Detector type: Shimadzu DSC-50
Acquisition date: 09/02/25
Acquisition time: 11:29:04
Sample name: Zaleplon
Weight: 4.000 (mg)
Cell: Aluminum
Atmosphere: Nitrogen
Rate Flow: 25.00 (ml/min)
Operator: Hany

Temp program
Rate (°C/min)   Hold temp (°C)   Hold time (min)
10.0   300.0   0.0

Peak 187.49C
Onset 181.35C
Endset 194.28C
Heat −363.17 mJ
−90.79 J/g

FIGURE 8.3 Differential scanning calorimetry thermogram of zaleplon, obtained at a heating rate of 10 °C/min.

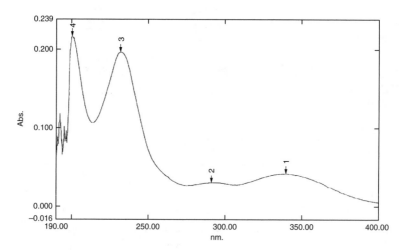

FIGURE 8.4 Ultraviolet spectrum of zaleplon.

1.2.5. Spectroscopy

1.2.5.1. UV–VIS spectroscopy [3]   The UV spectrum of zaleplon was obtained in ethanol, being scanned from 190 to 400 nm using Schimadzu UV-1600 spectrophotometer. The compound exhibits the characteristic UV spectrum shown in Fig. 8.4 characterized by absorption maxima at 232 nm.

1.2.5.2. Vibrational spectroscopy [3, 4]   The infrared absorption spectrum of zaleplon, obtained in KBr disk is shown in Fig. 8.5. The spectrum was recorded on Jasco FT/IR 460 plus Fourier transform infrared spectrometer model.

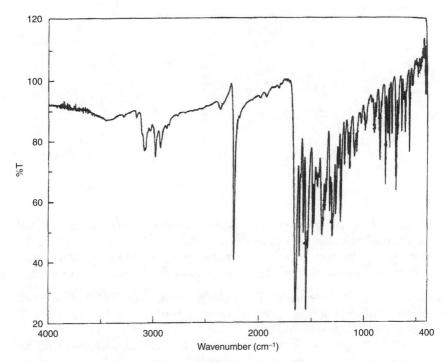

FIGURE 8.5 Infrared spectrum of zaleplon.

TABLE 8.3 Vibrational spectral assignment of zaleplon

Frequency (cm^{-1})	Assignment
3087, 3033	Weak aryl C–H stretch
2983, 2935	Weak aliphatic C–H stretching
2232.2	Strong C≡N stretch
1651.7	Strong amide C=O stretching
1614	Strong aryl C=C stretch and C≡N stretch
801, 697.10	Medium H out-of-plane bend

The major observed bands and their assignments are found in Table 8.3.

1.2.5.3. Nuclear magnetic resonance (NMR) spectrometry [4]   Both the ^{1}H NMR and ^{13}C NMR spectra of zaleplon have been obtained in DMSO-d_6 as a solvent and using tetramethylsilane as the internal standard (IS). The assignments for both the ^{1}H and ^{13}C NMR spectra make use of the following numbering scheme:

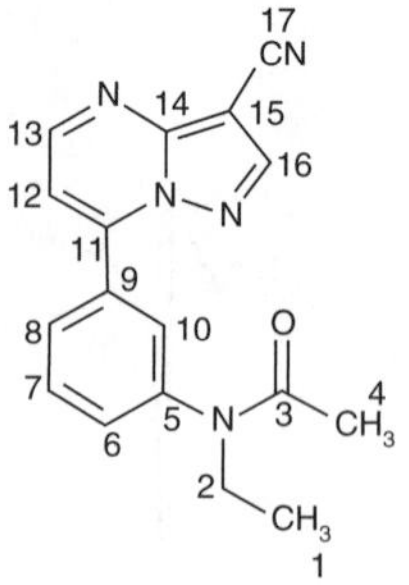

Zaleplon

1.2.5.3.1. ^{1}H NMR spectrum The ambient temperature 600 MHz ^{1}H NMR spectrum of zaleplon was obtained on a Bruker 600 MHz AVANCE III-cyoprobe spectrometer. The spectrum itself is shown in Fig. 8.6 and a summary of the chemical shifts and spectral assignments is provided in Table 8.4.

1.2.5.3.2. ^{13}C NMR spectrum The ambient temperature 75 MHz ^{13}C NMR spectrum of zaleplon was obtained on a Bruker 600 MHz AVANCE III-cyoprobe NMR spectrometer.

The spectrum is shown in Fig. 8.7, and a summary of the chemical shifts and spectral assignments is provided in Table 8.5.

The ^{13}C signal multiplicities were determined by DEPT experiments and shown in Fig. 8.8.

Evidence in support at the ^{1}H and ^{13}C spectral assignments was obtained from two-dimensional correlation spectroscopy (COSY) experiments. Proton–proton (H–H COSY) results are shown in Fig. 8.9.

1.2.6. Mass spectrometry [3]
The mass spectrum of zaleplon is shown in Fig. 8.10, and was recorded on Shimdzu GCMS/qp1000ex mass spectrometer.

The fragmentation assignments of the mass spectrum of zaleplon is shown in Table 8.6.

1.3. Stability

1.3.1. Solid-state stability
Store at control room temperature, 20–25 °C, protect from light.

1.3.2. Solution-phase stability
The physicochemical properties and dissolution profile of zaleplon β-cyclodextrin (βCD) inclusion complex were investigated. The phase solubility profile of Zaleplon with βCD was classified as A_L-type. Stability constant with 1:1 molar ratio was calculated from the phase solubility diagram and

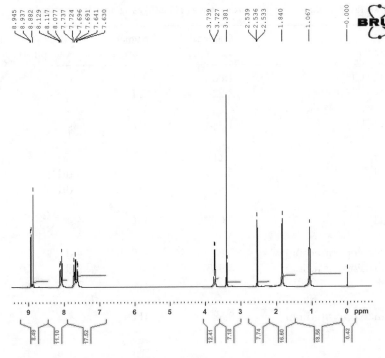

FIGURE 8.6 ^{1}H NMR spectrum of zaleplon.

the aqueous solubility of zaleplon was found to be enhanced by 714% ($p < 0.001$) for βCD. Binary systems of ZPN with βCD were prepared by kneading method. The solid-state properties of complex were characterized by DSC, Fourier transformation-infrared spectroscopy, and powder X-ray diffractometry. It could be concluded that zaleplon could form inclusion complex with βCD. The dissolution profile of inclusion complex was determined and compared with those of zaleplon alone and its physical mixture. The dissolution rate of ZPN was significantly increased by complexation with βCD, as compared with pure drug and physical mixture [5].

2. ANALYTICAL PROFILES OF DRUG SUBSTANCES AND EXCIPIENTS

2.1. Electrochemical methods of analysis

2.1.1. Polarography [6]
Gipsy *et al.* have developed a polarographic method for determination of zaleplon in ethanol–0.1 M Britton Robinson buffer solution (30–70) which showed two irreversible, well-defined cathodic responses in the pH range

TABLE 8.4 Proton nuclear magnetic resonance (^{1}H NMR) assignments for zaleplon

Assignment	Chemical shift δ (ppm)	Number of protons	Multiplicity
3H(CH$_2$·CH$_3$)	1.06	3H	t
2H(CH$_2$·CH$_3$)	3.727	2H	q
–	–	–	–
3H(COCH$_3$)	1.840	3H	s
–	–	–	–
H–C$_6$H$_4$	8.117	1H	dd
H–C$_6$H$_4$	7.724	1H	dd
H–C$_6$H$_4$	7.63	1H	dd
–	–	–	–
H–C$_6$H$_4$	8.077	1H	s
–	–	–	–
H(pyrazolopyrimidine)	7.64	–	d
H(pyrazolopyrimidine)	8.945	–	d
–	–	1H	–
–	–	–	–
H(pyrazolopyrimidine)	8.882	1H	s
–	–	–	–
–	–	–	–
–	–	–	–
–	–	–	–

s, singlet; d, doublet; dd, doublet of a doublet; t, triplet; m, multiplet; q, quartet.

of 2–12 using differential pulse polarography (DPP), tast polarography, and cyclic voltammetry.

The DPP technique working at pH 4.5 for peak I was selected, which exhibited adequate repeatability, reproducibility, and selectivity. The recovery was $99.97 \pm 1.5\%$ and the detection and quantitation limits were 5.13×10^{-7} and 1.11×10^{-6} M, respectively. The method was applied successfully to the individual assay of capsules to verify the content uniformity of zaleplon.

2.2. Spectroscopic methods of analysis

2.2.1. Spectrophotometry [7]

Zaleplon was determined in the presence of its alkaline degradation product namely N-[4-(3-cyano-pyrazolo[1,5-a]pyridin-7-yl)-phenyl]-N-ethyl-acetamide by measuring its D^2 (second derivative) and ^{1}DD (first

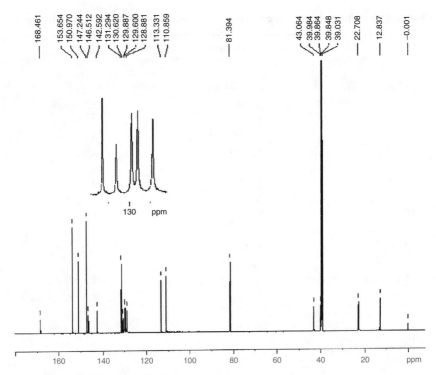

FIGURE 8.7 ^{13}C NMR spectrum of zaleplon.

derivative) of the ratio spectra of its solution in 0.01 N HCl at 235.2 and 241.8 nm, respectively.

Beer's law was obeyed over a concentration range of 1–10 μg/ml for both methods with mean percentage recovery of 100.24 ± 0.86% and 99.9 ± 1.07%, respectively.

2.2.2. Spectrofluorimetry [8]

A simple, rapid, sensitive, and selective spectrofluorimetric method ($\lambda_{ex}/\lambda_{em} = 345/455$ nm) has been developed for the determination of zaleplon. Tang *et al.* have studied the influence of micellar medium on the absorption, fluorescent excitation, and emission spectra character of zaleplon The nonionic surfactant of Triton X-100 showed a strong sensitizing effect for the fluorescence of zaleplon in a pH 5.0 buffer. The possible enhancement mechanism was discussed. Based on the optimum conditions, the linear range was 1.32×10^{-8}–1.00×10^{-5} mol/l. The detection limit was 4.0×10^{-9} mol/l with a relative standard deviation (RSD) of 0.06%. The proposed method was successfully applied to the determination of zaleplon in tablets, serum, and urine.

TABLE 8.5 ^{13}C (proton decoupled) and DEPT NMR assignments for zaleplon

Position	^{13}C δ (ppm)	DEPT
1	12.82	CH$_3$
2	43.05	CH$_2$
3	169.4	C-amide
4	22.69	CH$_3$
5	147.27	C-benzene
6	131.28	CH
7	131.32	CH
8	131.7	CH
9	143.5	C-benzene
10	130.1	CH
11	151.9	C-pyrimidine
12	111.8	CH
13	154.6	CH
14	131.5	C-pyrazole
15	82.3	C-pyrazole
16	147.2	CH
17	110.8	C-nitrine
18	–	–
19	–	–
20	–	–
21	–	–
22	–	–
23	–	–

2.3. Chromatographic methods of analysis

2.3.1. Thin-layer chromatography [7]

The spectrodensitometric analysis allows the separation and quantitation of zaleplon from its degradate on silica gel plates using chloroform–acetone–ammonia solution (9:1:0.2, v/v/v) as a mobile phase. This method depends on quantitative densitometric evaluation of thin-layer chromatogram of zaleplon at 338 nm over a concentration range of 0.2–1 μg/band, with mean percentage recovery 99.73 ± 1.35.7%.

2.3.2. High-performance liquid chromatography (HPLC)

Metwally et al. also developed a reversed-phase liquid chromatographic method using 5-C8 (22 cm × 4.6 mm i.d. × 5 μm particle size) column for quantitation of zaleplon using acetonitrile–deionized water (35:65, v/v) as a mobile phase using paracetamol as IS and a flow rate of 1.5 ml/min with UV detection of the effluent at 232 nm at ambient temperature over a

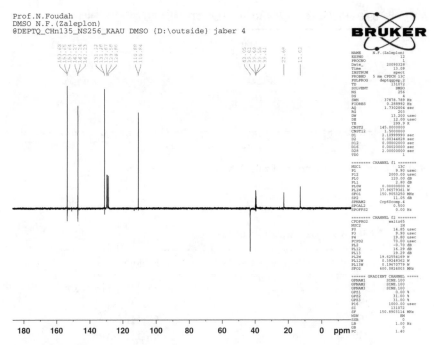

FIGURE 8.8 ^{13}C nuclear NMR spectrum of zaleplon with DEPT.

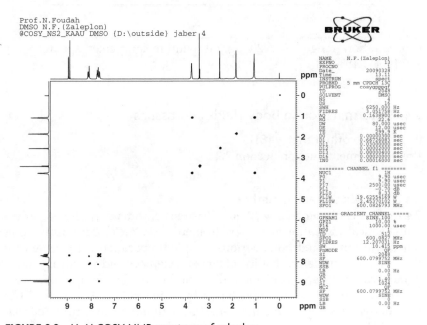

FIGURE 8.9   H–H COSY NMR spectrum of zaleplon.

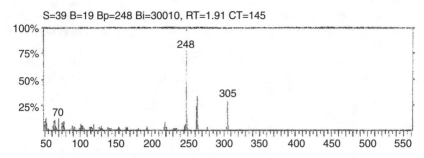

FIGURE 8.10   Mass spectrum of zaleplon.

TABLE 8.6   Mass spectrum of zaleplon

m/z	Relative intensity (%)	Fragment
305	27	$M + C_{17}H_{15}N_5O$
263	33	$C_{16}H_{15}N_4$
262	24	$C_{15}H_{12}N_5$
249	16	$C_{14}H_9N4O$
248	100	$C_{15}H_{12}N_4$
70	11.4	C_3H_4NO

M denotes the zaleplon unit.

concentration range of 2–20 μg/ml with mean percentage recovery 100.19 $\pm$ 1.15% [7].

2.4. Determination in body fluids and tissues

2.4.1. Spectrofluorimetry [8]
Previously mentioned in Section 2.2.2.

2.4.2. Chromatographic method
2.4.2.1. Gas chromatography [9, 10]   Fast gas chromatography/negative-ion chemical ionization mass spectrometric (GC/NICI-MS) assay combined with rapid and nonlaborious sample preparation is presented for the simultaneous determination of benzodiazepines and α-hydroxy metabolites, zaleplon, and zopiclone in whole blood. The compounds were extracted from 100 μl of whole blood by liquid–liquid extraction (LLE) and derivatized by *N*-methyl-*N*-(*tert*-butyldimethylsilyl)trifluoroacetamide (MTBSTFA).

Sensitive identification, screening, and quantitation of 18 compounds of interest were achieved in chromatographic separation in only 4.40 min. Accurate and reproducible results were obtained by using five different and carefully selected deuterated analogs on the basis of the chemical properties of the target analytes. The extraction efficiencies ranged from 74.3% to 105.7% and the limits of quantitation (LOQ) from 1 to 100 ng/ml. The method was fully validated and is applicable for sensitive, reliable, and quantitative determination of benzodiazepines, zaleplon, and zopiclone, for example, in clinical and forensic toxicology [9].

Gunnar *et al.* have developed comprehensively validated procedure for simultaneous semiquantitative/quantitative screening of 51 drugs of abuse (including zaleplon) or drugs potentially hazardous for traffic safety in serum, plasma or whole blood. The drugs were isolated by high-yield LLE. The dried extracts were derivatized by two-step silylation and analyzed by the combination of two different GC separations with both electron capture detection (ECD) and mass spectrometry (MS) operating in a selected ion-monitoring (SIM) mode. The method was fully validated. Intra- and interday precisions were within 2.5–21.8% and 6.0–22.5%, and square of correlation coefficients of linearity ranged from 0.9896 to 0.9999. The LOQ varied from 2 to 2000 ng/ml due to a variety of the relevant concentrations of the analyzed substances in blood. The method is feasible for highly sensitive, reliable, and possibly routinely performed clinical and forensic toxicological analysis [10].

2.4.2.2. High-performance liquid chromatography   A number of HPLC systems which are suitable for the identification and separation of zaleplon have been reported. A summary is represented in Table 8.7.

An atmospheric pressure chemical ionization (APCI-LC–MS) (Sciex API 150 EX) method was developed for the determination of zaleplon and zolpidem in the whole blood. After single-step LLE, the hypnotics were separated by gradient-elution with an ammonium formate buffer/acetonitrile eluent on an Inertsil ODS-3 column. Methaqualone was used as IS. The recovery was higher than 70% for both hypnotics and the IS. The method was successfully applied to forensic cases [11].

Another LC–APCI-MS method has been developed and validated by Zhang *et al.* for the identification and quantification of zaleplon in human plasma using estazolam as an IS. After the addition of estazolam and 2.0 M sodium hydroxide solution, plasma samples were extracted with ethyl acetate and then the organic layer was evaporated to dryness. The reconstituted solution of the residue was injected onto a prepacked Shim-pack VP-ODS C18 (250 mm × 2.0 mm i.d.) column and chromatographed with a mobile phase comprised methanol–water (70:30) at a flowrate of 0.2 ml/min.

Detection was performed on a single quadrupole mass spectrometer by SIM mode via APCI source. The intra- and interday precisions were

TABLE 8.7 HPLC assay for the analysis of zaleplon

Stationary phase	Mobile phase	Detector	Remarks	References
5-C8 (22 cm × 4.6 mm i.d. × 5 μm particle size) column	Acetonitrile–deionized water (35:65, v/v)	UV detection at 232 nm	Stability-indicating assay	[7]
Inertsil ODS-3 column	Ammonium formate buffer–acetonitrile	Mass spectrometry	Determination in whole blood applied to forensic cases	[11]
Shim-pack VP-ODS C18 (250 mm × 2.0 mm i.d.) column	Methanol–water (70:30)	Chemical ionization mass spectrometry	Determination in human plasma	[12]
XTerra MS C18 (3.5 μm, 100 × 2.1 mm i.d.) column	5% acetonitrile, 95% formic acid, 0.1% to a ratio 80–20%	MS/MS	Detection in oral fluids	[13]
Phenomenex Luna 5-μm C8(2) (250 mm × 4.6 mm i.d.) column	Methanol–water (75:25, v/v)	Electrospray ionization-mass spectrometry in selected ion-monitoring mode	Determination in plasma in phase 1 human PK study	[14]
C18 reversed-phase analytical column	Methanol–ammonium acetate buffer (50:50, v/v), pH 3.2	UV detection at 232 nm	Determination in human plasma	[15]

within 10% RSD and accuracy ranged from 85% to 115%. The limit of detection was 0.1 ng/ml. The validated LC–APCI-MS method has been used successfully to study zaleplon pharmacokinetic, bioavailability, and bioequivalence in 18 adult volunteers [12].

Pascal *et al.* have also developed a procedure for the screening of 17 benzodiazepines and hypnotics including zaleplon in oral fluid by LC–MS/MS. The method involves extraction of 0.5 ml of oral fluid (previously stored in the intercept blue buffer) treated with 0.5 ml of phosphate buffer (pH 8.4) in the presence of 5 ng diazepam-d_5 used as IS, with 3 ml of diethyl ether–methylene chloride (50:50) and separation using liquid chromatography-tandem mass spectrometry. The LOQ for all benzodiazepines and hypnotics range from 0.1 to 0.2 ng/ml. Linearity is observed from the LOQ of each compound to 20 ng/ml ($r^2 > 0.99$). Coefficients of variation at 2 ng/ml, measured on six points range from 4% to 8% for all drugs, except zopiclone (34%). Extraction recovery, measured at the same concentration, was higher than 90%. Ion suppression was evaluated for each compound and was lower than 10% for all drugs except zopiclone (93%). These results were found suitable to screen for 17 benzodiazepines in oral fluid and detect them at very low concentrations, making this method suitable for monitoring subjects under the influence [13].

A sensitive and rapid chromatographic procedure using a selective analytical detection method (electrospray ionization-mass spectrometry in SIM mode) in combination with a simple and efficient sample preparation step was presented for the determination of zaleplon in human plasma. The separation of the analyte, IS, and possible endogenous compounds are accomplished on a Phenomenex Luna 5-μm C8(2) column (250 mm × 4.6 mm i.d.) with methanol–water (75:25, v/v) as the mobile phase. To optimize the mass detection of zaleplon, several parameters such as ionization mode, fragmentor voltage, m/z ratios of ions monitored, type of organic modifier, and eluent additive in the mobile phase are discussed. Each analysis takes less than 6 min. The calibration curve of zaleplon in the range of 0.1–60.0 ng/ml in plasma is linear with a correlation coefficient of >0.9992, and the detection limit ($S/N = 3$) is 0.1 ng/ml. The within- and between-day variations (RSD) in the zaleplon plasma analysis are less than 2.4% ($n = 15$) and 4.7% ($n = 15$), respectively. The application of this method is demonstrated for the analysis of zeleplon plasma samples [14].

Foda *et al.* have developed a reversed-phase chromatography method using C18 reversed-phase analytical column.

The method was described and validated for intra- and interday variation for the quantitation of zaleplon in plasma and for pharmacokinetic application, using methanol–ammonium acetate buffer (50:50, v/v) as a mobile phase and drotaverine hydrochloride as the IS and a flow rate of 1.4 ml/min with UV detection of the effluent at 232 nm at ambient temperature over a concentration range 0.005–0.2 mg/ml with mean percentage recovery of 93.29%, and its respective CV% was 2.557% [15].

2.4.2.3. Capillary electrophoresis plasma   A capillary electrophoresis (CE) method using laser-induced fluorescence (LIF) detection for the determination of the hypnotic drug zaleplon and its metabolites in human urine was developed by Horstkotter *et al.* using carboxymethyl-β-cyclodextrin as a charged carrier. By the help of a complementary HPLC method coupled to MS, three metabolites present in human urine could be identified as 5-oxozaleplon, 5-oxo-N-deethylzaleplon, and 5-oxozaleplon glucuronide. N-Deethylzaleplon, a previously described zaleplon metabolite, as well as zaleplon itself could not be detected in human urine by the CE–LIF assay. The results were confirmed by spiking with reference compounds of the phase I metabolites. The metabolites differed very much concerning their fluorescence intensities, thus the 5-oxo metabolites present as lactam tautomer fluoresced 10-fold lower than the unchanged drug zaleplon and its N-deethylated metabolite. The glucuronide of the 5-oxozaleplon, however, showed high fluorescence due to its lactim structure. LOQ yielded by the CE–LIF assay including a 10-fold preconcentration step by solid-phase extraction were 10 ng/ml for zaleplon and N-deethylzaleplon and 100 ng/ml for 5-oxozaleplon and 5-oxo-N-deethylzaleplon [16].

3. ADME PROFILES OF DRUG SUBSTANCES AND EXCIPIENTS

3.1. Uses and applications

Zaleplon (N-[3-(3-cyanopyrazolo[1,5-a] pyrimidin-7-yl)phenyl]-N-ethylacetamide) is a nonbenzodiazepine recently introduced for clinical use [17]. This agent is indicated for the short-term treatment of insomnia [18]. There is evidence that zaleplon has sustained hypnotic efficacy without the occurrence of rebound insomnia or withdrawal symptoms on disrupt discontinuation [19].

The drug appears to be effective for reducing sleep latency in the elderly without evidence of any undesired effects [20].

Zaleplon is also efficacious in the treatment of middle of the night insomnia without causing residual hangover effects [21, 22].

3.1.1. Administration in hepatic impairment [23]

The dose of zaleplon should be reduced to 5 mg at bedtime in patients with mild to moderate hepatic impairment; it should not be given to those with severe impairment.

3.2. Absorption

Zaleplon is rapidly and almost completely absorbed following oral administration. Peak plasma concentrations are attained within approximately 1 h after oral administration. Although zaleplon is well absorbed, its absolute bioavailability is approximately 30% because it undergoes significant presystemic metabolism [24].

3.2.1. Effect of food [23]

In healthy adults a high-fat/heavy meal prolonged the absorption of zaleplon compared to the fasted state, delaying t_{max} by approximately 2 h and reducing C_{max} by approximately 35%. Zaleplon AUC and elimination half-life were not significantly affected. These results suggest that the effects of zaleplon on sleep onset may be reduced if it is taken with or immediately after a high-fat meal.

3.3. Distribution [22]

The volume of distribution of zaleplon has been determined to be 1.27 l/kg, or 90 l for a 70-kg person. This large volume of distribution indicates extensive distribution to extravascular tissue, which is not uncommon for lipophilic compounds such as zaleplon.

3.4. Metabolism

Zaleplon is extensively metabolized in the liver after oral administration. All metabolites are inactive. Less than 0.1% of unchanged zaleplon is recovered in the urine [24].

Zaleplon is metabolized primarily by aldehyde oxidase to form 5-oxo-zaleplon and, to a lesser extent, by the cytochrome P450 isoenzyme CYP3A4 to desethylzaleplon, which is further metabolized by aldehyde oxidase to 5-oxo-desethylzaleplon [25].

A marked difference in hepatic activity of aldehyde oxidase between rats and monkeys was found to be responsible for the reported marked species difference in the metabolism of Zaleplon *in vivo*. In the postmitochondrial fractions, S-9s, from liver homogenates of these animals, zaleplon was transformed in the presence of NADPH into the side chain oxidation product, N-desethyl-zaleplon, and the aromatic ring oxidation product, 5-oxo-zaleplon. In the rat S-9, N-desethyl-zaleplon and 5-oxo-zaleplon were a major and a very minor metabolites, respectively.

The metabolic pathways of zaleplon in humans and in monkeys and rats are shown in Figs. 8.11 and 8.12, respectively [26].

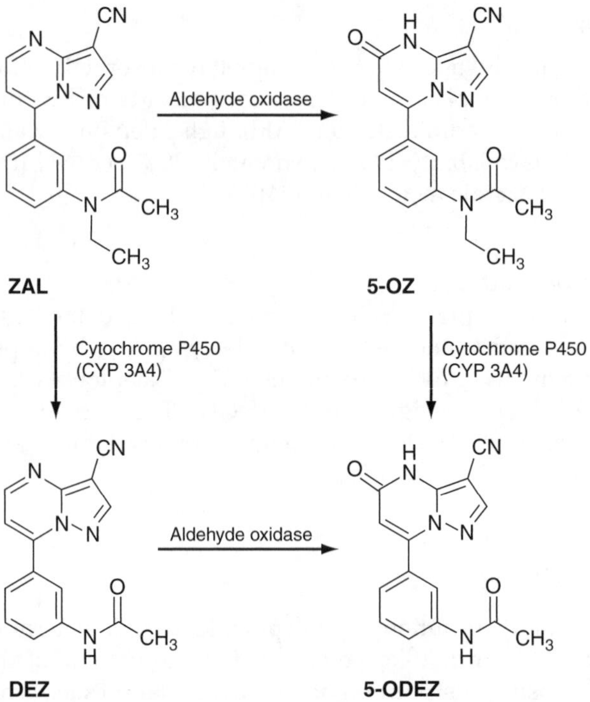

FIGURE 8.11   Metabolic pathway of zaleplon in humans.

3.5. Elimination [27]

The elimination half-life of zaleplon is 1 h after both oral and IV administration. The elimination half-life is not related to dose.

3.6. Pharmacological effects

3.6.1. Abuse

In a controlled study in healthy patients with a history of drug abuse, zaleplon was shown to have a comparable abuse potential to that of the benzodiazepine, triazolam [28].

The underlying mechanism of zaleplon abuse and stimulating effect is unknown. However, following chronic exposure to benzodiazepines or benzodiazepine-like agents, alterations in $GABA_A$ receptor sensitivity occur, which contribute to the development of tolerance, dependence, and withdrawal [29].

3.6.2. Breast feeding [30]

The manufacturers of zaleplon advise that it should not be given to breast-feeding mothers since, although only a small amount is excreted into breast milk, the effect on the nursing infant is not known. Zaleplon was

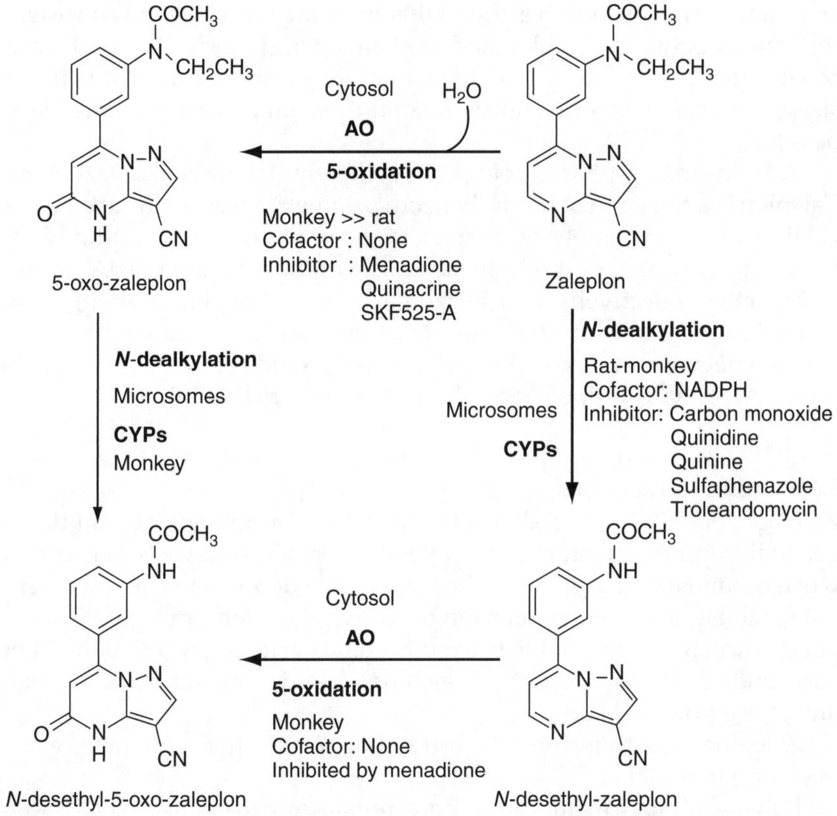

FIGURE 8.12 Metabolic pathway of zaleplon in monkeys and rats.

detected in the breast milk of five lactating women who had been given a 10-mg dose. The milk-to-plasma concentration ratio for zaleplon was about 0.5.

3.6.3. Mechanism of action

Zaleplon selectively binds with high efficacy to the benzodiazepine site (ω_1) on the α_1 containing GABA$_A$ receptors which produces its therapeutic hypnotic properties. The ultra short half-life gives zaleplon a unique advantage over other hypnotics because of its lack of next-day residual effects on driving and other performance related skills [28]. Unlike nonselective benzodiazepine drugs and zopiclone which distort the sleep pattern, zaleplon appears to induce sleep without disrupting the natural sleep architecture [31].

A meta-analysis of randomized controlled clinical trials, which compared benzodiazepines against Zaleplon or other Z drugs such as zolpidem and zopiclone, has found that there are few clear and consistent differences between zaleplon and the benzodiazepines in terms of sleep onset latency, total sleep duration, number of awakenings, quality of sleep, adverse events, tolerance, rebound insomnia, and daytime alertness [32].

Zaleplon has a pharmacological profile similar to benzodiazepines. Zaleplon is a full agonist for the benzodiazepine α_1 receptor located on the $GABA_A$ receptor ionophore complex in the brain, with lower affinity for the α_2 and α_3 subtypes. It selectively enhances the action of GABA similar to but more selectively than benzodiazepines. Zaleplon, although not benzodiazepine-like in chemical structure, induces sedative-hypnotic, anticonvulsant, and anticonflict effects via its binding to the central nervous system (CNS)-type benzodiazepine receptors [33–36].

3.6.4. Adverse reactions

Zaleplon possesses several of the clinical characteristics of traditional benzodiazepines, including the potential for additive CNS depression when administered with alcohol or other CNS depressants, a low potential for abuse, and relative safety in overdose. Zaleplon exhibited sedative effects similar to those of the benzodiazepines, with a lower likelihood of such undesirable side effects as memory loss, interaction with alcohol, and abuse potential [22].

Zaleplon was found to adversely affect respiration, attention, alertness, or mood [37].

Driving impairment and perceptual impairments have been reported [38].

In general, the better safety profile of the newer generation nonbenzodiazepines, for example, zaleplon compared to benzodiazepines, makes them better first-line choices for long-term treatment of chronic insomnia [39].

3.6.5. Interactions

Zaleplon is primarily metabolized by aldehyde oxidase and use with inhibitors of this enzyme, such as cimetidine, may result in increased plasma concentrations of zaleplon. Zaleplon is also partly metabolized by the cytochrome P450 isoenzyme CYP3A4 and, consequently, caution is advised when zaleplon is given with drugs that are substrates for, or potent inhibitors of, this isoenzyme. Cimetidine is also an inhibitor of CYP3A4 and thus inhibits both the primary and secondary metabolic pathways of zaleplon. Use with rifampicin or other potent enzyme-inducing drugs may accelerate the metabolism of zaleplon and reduce its plasma concentrations [40].

4. METHODS OF CHEMICAL SYNTHESIS

4.1. Preparative chemical methods

Several methods have been reported for the preparation of zaleplon in the reaction of 3-dimethylamino-1-(3-N-ethyl-N-acetylaminophenyl)-2-propen-1-one with 3-aminopyrazole-4-carbonitrile, which comprises carrying out said reaction in an aqueous solution of formic acid (as shown in Fig. 8.13, Scheme 1) at formic acid concentrations in the range of 20–80% (w/w) [41].

Another methods were reported either by heating in acetic acid [42] or by heating in an aqueous solution of acetic acid [43]. According to the teachings of (EP 0776898), carrying out the reaction in aqueous acetic acid would make it possible to obtain the product free from color impurities, in a much higher yield (ca. 90%) and of much better purity (>98.77%), compared to the reaction carried out in neat acetic acid. Such improved approach would also allow one to shorten the reaction time and to lower the reaction temperature.

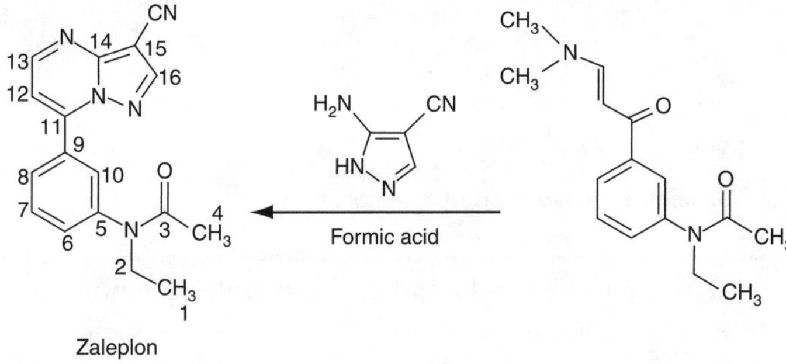

FIGURE 8.13   Scheme 1 for synthesis of zaleplon.

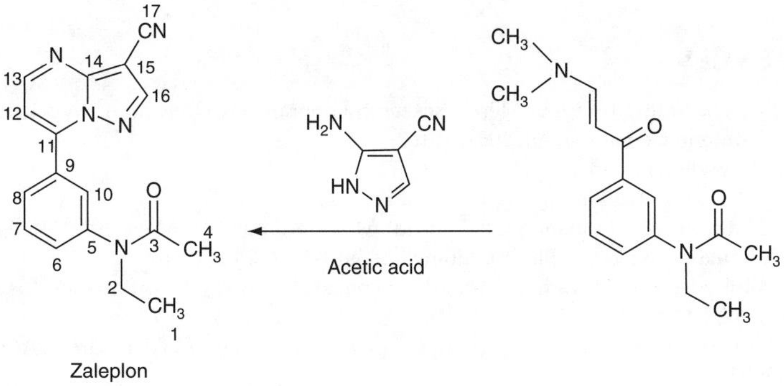

FIGURE 8.14   Scheme 2 for synthsis of zaleplon.

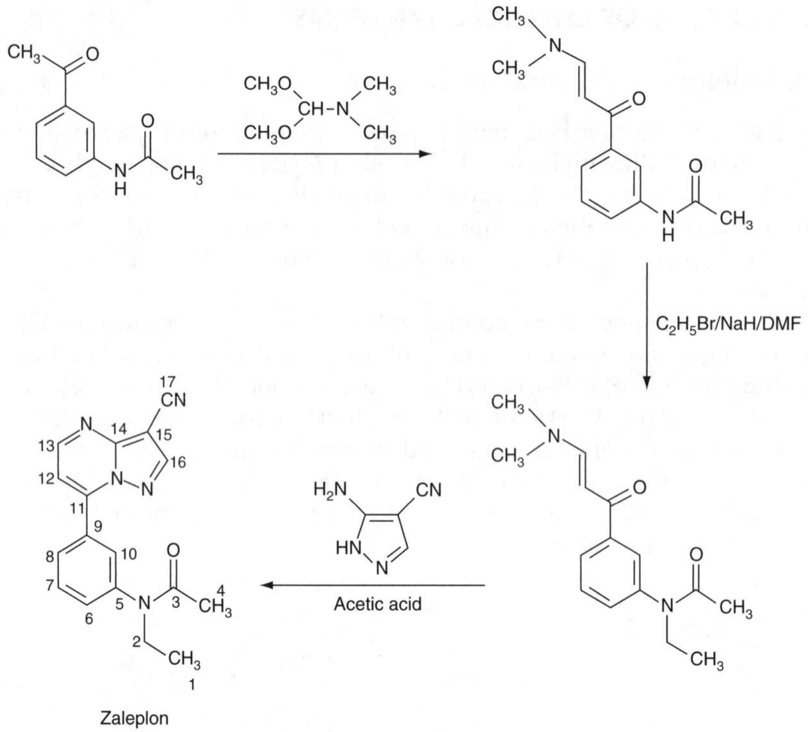

FIGURE 8.15   Scheme 3 for synthesis of zaleplon.

Schemes 2 and 3 in Figs. 8.14 and 8.15 illustrate zaleplon second and third methods of preparations, respectively.

Impurities: The HPLC analysis of zaleplon bulk drug revealed the presence of nine impurities, which were up to 0.1% [4].

REFERENCES

[1] The Merck Index, 13th ed., Merck Research Laboratories Division of Merck and Co., Inc., Whitehouse Station, NJ, 2001, 10162.
[2] http://www.wikipedia.org.
[3] N.H. Foda, R.B. Bakhaidar, unpublished results.
[4] Ch. Bharathi, K.J. Prabahar, Ch. Prasad, M. Sa Kumara, S. Magesha, V.K. Handaa, R. Dandala, A. Naidu, J. Pharm. Biomed. Anal. 44 (2007) 101–109.
[5] D. Doiphode, S. Gaikwad, Y. Pore, B. Kuchekar, S. Late, J. Incl. Phenom. Macrocycl. Chem. 62 (2008) 43–50.
[6] L. Gipsy, B. Soledad, R. Marcelo, L. Igor, J.N. Luis, A.S. Juan, A. Alejandro, J. AOAC Int. 88 (2005) 1135–1141.
[7] F. Metwally, M. Abdelkawy, N. Abdelwahab, Spectrochim. Acta Part A 68 (2007) 1220–1230.

[8] B. Tang, X. Wang, B. Jia, J. Niu, Y. Wei, Z. Chen, Y. Wang, Anal. Lett. 36 (2003) 2985–2997.

[9] T. Gunnar, K. Ariniemi, P. Lillsunde, Mass Spectrom. 41 (2006) 741–754.

[10] T. Gunnar, S. Mykkänen, K. Ariniemi, Lillsunde, Chromatogr. B Anal. Technol. Biomed. Life Sci. 806 (2004) 205–219.

[11] C. Giroud, M. Augsburger, A. Menetrey, P. Mangin, J. Chromatogr. B 789 (2003) 131–138.

[12] B. Zhang, Z. Zhang, Y. Tian, F. Xu, Y. Chen, J. Pharm. Biomed. Anal. 24 (2005) 707–714.

[13] P. Kintz, M. Villain, M. Concheiro, V. Cirimele, Forensic Sci. Int. 150 (2005) 213–220.

[14] F. Feng, J. Jiang, H. Dai, J. Wu, J. Chromatogr. Sci. 41 (2003) 17–21.

[15] N. Foda, A. Abdelbary, O. El Gazayerly, Anal. Lett. 39 (2006) 1891–1905.

[16] C. Horstkötter, D. Schepmann, G. Blaschke, J. Chromatogr. A 1014 (2003) 71–81.

[17] D.S. Charney, S.J. Mihic, R.A. Harris, Hypnotics and sedatives, in: J.G. Hardman, L.E. Limbird, A.G. Gilman (Eds.), Goodman & Gilman's The Pharmacological Basis of Therapeutics, 10th ed., McGraw-Hill, New York, USA, 2001.

[18] M. Dooley, G.L. Plosker, Drugs 60 (2000) 413–445.

[19] J. Fry, M. Scharf, R. Mangano, M. Fujimori, Int. Clin. Psychopharmacol. 15 (2000) 141–152.

[20] S. Ancoli-Israel, J.K. Walsh, R.M. Mangano, M. Fujimori, J. Clin. Psychiatry 1 (1999) 114–120.

[21] J.K. Walsh, C.P. Pollak, M.B. Scharf, P.K. Schweitzer, G.W. Vogel, Clin. Neuropharmacol. 23 (2000) 17–21.

[22] J.C. Verster, D.S. Veldhuijzen, E.R. Volkerts, Sleep Med. Rev. 8 (2004) 309–325.

[23] S.C. Sweetman (Ed.), Martindale: The Complete Drug Reference, 35th ed., Pharmaceutical Press, London, 2008 (electronic version).

[24] D.R. Drover, Clin. Pharmacokinet. 43 (2004) 227–238.

[25] D.J. Greenblatt, et al., Clin. Pharmacol. Ther. 64 (1998) 553–561.

[26] K. Kawashima, K. Hosoi, T. Naruke, T. Shiba, M. Kitamura, T. Tadashi Watabe, Drug Metabol. Dispos. 27 (1999) 422–428.

[27] K. Weikel, J. Wickman, S. Augustin, J. Strom, Clin. Ther. 22 (2000) 1254–1267.

[28] A.S. Rosen, P. Fournie, M. Darwish, Biopharm. Drug Dispos. 20 (1999) 171–175.

[29] B. Beer, J.R. Ieni, W.-H. Wu, et al., J. Clin. Pharmacol. 13 (1994) 335–344.

[30] C.R. Rush, et al., Psychopharmacology 145 (1999) 39–51.

[31] T. Paparrigopoulos, E. Tzavellas, D. Karaiskos, I. Liappas, Am. J. Psychiatry 165 (2008) 1489–1490.

[32] M. Darwish, et al., J. Clin. Pharmacol. 39 (1999) 670–674.

[33] M. Leah, L. Lisa, J. David, CNS Drugs 17 (2004) 513–532.

[34] A. Patat, I. Paty, I. Hindmarch, Hum. Psychopharmacol. 16 (2001) 369–392.

[35] J.K. Rowlett, R.D. Spealman, S. Lelas, J.M. Cook, W. Yin, Psychopharmacology (Berl.) 165 (2003) 209–215.

[36] H. Noguchi, K. Kitazumi, M. Mori, T. Shiba, Eur. J. Pharmacol. 434 (2002) 21–28.

[37] A.B. Renwick, S.E. Ball, J.M. Tredger, R.J. Price, D.G. Walters, et al., Xenobiotica 32 (2002) 849–862.

[38] M. Beaumont, D. Batéjat, C. Piérard, P. Van Beers, M. Philippe, D. Léger, G. Savourey, J. C. Jouanin, Sleep 30 (2007) 1527–1533.

[39] A. De souse, JPPS 5 (2008) 34–35.

[40] J.S. Wang, C.L. DeVane, Psychopharmacol. Bull. 37 (2003) 10–29.

[41] K. Ramakrishnan, D.C. Scheid, Am. Fam. Physician 15 (2007) 517–526.

[42] M. Korycinska, T. Stawinski, M. Wieczorek, US Patent 7057041.

[43] Padmanathan, Thurairajah, EP 0776898.

CHAPTER **9**

Cocrystal Systems of Pharmaceutical Interest: 2007–2008

Harry G. Brittain

1. INTRODUCTION

The substantial research focus on the crystal structures of drug substances that has developed over the past decade [1–5a,b] has led to extensive interest in any modification of the properties of drug substances whose physical properties are less than desirable. Within this context, the advances being made in the areas of polymorphism and solvatomorphism have been amply documented [6–9].

Center for Pharmaceutical Physics, Milford, New Jersey, USA

Profiles of Drug Substances, Excipients, and Related Methodology, Volume 35
ISSN 1871-5125, DOI: 10.1016/S1871-5125(10)35009-6

With the recognition that many substances may cocrystallize in a single continuous lattice structure, scientists have more recently initiated intense studies of the mixed molecular crystal systems that have become known as cocrystals [10]. This particular area of solid-state research has led pharmaceutical scientists into the areas of crystal engineering and assembly of appropriate supramolecular synthons, with particular emphasis on understanding the origins of the molecular self-assembly that takes place in the formation of cocrystal systems.

Without stepping into any nomenclature controversies, for the purposes of this review, cocrystal systems will be regarded as those mixed crystal systems where the individual components exist as solids under ambient conditions [11]. Aakeröy has summarized guidelines for cocrystal formation from supramolecular synthons as being constructed from discrete neutral molecular species that are solids at ambient temperatures, and where the cocrystal is a structurally homogeneous crystalline material that contains the building blocks in definite stoichiometric amounts [12].

Cocrystal systems are assembled through the association of individual molecules into fundamental building block units that are known as supramolecular synthons [13]. For example, one such synthon would be formed by hydrogen-bond interactions between a phenyl-carboxylic acid and a phenyl-amide, with the molecules being linked into a dimeric species through $O \cdots H–N$ and $O–H \cdots O$ hydrogen bonds [14]. This mode of interaction can be illustrated using the synthon that would result from the dimerization of benzoic acid and benzamide:

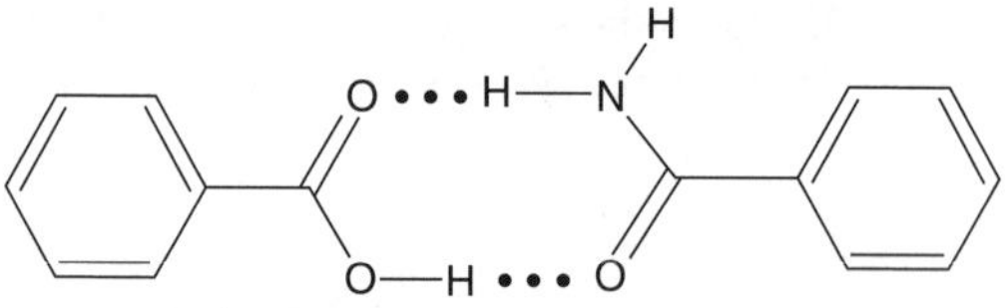

This mode of interaction has been empirically observed in the crystal structures of the cocrystals formed by benzamide and pentafluorobenzoic acid [15, 16].

Since both cocrystals and salts are multicomponent crystalline forms, it is clear that the distinction between the two depends on the degree of proton transfer between the donor and the acceptor. In this view, salts would be characterized by effectively complete proton transfer, while cocrystals would exhibit proton sharing with little or no transfer [17, 18]. In a survey study of over 80 salts and cocrystals prepared by the interaction of carboxylic acids and N-heterocyclic compounds, it was reported that structure prediction and targeted synthesis appeared to be more difficult for salts than for cocrystals [19].

The occurrence of salt forms and cocrystal systems between six sulfonamide drug substances and carboxylic acids was studied, with the nature of the product being determined by the relative acidities and basicities of the components [20]. For example, trimethoprim formed a salt with acidic sulfamethoxazole, and a cocrystal with the less acidic sulfadimidine. The principle has also been demonstrated for mixed crystals of norfloxacin, which forms a cocrystal system with isonicotinamide and salts with dicarboxylic acids [21].

Aakeröy and coworkers used the seven new cocrystals of 1,4-diiodotetrafluorobenzene they discovered to critically discuss the progress made in the field, and to comment on challenges that remain unsolved [22]. The goal of the work was to match the desired supramolecular outcome with a strategy for synthesis of the intended system, and the investigators were able to obtain the anticipated intermolecular interactions in several cases. The crystallization of one intended system was complicated by the production of a solvated crystal form, and it was concluded that the serendipitous inclusion of solvent molecules in a crystal lattice was not always predictable, and the outcome not always favorable.

In a recent review, Shan and Zaworotko have discussed cocrystals having pharmaceutical interest, and presented several case studies that they used to demonstrate how one could enhance the solubility, bioavailability, and/or stability of drug substances [23]. The systems considered were the cocrystals of fluoxetine hydrochloride with carboxylic acids, itraconazole with dicarboxylic acids, sidenafil with acetylsalicylic acid, and melamine with cyanuric acid. One main conclusion advanced by the authors was that the use of cocrystal systems in pharmaceutical dosage forms was inevitable, and that the main questions were who would benefit and how drastic the influence on development would ultimately turn out to be.

The present review will examine literature published primarily during 2007 and 2008 regarding cocrystal systems that have a pharmaceutical interest, although some earlier papers will also be mentioned when they bring a particular enlightenment to a given topic. The literature cited in the present review has been drawn primarily from the major physical, crystallographic, and pharmaceutical journals, and consequently the coverage cannot be represented as being encyclopedic or comprehensive. Apologies are presented in advance to any scientist in the field whose works have been inadvertently omitted.

2. SCREENING FOR AND PREPARATION OF COCRYSTAL SYSTEMS

Historically, the typical method to produce either new crystal forms of mixed crystals was through dissolution of the reactants in a suitable solvent system, followed by generation of a supersaturated solution that could be perturbed so as to yield a crystalline product. However, it became recognized that through the use of solid-state grinding of the reactants in the presence of small quantities of solvent, one could use the mechanical forces to produce the supramolecular synthons that formed cocrystal systems in a much more efficient manner [24–27]. The use of solid-state grinding (sometimes referred to as mechanochemistry) to generate cocrystals of organic molecules has been reviewed in detail [28].

The plasticizing effect of water during the grinding process has been found to facilitate conversion of reactants into cocrystal products. During a study of the 1:1 carbamazepine–saccharin cocrystal system, products were prepared using both the anhydrate and dihydrate solvatomorphs of carbamazepine [29]. It was found that the presence of either water or an amorphous content would increase the rate of cocrystal formation owing to increased molecular mobility in the solids. The role of water in cocrystallization phenomena was investigated further for compounds known to form hydrates, and it was found that moisture uptake generated cocrystals for carbamazepine–nicotinamide, carbamazepine–saccharin, as well as cocrystals of caffeine or theophylline with dicarboxylic acids when the reactants underwent deliquescence [30].

Both neat and liquid-assisted grinding has been used to screen for hydrates of cocrystals, where bulk water is present in the reaction mixture during liquid-assisted grinding and crystalline hydrates were used as reactants in the case of neat grinding [31]. It was determined that liquid-assisted grinding was less sensitive to the solvatomorphic nature of the reactant relative to neat grinding, and hence it was concluded that liquid-assisted grinding would be a more efficient method of screening for hydrated forms of cocrystals. Carrying the liquid-assisted route of cocrystal formation further, the concept of solution-mediate phase transformations has been used to screen for cocrystal systems, and was shown to yield the anticipated products in 16 systems having pharmaceutical interest [32].

Since the presence of solvent during cocrystal formation is of extreme importance, it is not surprising that workers have attempted to use solubility concepts as a means to understand the processes involved. A mathematical model has been developed that describes the solubility of cocrystals by considering equilibria existing between cocrystal, cocrystal components, and solution complexes, and was applied to the phase diagrams of the carbamazepine–nicotinamide cocrystal system in organic solvents [33]. Among the conclusions was that the solubility of a cocrystal is described by the solubility product of cocrystal components and by solution complexation constants, and that these equilibrium constants can be determined from solubility methods. In another work, 27 unique cocrystals of carbamazepine were generated during experiments that employed 18 carboxylic acid coformers, where the screening methods were designed through the use of phase solubility diagrams and triangular phase diagrams [34].

Of course, not all methods of cocrystal production require the use of auxiliary solvents. Thermal microscopy was used to determine if a particular carboxylic acid could cocrystallize with 2-[4-(4-chloro-2-fluorophenoxy)phenyl]pyrimidine-4-carboxamide, with positive interactions being detected as crystalline material being produced at the binary interface [35]. Once identified, authentic cocrystal systems were prepared on a larger scale using solution-phase methods. In a similar study, hot-state microscopy was used to screen the possible interactions of nicotinamide with seven compounds of pharmaceutical interest that contained carboxylic acid groups [36]. A screening method for cocrystal formation based on differential scanning calorimetry has also been described, and used to demonstrate cocrystal formation in 16 out of 20 tested binary systems [37].

In an extraordinarily comprehensive review of polymorph screening procedures, it has been reported that during the conduct of 245 polymorph screens, about 90% of the systems studied exhibited multiple crystalline and noncrystalline forms, and about 50% exhibited polymorphism [38]. As to cocrystal screening, it was concluded that it is most efficient to use a combination of structural and physical property evaluation methods in conjunction with screening protocols similar to those used to detect new polymorphic forms. Data from 64 cocrystal screening studies were considered, and it was shown that cocrystals were found in 61% of the studied systems.

Additional understanding of the formation of the supramolecular synthons that form in solution and which result in the crystallization of cocrystal systems was obtained using pulsed gradient spin–echo nuclear magnetic resonance as a means to record the long-time self-diffusivities associated with intermolecular interactions that existed in solutions of cocrystallizing systems [39]. It was concluded that for reactants that went on to produce cocrystals, the heteromeric attractions between unlike molecules were stronger than the homomeric attractions between like molecules, which was taken to indicate that an energetically favorable condition is required for the formation of cocrystals.

3. COCRYSTAL SYSTEMS HAVING PHARMACEUTICAL INTEREST

As cocrystal systems achieve more and more interest as new solid-state forms of active pharmaceutical ingredients, the research conducted on these systems continues to grow. As one might expect, a considerable amount of work has also been conducted on cocrystal systems of less interest to pharmaceutical scientists. The following sections will summarize works where drug substances have been obtained as cocrystals, and hence are of pharmaceutical interest.

3.1. Cocrystal systems formed by carbamazepine-type molecules

As the prototypical amide-containing compound, carbamazepine (5*H*-dibenz[*b,f*]azepine-5-carboxamide):

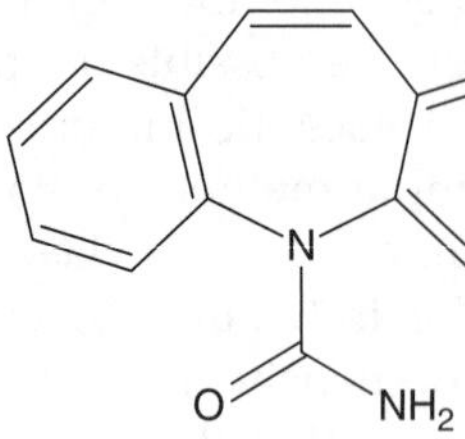

has been found to form cocrystal products with a variety of compounds, studies within the system have been used to understand a great deal about cocrystal formation. For example, in one pioneering work the crystallography of carbamazepine cocrystals with formamide, adamantane-1,3,5,7-tetracarboxylic acid, 5-nitroisophthalic acid, trimesic acid, butyric acid, formic acid, acetic acid, nicotinamide, saccharin, terephthaladehyde, benzoquinone, dimethyl sulfoxide, and acetone were studied in detail [40].

Using the carbamazepine–nicotinamide cocrystal system, a mathematical model has been developed to predict the solubility of cocrystals [41]. The model predicted that the solubility of a solid cocrystal is determined by the solubility products of the reactant species and solution complexation constants that could be obtained from the performance of solubility studies. In addition, graphical methods were developed to use the dependence of cocrystal solubility on ligand concentration for evaluation of the stoichiometry of the solution-phase complexes that are the precursor to the crystalline cocrystal itself. It was proposed that the dependence of cocrystal solubility on solubility product and complexation constants would aid in the design of screening protocols, and would provide guidance for systems where crystallization of the cocrystal did not take place.

The carbamazepine–nicotinamide cocrystal system has been used to illustrate a mechanism for the formation of cocrystals, for which nucleation and growth of solid products are determined by the combination of the reactant components to reduce the solubility of the intermolecular complex that eventually becomes crystallized [42]. The principles were studied through the use of *in situ* monitoring of the cocrystallization process in solutions, suspensions, slurries, and wet solid phases of the cocrystal reactants. In this work, it was noted that consideration of the solubility product behavior could be used to widen the range of solvents used in cocrystallization screening studies.

In another study involving the carbamazepine–nicotinamide cocrystal system, the effect of the amorphous state on the process of cocrystal formation was investigated [43]. The hypothesis was that the degree of molecular mobility associated with amorphous materials could promote intermolecular interactions between reactants, resulting in the formation of a single crystalline phase of multiple components (i.e., a cocrystal). Beginning with amorphous starting materials, low heating rates caused formation of a metastable cocrystal product that underwent a phase transformation to the stable cocrystal phase. At higher heating rates, it was found that the reactants initially crystallized, and after melting formed the stable cocrystal phase that grew from the melt. It was concluded that the amorphous phase could play an important role in determining the outcome of a cocrystal preparation.

It was observed that needle-like crystals of the metastable Form-II of the carbamazepine–isonicotinamide cocrystal formed when the ingredients were initially dissolved in ethanol, but that over time plate-like crystals of the stable Form-I grew in the suspension [44]. It was concluded that the molecular packing in Form-II consisted of carbamazepine dimers hydrogen-bonded to nicotinamide chains, and where the pyridine nitrogen of isonicotinamide did not participate in the hydrogen bonding.

Two different of supramolecular synthons have been detected in the two polymorphs of the carbamazepine–saccharin cocrystal system [45]. In the Form-I structure, the carbamazepine molecules formed a homosynthon, with the saccharine molecules also forming a hydrogen-bonded homodimer. The interaction of these two synthons resulted in formation of a one-dimensional array of molecules in a crinkled tape motif. In the Form-II structure, a heterosynthon is formed by the interaction of a carbamazepine and a saccharin molecule. This latter synthon packs in one-dimensional chains that extended along the crystallographic c-axis.

For pharmaceutical scientists, the value in cocrystals would be if such materials would be superior active pharmaceutical ingredients relative to the drug substance itself. This possibility has been studied for the carbamazepine–saccharin cocrystal, where its performance characteristics were compared with the marketed form of carbamazepine [46]. It was learned that the physical and chemical stability of formulations containing the carbamazepine–saccharin cocrystal product were similar to those of carbamazepine in the marketed product, and comparative bioavailability studies demonstrated that the cocrystal was a viable alternative drug substance to the anhydrous drug form used in the conventional solid dose forms.

While the polymorphic forms of carbamazepine all exhibit an anticarbox-amide hydrogen-bond dimer motif, the related compounds oxcarbazepine (10-oxo-10,11-dihydro-5H-dibenzo[b,f]azepine-5-carboxamide) and dihy-drocarbamazepine (10,11-dihydro-5H-dibenzo[b,f]azepine-5-carboxamide) adopt hydrogen-bond chain motifs in their crystal structures [47]. The struc-tures of several cocrystals of the structurally related compound cytenamide (5H-dibenzo[a,d][7]annulene-5-carboxamide) have been reported, with the details of the hydrogen-bonding patterns being discussed [48–51].

3.2. Cocrystal systems formed by nicotinamide with carboxylic acids

The amide–carboxylic acid interaction forms one of the most studied supramolecular synthons. For example, the binary synthon formed by benzamide and benzoic acid arises by the hydrogen-bond interactions that cause the molecules to become linked into a dimeric species through $O \cdots H-N$ and $O-H \cdots O$ hydrogen bonds:

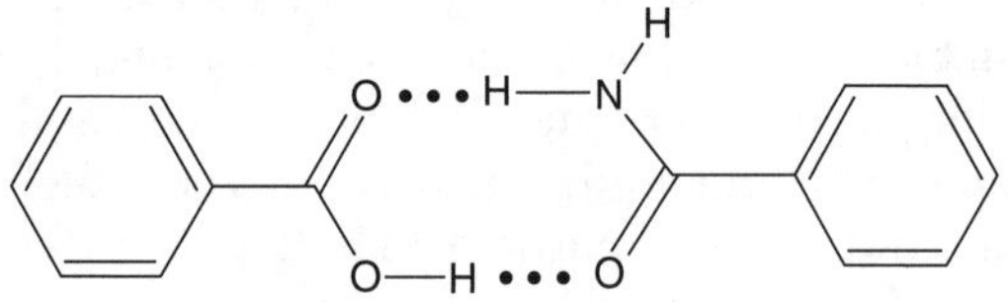

This mode of interaction has been empirically observed in the crystal structures of the cocrystals formed by benzamide and pentafluorobenzoic acid [52].

A great deal of understanding into the formation of cocrystals has been deduced through studies in which mixed crystals were prepared by the interaction of carboxylic acids with either nicotinamide or isonicoti-namide:

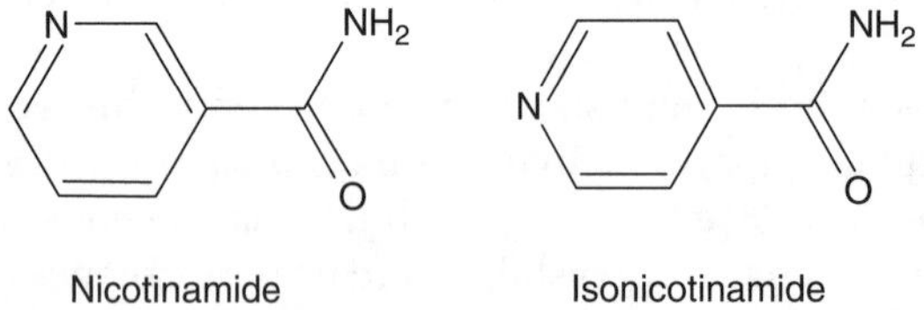

Nicotinamide   Isonicotinamide

In one study, a solvent evaporation method was used to produce cocrystals of nicotinamide with ibuprofen (2-(4-isobutylphenyl)propanoic acid) [53]. The properties of the cocrystal could be studied in the solid state, but the synthon proved to be too weak to survive in fluid solutions. Nevertheless, the solubility of ibuprofen was enhanced by 62 times when the nicotinamide concentration was 13.3 mg/ml, suggesting that the cocrystal had the potential of being a better medication for pain relief in patients with osteoarthritis.

The crystal structure of the cocrystal formed by celecoxib (4-[5-(4-methylphenyl)-3-(trifluoromethyl)-1H-pyrazol-1-yl]benzenesulfonamide) with nicotinamide has been solved from powder X-ray diffraction data [54]. The dissolution and solubility of the cocrystal product were found to depend on the medium involved, and a number of the observed phenomena were shown to originate from differences in conversion of the cocrystal celecoxib polymorphic forms I and III. However, through the judicious use of choice excipients, a formulation was developed that took advantage of the crystalline conversion to be up to fourfold more bioavailable than the celecoxib Form-III marketed product.

It has been reported that a 1:1 nicotinamide and ethyl paraben (ethyl 4-hydroxybenzoate) cocrystal can be crystallized from ethanol, with the crystal structure being used to demonstrate that both reactant molecules form corrugated layers within the cocrystal that are linked by extensive hydrogen bonding [55]. While excess amounts of nicotinamide serve to enhance the solubility of ethyl paraben somewhat, the cocrystal exhibits lower solubility in water with respect to the reactant species, and will precipitate from aqueous solutions at ambient temperature.

To understand why solution experiments sometimes fail to produce cocrystal products, and why solvent-drop grinding experiments can work when performed on the same system, the 1:1 cocrystal formed by nicotinamide and trans-cinnamic acid (trans-(2E)-3-phenylacrylic acid) has been studied [56]. In this work ternary isothermal phase diagrams of the cocrystal system was used to understand the crystallization phenomena, and to deduce methodologies and for the experimental design of cocrystal preparation. Cocrystals are most likely to form from solutions in which the two reactants have similar degrees of solubility, and the success of solvent-drop grinding was explained in that crystallization took place in the region of low solvent mole fractions where the cocrystal would be more stable relative to the separated reactants.

The crystal structures of cocrystals of isonicotinamide with two chiral carboxylic acids (optically active and racemic 2-phenylpropionic acid and 2-phenylbutyric acid) were studied with the aim of understanding the effects of the chirality on the melting point of the cocrystal products [57]. It was reported that the cocrystals formed from the racemic carboxylic acid reactants had a higher melting point relative to the melting points of cocrystals prepared from the optically active acid reactants. This finding was correlated with the more efficient crystal packing and associated higher density of the cocrystals prepared from the racemic acids, a property that would lead to the observation of greater thermal stability.

While a 1:1 isonicotinamide cocrystal with benzoic is the product formed during solution-phase preparation, the formation of a 2:1

isonicotinamide–benzoic acid cocrystal on a benzoic acid subphase has been observed as a result of epitaxial growth [58]. During the template studies, sufficient supersaturation excess of benzoic was required to ensure the integrity of the template surface, a property that was crucial to growth of the 2:1 cocrystal. It was further demonstrated that templated growth permitted isolation of the new 2:1 cocrystal directly from a solution containing a 1:1 mole ratio of the reactants.

As part of a more extensive study of cocrystals formed by isonicotinamide with carboxylic acids, 1:1 products containing the dicarboxylic fumaric or succinic acids [59]. In the structures of these particular cocrystals, the typical discrete dimeric synthon was not observed, but instead effectively infinite assemblies of one-dimensional chains were found instead. In a subsequent work, cocrystals of isonicotinamide containing mixed fumaric/succinic acids were prepared using both solid-state grinding and solution crystallization [60]. A full physical characterization of the products demonstrated that the products consisted of a single cocrystal phase, and were not simple physical mixtures of two cocrystal components. Such solid solutions were proposed as yet another method whereby one might obtain even finer control over the physical properties of cocrystal systems proposed as drug substances.

3.3. Cocrystal systems formed by caffeine and theophylline

Hydration and dehydration studies of the solvatomorphs of caffeine (1,3,7-trimethyl-3,7-dihydro-1H-purine-2,6-dione) and theophylline (1,3-dimethyl-3,7-dihydro-1H-purine-2,6-dione):

Caffeine   Theophylline

represent a significant body of work conducted to understand the mechanisms of such processes. These substances also possess a number of functional groups capable of hydrogen bonding with other molecules, and hence are able to form cocrystals with a variety of interactive molecules.

In a study of the cocrystal systems formed by caffeine with formic acid, acetic acid, and trifluoroacetic acid, it was found that products prepared by means of solid-state grinding and solution crystallization were not always the same [61]. Three possible outcomes from the two synthetic procedures were noted, namely that the same cocrystal product could be obtained from both methods, different cocrystal stoichiometries could result from the application of each method, or different cocrystal polymorphs could result when using the different methods. The utility of using solid-state grinding as the first procedure was demonstrated for those instances where products that could be initially obtained only by grinding were subsequently be crystallized from solution by seeding with crystals of the ground product.

The structures of three cocrystals of caffeine having a 1:1 stoichiometry with various hydroxy-2-naphthoic acids have been reported [62]. The anticipated imidazole-carboxylic acid supramolecular synthon was observed in caffeine cocrystals containing 1-hydroxy-2-naphthoic acid and 3-hydroxy-2-naphthoic acid, while a hydrogen-bonded carboxylic acid dimer (and no hydroxyl-caffeine heterosynthon) was observed in the caffeine cocrystal with 6-hydroxy-2-naphthoic acid.

The cocrystallization of caffeine and succinic acid in the presence of an appropriate guest was found to result in inclusion host lattices in which the guest molecule was entrapped [63]. The host frameworks were found to be built up from a dumbbell-shaped heteromolecular building block involving succinic acid and caffeine molecules. The selectivity of the system to form a particular host structure was determined to be directed by the molecular recognition properties of the guest, with the formation of $C-H\cdots O$ hydrogen bonds and halogen bonds being the key factor in controlling the nature of the inclusion matrix. Variations in the bonding patterns allow variations in the stoichiometric ratio of caffeine and succinic acid by the properties of the guest molecule.

In a demonstration of the pharmaceutical advantage that can be realized through the use of a cocrystal form of a substance, it was shown that the 1:1 cocrystal of caffeine and methyl gallate exhibited significantly improved powder compaction properties [64]. The compression characteristics of the cocrystal were reported to be excellent over the entire pressure range studied, with the tablet tensile strength of the cocrystal being twice that of caffeine at pressures less than 200 MPa. The superior compaction properties of the cocrystal product were attributed to the presence of slip planes in crystal structure.

The tendency of theophylline to convert to its monohydrate crystal form when exposed to either high relative humidity or bulk water was investigated in its cocrystal products formed with oxalic, malonic, maleic, and glutaric acids [65]. It was found that the theophylline–oxalic acid cocrystal demonstrated superior humidity stability relative to theophylline anhydrate under the conditions studied, while the other cocrystal products appeared to offer comparable stability to that of theophylline anhydrate. These workers concluded that one could demonstrate the feasibility of pharmaceutical cocrystal design based on the crystallization preferences of a molecular analog (caffeine, in this case), and that avoidance of hydrate formation and improvement in physical stability could be possible cocrystallization.

The crystal structure of the cocrystal formed by theophylline with 2,4-dihydroxybenzoic acid has been reported, with the product actually being obtained as a monohydrate [66]. In the structure, the three molecules were observed to form S(6) and $C_2^2(8)$ ring motifs, making use of both intramolecular and intermolecular hydrogen bonds, respectively. These molecules generate one-dimensional chains via $O-H \cdots O$ and $N-H \cdots O$ hydrogen bonds, with these chains being organized into a two-dimensional network through additional hydrogen bonds.

3.4. Cocrystal systems formed by saccharin

Owing to the presence of an acidic –NH group, saccharin (1,2-benzisothiazol-3(2H)-one–1,1-dioxide):

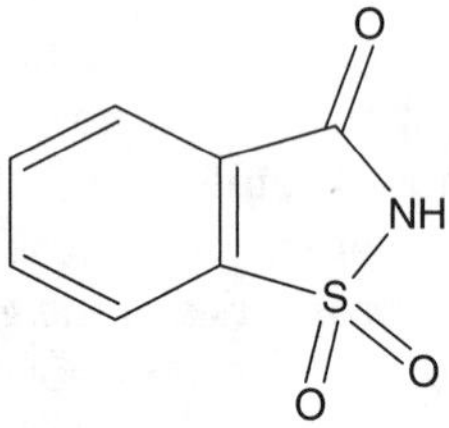

is known to form salts with sufficiently basic compounds [67]. When the difference in ionization constants between the saccharin proton donor and a potential proton acceptor is not sufficiently large, formation of cocrystal products (such as the carbamazepine–saccharin cocrystal of reference [40]) are also known.

During the course of a study of the salts formed by saccharin with quinine, haloperidol, mirtazapine, pseudoephedrine, lamivudine, risperidone, sertraline, venlafaxine, zolpidem, and amlodipine, a 1:1 cocrystal of saccharin and piroxicam was detected [68]. In the crystal structure, the asymmetric unit was found to consist of one saccharin molecule and one zwitterionic piroxicam molecule that were linked by two sets of $N-H \cdots O$ hydrogen bonds. The piroxicam–saccharin synthons were in turn linked through bridging $C-H \cdots O$ hydrogen bonds. Interestingly, the drug substance solubility out of the cocrystal was found to be comparable to that of the marketed piroxicam product.

Given the insolubility of indomethacin, a number of formulation approaches have been taken to improve the physical properties of this nonsteroidal anti-inflammatory drug substance. A 1:1 cocrystal was successfully prepared with saccharin using both solid-state dry grinding and liquid-assisted grinding methods, and the products characterized by a full range of physical analytical techniques [69]. In the crystal structure, the presence of acid–imide dimer synthons were detected, as well as acid dimer synthons weakly linked to individual saccharin molecules via a N−H$\cdots$O hydrogen bond. The pharmaceutical utility of the cocrystal products was evident in its nonhygroscopic nature, and with its significantly faster dissolution rate relative to the indomethacin γ-form.

Cocrystal products were obtained from the interaction of exemestane with maleic acid, and from the interaction of megestrol acetate with saccharin [70]. While both cocrystal products exhibited improved initial dissolution rates relative to those of respective initial reactants, the rationale for the dissolution enhancement varied. For the exemestane–maleic acid cocrystal, the formation of fine particles explained the observed enhancement in dissolution rate. On the other hand, for the megestrol acetate–saccharin cocrystal produce, the dissolution enhancement was attributed to maintenance of the cocrystal form and its rapid dissolution.

3.5. Cocrystal systems formed by carboxylic acids

The carboxylic acid functionality is present in a large number of compounds having pharmaceutical interest, and frequently represents one of the points of interaction between reactants in a supramolecular synthon. For instance, an analysis of structures published in the Cambridge Structural Database was conducted to evaluate the hierarchy of carboxylic acid and alcohol supramolecular heterosynthons in the context of active pharmaceutical ingredients [71]. It was found that 34% of the 5690 molecular carboxylic acid entries and 26% of the 25,035 molecular alcohol entries formed supramolecular homosynthons, whereas the remaining entries form supramolecular heterosynthons with other functional groups. The analysis indicated that the heterosynthons were strongly favored over their respective COOH$\cdots$COOH and OH$\cdots$OH supramolecular homosynthons.

Since the formation of cocrystals of piroxicam could result in drug substances having increased bioavailability, a total of 50 unique cocrystals containing piroxicam and a guest carboxylic acid were identified in screening experiments [72]. From the screening study, it was learned that all 23 guest molecules tried formed at least one cocrystal, and other isolation conditions led to formation of the three known polymorphs of piroxicam. The cocrystal products could be differentiated into three groupings based on their Raman spectra, which reflected the piroxicam tautomer present in the cocrystal and the presence or absence of a strong hydrogen-bond donor interacting with the piroxicam amide carbonyl group. Crystal structure analysis was used to demonstrate that the carboxylic acids were not ionized in the cocrystals, and piroxicam was present as the zwitterionic tautomer.

Although one might have anticipated that of gabapentin and oxalic acid would form a salt species, a cocrystal has been prepared through the use of a strategy based on the hydrogen-bonded synthon described by the graph set notation $R_4^2(8)$ [73]. Along with traditional crystallization methods (including crystallization in aqueous solutions at various pH values), high-throughput screening techniques and solvent drop grinding procedures were used to obtain the cocrystal product. The gabapentin–oxalic acid cocrystal was fully characterized by a variety of physical means, including the single crystal structure analysis that demonstrated the presence of the sought-after $R_4^2(8)$ synthon.

As part of a crystallization study to investigate the structural similarities between 2-methylbenzoic acid and 2-chlorobenzoic acid, a 1:1 cocrystal was obtained in which the chlorine atom and methyl group were disordered [74]. The crystal structures of the free acids and the cocrystal were characterized by the same type of hydrogen-bonded ribbons, but the solids were not isostructural owing to significant differences in the packing of the ribbons. Computationally generated structures generated by interchanging the chloro and methyl groups were found to be slightly less stable than were the experimentally observed forms. The small energy differences associated with interchanging the methyl and chloro substituents, and between alternative packing arrangements of the hydrogen-bonded ribbons, was judged to be consistent with the polymorphism of 2-methylbenzoic acid and the observed disorder in the cocrystal.

REFERENCES

[1] S.R. Byrn, R.R. Pfeiffer, J.G. Stowell, Solid State Chemistry of Drugs, third ed., SSCI Inc., West Lafayette, 1999.

[2] H.G. Brittain, Polymorphism in Pharmaceutical Solids, first ed., Marcel Dekker, New York, 1999.

[3] J. Bernstein, Polymorphism in Molecular Crystals, Clarendon Press, Oxford, 2002.

[4] R. Hilfiker, Polymorphism in the Pharmaceutical Industry, Wiley-VCH, Weinheim, 2006.

[5a] H.G. Brittain, Polymorphism in Pharmaceutical Solids, second ed., Informa Healthcare Press, New York, 2009.

[5b] H.G. Brittain, Polymorphism and Solvatomorphism 2004, in: H.G. Brittain (Ed.), Profiles of Drug Substances, Excipients, and Related Methodology, vol. 32, Elsevier Academic Press, Amsterdam, 2005, pp. 263–283.

[6] H.G. Brittain, Polymorphism and solvatomorphism 2005, J. Pharm. Sci. 96 (2007) 705–728.

[7] H.G. Brittain, Polymorphism and solvatomorphism 2006, J. Pharm. Sci. 97 (2008) 3611–3636.

[8] H.G. Brittain, Polymorphism and solvatomorphism 2007, J. Pharm. Sci. 98 (2009) 1617–1642.

[9] H.G. Brittain, Polymorphism and solvatomorphism 2008, J. Pharm. Sci. 99 (2010) accepted for publication (2010).

[10] P. Vishweshwar, J.A. McMahon, J.A. Bis, M.J. Zaworotko, Pharmaceutical cocrystals, J. Pharm. Sci. 95 (2006) 499–516.

[11] M.J. Zaworotko, Molecules to crystals, crystals to molecules . . . and back again? Cryst. Growth Des. 7 (2007) 4–9.

[12] C.B. Aakeröy, D.J. Salmon, Building co-crystals with molecular sense and supramolecular sensibility, Cryst. Eng. Comm. 7 (2005) 439–448.

[13] G.R. Desiraju, Supramolecular synthons in crystal engineering: a new organic synthesis, Angew. Chem. Int. Ed. 34 (1995) 2311–2327.

[14] L. Leiserowitz, F. Nader, The molecular packing modes and the hydrogen-bonding properties of amide:dicarboxylic acid complexes, Acta Cryst. B33 (1977) 2719–2733.

[15] L.S. Reddy, A. Nangia, V.M. Lynch, Phenyl-perfluorophenyl synthon mediated cocrystallization of carboxylic acids and amides, Cryst. Growth Des. 4 (2004) 89–94.

[16] W. Janowski, M. Gdaniec, T. Połoński, The 2:1 cocrystal of benzamide and pentafluorobenzoic acid, Acta Cryst. C62 (2006) o492–o494.

[17] S.L. Childs, G.P. Stahly, A. Park, The salt–cocrystal continuum: the influence of crystal structure on ionization state, Mol. Pharm. 4 (2007) 323–338.

[18] G.P. Stahly, Diversity in single- and multiple-component crystals: the search for and prevalence of polymorphs and cocrystals, Cryst. Growth Des. 7 (2007) 1007–1026.

[19] C.B. Aakeröy, M.E. Fasulo, J. Desper, Cocrystal or salt: does it really matter? Mol. Pharm. 4 (2007) 317–322.

[20] M.R. Caira, Sulfa drugs as model cocrystal formers, Mol. Pharm. 4 (2007) 310–316.

[21] S. Basavoju, D. Boström, S.P. Velga, Pharmaceutical cocrystal and salts of norfloxacin, Cryst. Growth Des. 6 (2006) 2699–2708.

[22] C.B. Aakeröy, J. Desper, M. Fasulo, I. Hussain, B. Levin, N. Schultheiss, Ten years of cocrystalsynthesis: the good, the bad, and the ugly, CrystEngComm 10 (2008) 1816–1821.

[23] N. Shan, M.J. Zaworotko, The role of cocrystals in pharmaceutical science, Drug Discov. Today 13 (2008) 440–446.

[24] V.R. Pedireddi, W. Jones, A.P. Chorlton, R. Docherty, Creation of crystalline supramolecular arrays: a comparison of cocrystal formation from solution and by solid-state grinding, Chem. Commun. (1996) 987–988.

[25] N. Shan, F. Toda, W. Jones, Mechanochemistry and co-crystal formation: effect of solvent on reaction kinetics, Chem. Commun. (2002) 2372–2373.

[26] A.V. Trask, W.D.S. Motherwell, W. Jones, Solvent-drop grinding: green polymorph control of cocrystallization, Chem. Commun. (2004) 890–891.

[27] A.V. Trask, N. Shan, W.D.S. Motherwell, W. Jones, S. Feng, R.B.H. Tan, K.J. Carpenter, Selective polymorph transformation via solvent-drop grinding, Chem. Commun. (2005) 880–882.

[28] A.V. Trask, W.D.S. Motherwell, W. Jones, Crystal engineering of organic cocrystals by the solid-state grinding approach, Top. Curr. Chem. 254 (2005) 41–70.

[29] A. Jayasankar, A. Somwangthanaroj, Z.J. Shao, N. Rodríguez-Hornedo, Cocrystal formation during cogrinding and storage is mediated by amorphous phase, Pharm. Res. 23 (2006) 2381–2392.

[30] A. Jayasankar, D.J. Good, N. Rodríguez-Hornedo, Mechanism by which moisture generates cocrystals, Mol. Pharm. 4 (2007) 360–372.

[31] S. Karki, T. Frišić, W. Jones, W.D.S. Motherwell, Screening for pharmaceutical cocrystal hydrates via neat and liquid-assisted grinding, Mol. Pharm. 4 (2007) 347–354.

[32] G.G.Z. Zhang, R.F. Henry, T.B. Borchardt, X. Lou, Efficient co-crystal screening using solution-mediated phase transformation, J. Pharm. Sci. 96 (2007) 990–995.

[33] S.J. Nehm, B. Rodríguez-Spong, N. Rodríguez-Hornedo, Phase solubility diagrams of cocrystals are explained by solubility product and solution complexation, Cryst. Growth Des. 6 (2006) 592–600.

[34] S.L. Childs, N. Rodríguez-Hornedo, L.S. Reddy, A. Jayasankar, C. Maheshwari, L. McCausland, R. Shipplett, B.C. Stahly, Screening strategies based on solubility and solution composition generate pharmaceutically acceptable cocrystals of carbamazepine, CrystEngComm 10 (2008) 856–864.

[35] D.P. McNamara, S.L. Childs, J. Giordano, A. Iarriccio, J. Cassidy, M.S. Shet, R. Mannion, E. O'Donnell, A. Park, Use of a glutaric acid cocrystal to improve oral bioavailability of a low solubility API, Pharm. Res. 23 (2006) 1888–1897.

[36] D.J. Berry, C.C. Seaton, W. Clegg, R.W. Harrington, S.J. Coles, P.N. Horton, M. B. Hursthouse, R. Storey, W. Jones, T. Frii, N. Blagden, Applying hot-stage microscopy to co-crystal screening: a study of nicotinamide with seven active pharmaceutical ingredients, Cryst. Growth Des. 8 (2008) 1697–1712.

[37] E. Lu, N. Rodríguez-Hornedo, R. Suryanarayanan, A rapid thermal method for cocrystal screening, CrystEngComm 10 (2008) 665–668.

[38] G.P. Stahly, Diversity in single- and multiple-component crystals. The search for and prevalence of polymorphs and cocrystals, Cryst. Growth Des. 7 (2007) 1007–1026.

[39] G. He, C. Jacob, L. Guo, P.S. Chow, R.B.H. Tan, Screening for cocrystallization tendency: the role of intermolecular interactions, J. Phys. Chem. B 112 (2008) 9890–9895.

[40] S.G. Fleischman, S.S. Kuduva, J.A. McMahon, B. Moullton, R.D.B. Walah, N. Rodríguez-Hornedo, M.J. Zaworotko, Crystal engineering of the composition of pharmaceutical phases: multiple-component crystalline solids involving carbamazepine, Cryst. Growth Des. 3 (2003) 909–919.

[41] S.J. Nehm, B. Rodríguez-Spong, N. Rodríguez-Hornedo, Phase solubility diagrams of cocrystals are explained by solubility product and solution complexation, Cryst. Growth Des. 6 (2006) 592–600.

[42] N. Rodríguez-Hornedo, S.J. Nehm, K.F. Seefeldt, Y. Pagán-Torres, C.J. Falkiewicz, Reaction crystallization of pharmaceutical molecular complexes, Mol. Pharm. 3 (2006) 362–367.

[43] K. Seefeldt, J. Miller, F. Alvarez-Núñez, N. Rodríguez-Hornedo, Crystallization pathways and kinetics of carbamazepine–nicotinamide cocrystals from the amorphous state by in situ thermomicroscopy, spectroscopy, and calorimetry studies, J. Pharm. Sci. 96 (2007) 1147–1158.

[44] J.H. ter Horst, P.W. Cains, Cocrystal polymorphs from a solvent-mediated transformation, Cryst. Growth Des. 8 (2008) 2537–2542.

[45] W.W. Porter III, S.C. Elie, A.J. Matzger, Polymorphism in carbamazepine cocrystals, Cryst. Growth Des. 8 (2008) 14–16.

[46] M.B. Hickey, M.L. Peterson, L.A. Scoppettuolo, S.L. Morrisette, A. Vetter, H. Guzmán, J.F. Remenar, Z. Zhang, M.D. Tawa, S. Haley, M.J. Zaworotko, Ö. Almarsson, Performance comparison of a co-crystal of carbamazepine with marketed product, Eur. J. Pharm. Biopharm. 67 (2007) 112–119.

[47] A.J.C. Cabeza, G.M. Day, W.D.S. Motherwell, W. Jones, Importance of molecular shape for the overall stability of hydrogen bond motifs in the crystal structures of various carbamazepine-type drug molecules, Cryst. Growth Des. 7 (2007) 100–107.

[48] A. Johnston, A.J. Florence, F.J.A. Fabbiani, K. Shankland, C.T. Bedford, J. Bardin, Cytenamide trifluoroacetic acid solvate, Acta Cryst. E64 (2008) o1215–o1216.

[49] A. Johnston, A.J. Florence, F.J.A. Fabbiani, K. Shankland, C.T. Bedford, Cytenamide-butyric acid (1/1), Acta Cryst. E64 (2008) o1295–o1296.

[50] A. Johnston, A.J. Florence, F.J.A. Fabbiani, K. Shankland, C.T. Bedford, Cytenamide-1, 4-dioxane (2/1), Acta Cryst. E64 (2008) o1345–o1346.

[51] A. Johnston, A.J. Florence, F.J. Miller, A.R. Kennedy, C.T. Bedford, Cytenamide–formic acid (1/1), Acta Cryst. E64 (2008) o1379–o1380.

[52] L.S. Reddy, A. Nangia, V.M. Lynch, Phenyl-perfluorophenyl synthon mediated cocrystallization of carboxylic acids and amides, Cryst. Growth Des. 4 (2004) 89–94.

[53] L.M. Oberoi, K.S. Alexander, A.T. Riga, Study of interaction between ibuprofen and nicotinamide using differential scanning calorimetry, spectroscopy, and microscopy and formulation of a fast-acting and possibly better ibuprofen suspension for osteoarthritis patients, J. Pharm. Sci. 94 (2005) 93–101.

[54] J.F. Remenar, M.L. Peterson, P.W. Stephens, Z. Zhang, Y. Zimenkov, M.B. Hickey, Celecoxib:nicotinamide dissociation: using excipients to capture the cocrystal's potential, Mol. Pharm. 4 (2007) 386–400.

[55] S. Nicoli, S. Bilzi, P. Santi, M.R. Caira, J. Li, R. Bettini, Ethyl-paraben and nicotinamide mixtures: apparent solubility, thermal behavior and X-ray structure of the 1:1 cocrystal, J. Pharm. Sci. 97 (2008) 4830–4839.

[56] R.A. Chiarella, R.J. Davey, M.L. Peterson, Making cocrystals—the utility of ternary phase diagrams, Cryst. Growth Des. 7 (2007) 1223–1226.

[57] A. Lemmerer, N.B. Báthori, S.A. Bourne, Chiral carboxylic acids and their effects on melting-point behavior in cocrystals with isonicotinamide, Acta Cryst. B64 (2008) 780–790.

[58] Colin C. Seaton, Andrew Parkin, Chick C. Wilson, Nicholas Blagden, Growth of an organic cocrystal upon a component subphase, Cryst Growth Des. 8 (2008) 363–368.

[59] C.B. Aakeröy, A.M. Beatty, B.A. Helfrich, A high-yielding supramolecular reaction, J. Am. Chem. Soc. 124 (2002) 14425–14432.

[60] M.A. Oliveira, M.L. Peterson, D. Klein, Continuously substituted solid solutions of organic cocrystals, Cryst. Growth Des. 8 (2008) 4487–4493.

[61] A.V. Trask, J. van de Streek, W.D.S. Motherwell, W. Jones, Achieving polymorphic and stoichiometric diversity in cocrystal formation: importance of solid-state grinding, powder X-ray structure determination, and seeding, Cryst. Growth Des. 5 (2005) 2233–2241.

[62] D.-K. Buar, R.F. Henry, X. Lou, R.W. Duerst, T.B. Borchardt, L.R. MacGillivray, G.G.Z. Zhang, Cocrystals of caffeine and hydroxy-2-naphthoic acids: unusual formation of the carboxylic acid dimer in the presence of a heterosynthon, Mol. Pharm. 4 (2007) 339–346.

[63] T. Frii, A.V. Trask, W.D.S. Motherwell, W. Jones, Guest-directed assembly of caffeine and succinic acid into topologically different heteromolecular host networks upon grinding, Cryst. Growth Des. 8 (2008) 1605–1609.

[64] C.C. Sun, H. Hou, Improving mechanical properties of caffeine and methyl gallate crystals by cocrystallization, Cryst. Growth Des. 8 (2008) 1575–1579.

[65] A.V. Trask, W.D.S. Motherwell, W. Jones, Physical stability enhancement of theophylline via cocrystallization, Int. J. Pharm. 320 (2006) 114–123.

[66] Z.-L. Wang, L.-H. Wei, Theophylline-2,4-dihydroxybenzoic acid–water (1/1/1), Acta Cryst. E63 (2007) o1681–o1682.

[67] P.M. Bhatt, N.V. Ravindra, R. Banerjee, G.R. Desiraju, Saccharin as a salt former. Enhanced solubilities of saccharinates of active pharmaceutical ingredients, Chem. Comm. (2005) 1073–1075.

[68] R. Banerjee, P.M. Bhatt, N.V. Ravindra, G.R. Desiraju, Saccharin salts of active pharmaceutical ingredients, their crystal structures, and increased water solubilities, Cryst. Growth Des. 5 (2005) 2299–2309.

[69] S. Basavoju, D. Boström, S.P. Velaga, Indomethacin–saccharin cocrystal: design, synthesis and preliminary pharmaceutical characterization, Pharm. Res. 25 (2008) 530–541.

[70] K. Shiraki, N. Takata, R. Takano, Y. Hayashi, K. Terada, Dissolution improvement and the mechanism of the improvement from cocrystallization of poorly water-soluble compounds, Pharm. Res. 25 (2008) 2581–2592.

[71] T.R. Shattock, K.K. Arora, P. Vishweshwar, M.J. Zaworotko, Hierarchy of supramolecular synthons: persistent carboxylic acid pyridine hydrogen bonds in cocrystals that also contain a hydroxyl moiety, Cryst. Growth Des. 8 (2008) 4533–4545.

[72] S.L. Childs, K.I. Hardcastle, Cocrystals of piroxicam with carboxylic acids, Cryst. Growth Des. 7 (2007) 1291–1304.

[73] M. Wenger, J. Bernstein, An alternate crystal form of gabapentin: a cocrystal with oxalic acid, Cryst. Growth Des. 8 (2008) 1595–1598.

[74] M. Polito, E. D'Oria, L. Maini, P.G. Karamertzanis, F. Grepioni, D. Braga, S.L. Price, The crystal structures of chloro and methyl ortho-benzoic acids and their co-crystal: rationalizing similarities and differences, CrystEngComm 10 (2008) 1848–1854.

Cumulative Index

Bold numerals refer to volume numbers.